ANALOG CIRCUIT DESIGN

Analog Circuit Design

High-Speed Analog-to-Digital Converters; Mixed-Signal Design; PLL's and Synthesizers

Edited by

Rudy J. van de Plassche
Broadcom Netherlands B.V., Bunnik

Johan H. Huijsing
Delft University of Technology

and

Willy Sansen
Katholieke Universiteit Leuven

KLUWER ACADEMIC PUBLISHERS
BOSTON / DORDRECHT / LONDON

A C.I.P. Catalogue record for this book is available from the Library of Congress.

ISBN 0-7923-7956-X

Published by Kluwer Academic Publishers,
P.O. Box 17, 3300 AA Dordrecht, The Netherlands.

Sold and distributed in North, Central and South America
by Kluwer Academic Publishers,
101 Philip Drive, Norwell, MA 02061, U.S.A.

In all other countries, sold and distributed
by Kluwer Academic Publishers,
P.O. Box 322, 3300 AH Dordrecht, The Netherlands.

Printed on acid-free paper

Printed in the Netherlands.

Table of Contents

Preface

This book contains the extended and revised editions of all the talks of the ninth AACD Workshop held in Hotel Bachmair, April 11 – 13 2000 in Rottach-Egern, Germany. The local organization was managed by Rudolf Koch of Infineon Technologies AG, Munich, Germany.
The program consisted of six tutorials per day during three days. Experts in the field presented these tutorials and state of the art information is communicated.
The audience at the end of the workshop selects program topics for the following workshop.
The program committee, consisting of Johan Huijsing of Delft University of Technology, Willy Sansen of Katholieke Universiteit Leuven and Rudy van de Plassche of Broadcom Netherlands BV Bunnik elaborates the selected topics into a three-day program and selects experts in the field for presentation.
Each AACD Workshop has given rise to publication of a book by Kluwer entitled "Analog Circuit Design". A series of nine books in a row provides valuable information and good overviews of all analog circuit techniques concerning design, CAD, simulation and device modeling. These books can be seen as a reference to those people involved in analog and mixed signal design.
The aim of the workshop is to brainstorm on new and valuable design ideas in the area of analog circuit design. It is the hope of the program committee that this ninth book continues the tradition of emerging contributions to the design of analog and mixed signal systems in Europe and the rest of the world.

Rudy J. van de Plassche
Broadcom Netherlands BV.

High-Speed Analog-to-Digital Converters

R.J. van de Plassche

The application of digital techniques to process analog signals in systems on a chip continues to be a very a crucial part of such a system. Analog-to-digital converters with analog preprocessing circuitry are combined with digital circuitry on the same chip using standard digital CMOS process technology. The scaling of CMOS technology to reduce power and increase the system size results in a reduction of the available supply voltage. Soon this supply voltage will reach 1 V or below. To design an analog-to-digital converter using such a small supply voltage will be a tremendous challenge. Furthermore the susceptibility of the analog circuit part to ground bounces and (digital) substrate noise injection has to be taken into account. In this session the progress in high-speed analog-to-digital converter design will be reported.

In the first paper by Steyaert et al. The influence of technology scaling on the performance of converters is discussed. This paper gives insight in future problems and possibilities to design high-performance converters.

In the second paper by Nagaraj et al, examples of a 6-bit and a 7-bit high-speed analog-to-digital converter are shown. These converters use a standard 0.25 micron digital CMOS technology.

The third paper describes a 12-bit CMOS analog-to-digital converter using folding an interpolation. This converter has been optimized to obtain a large Spurious Free Dynamic Range of 80 dB suitable for radio receiver applications.

The fourth paper by Brandt et al, shows what can be obtained using a subranging architecture to optimize performance, power and dies size. A 10-bit converter with 9.5 effective bits is reported.

In the field of sigma-delta conversion Fujimori et al, shows what techniques can be used to optimize the performance of sigma-delta modulator with a large signal bandwidth and a limited oversampling factor. Over a bandwidth of 2.5 MHz 90 dB signal-to-noise ratio is obtained with and SFDR of 102 dB.

R. J. van de Plassche et al. (eds.), Analog Circuit Design, 1–2.

The last contribution by van Engelen discusses stability and design criteria for a bandpass sigma-delta modulator to be used in AM/FM radio systems. This continuous time bandpass modulator obtains at 10.7 MHz Intermediate Frequency input signal a resolution of 13-bit with a sampling frequency between 40 and 80 MHz.

Speed-Power-Accuracy Trade-off In high-speed Analog-to-digital converters: Now and in the future ...

M. Steyaert and K. Uyttenhove
K.U.Leuven, ESAT-MICAS
Kardinaal Mercierlaan 94, 3001 Heverlee, Belgium

Abstract

High-speed analog-to-digital converters (ADC's) are an essential part in a signal processing system. Radar applications and hard disk drive read channels require very high conversion speeds and relatively low resolutions (6-8 bits) [1][4]. Since several ADC's may be needed in a "system-on-chip", the ADC should only consume a small fraction of the total power budget [15]. In this article, a fundamental trade-off between speed, power and accuracy for high-speed converters is shown. This trade-off only depends on the matching data of the used process. Technology-scaling issues influencing this trade-off will be discussed. An important factor is the supply voltage; the never-ending story of technology trends towards smaller transistor dimensions has resulted to date in deep sub-micron transistors. The consequence is the downscaling of the power supply voltages, to date even lower than 2V, with almost the same threshold voltages of the CMOS transistors (in order to keep the leakage current in digital circuits small enough). This voltage scaling will have an impact on the previous mentioned trade-off between speed, power and accuracy. In the first section, high-speed ADC's architectures are presented.

In the second section the impact of mismatch or accuracy in analog circuits (especially in high-speed ADC's) and the impact on power drain is discussed. Secondly in section three some fundamental limitations of analog integrated circuit design in the trade-off between speed, accuracy and power drain are analysed. In the following section the impact of the supply-voltage scaling on this trade-off is studied. After this, some modifications are presented to circumvent this trade-off, and the article is ended with a conclusion.

R. J. van de Plassche et al. (eds.), Analog Circuit Design, 3–24.

High-speed ADC architecture

An A/D conversion algorithm is a description of the functional operation of the ADC. The architecture of the ADC is the translation of this algorithm in hardware. The choice of architecture is strongly related to the system design. This system design is ruled by the trade-off between performance and hardware cost of its building blocks. From literature, high-speed, low/medium resolution ADC architectures can be roughly divided in three groups: full flash, folding/interpolating and pipelined architectures. Each of these architectures has its own place in the resolution-bandwidth picture shown in Figure 1 .
Flash-type architectures are typically the fastest structures that can be used to implement low resolution ADC's. Figure 2 presents a typical block diagram of a N-bit flash converter. The resistive ladder subdivides the converter external reference voltages ($+V_{ref}$ $-V_{ref}$) in a set of 2^N reference voltages on chip, which are compared in parallel with the analog input signal.

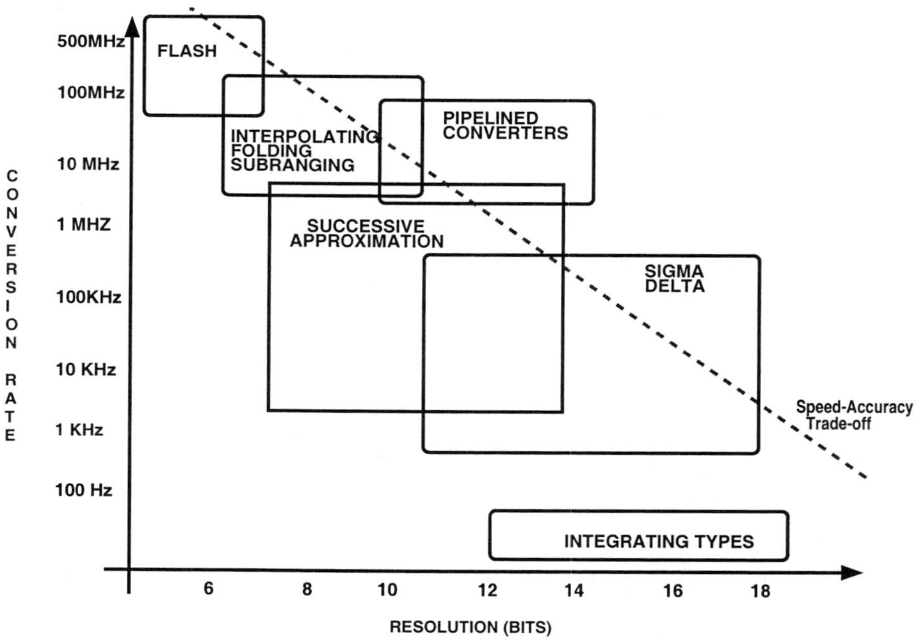

Figure 1: Speed-Resolution Trade-off

A logic encoder converts the thermometer code generated by all the comparators into a binary code that approximates the input signal every clock cycle. The comparators are preceded by an array of pre-amplifiers to reduce the input-referred offset of the regenerative comparator and to reduce also the kickback noise.
Note that the major advantages (simplicity and parallelism) of flash architectures also present their main problem: the number of comparators increases exponentially with the resolution specification, leading typically to a large die area and a high power consumption. As

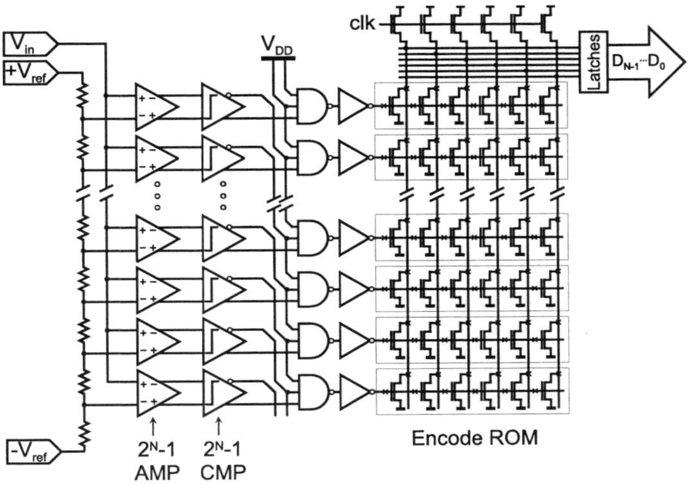

Figure 2: Typical Flash ADC architecture

can be seen from the block diagram of the flash converter, the correct operation of a flash converter depends on the accurate definition of the reference voltages sensed by each comparator.
Since the comparator offset voltage is a random variable (which depends on the matching properties of the used technology [10]), it directly influences the differential and integral nonlinearity (DNL/INL) characteristics of the A/D converter.
Therefore, this flash-architecture is only used for $N < 8$. For higher resolutions, analog preprocessing steps, like folding/interpolating or pipelining, are used to break the exponential relationship between resolution and area (power).
Therefore, the first step in the design of a flash converter consists in deriving an offset voltage standard deviation that garanties with a high

probability that the design complies with a certain performance specification (high yield). As shown in figure 3, the yield of the analog part (e.g. ADC) of a mixed-mode chip must be much higher than the overall yield, because of the relatively small area contribution of the analog part.

Consider that the offset voltage of all the comparators are independent

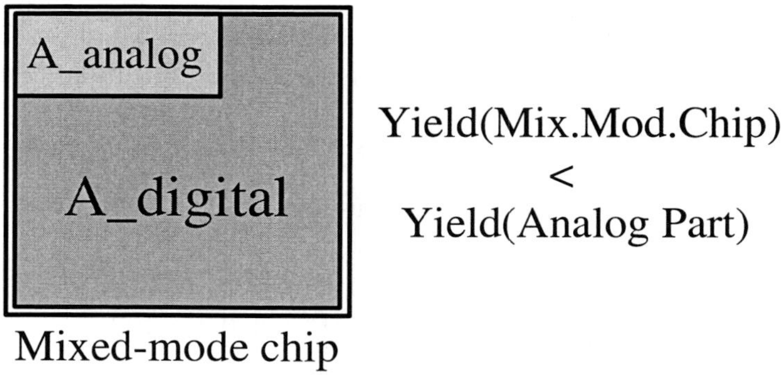

Figure 3: Yield of mixed-mode chip

variables that follow a normal distribution. Monte Carlo simulations have been used to estimate the design yield as a function of the offset voltage standard deviation (Closed form expressions also exist for calculating the yield as a function of offset standard deviation [19])

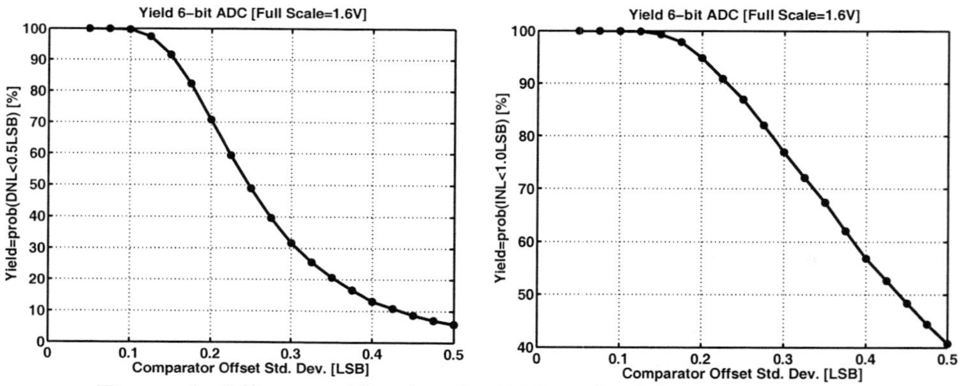

Figure 4: Offset specification for DNL < 0.5 lsb and INL < 1 lsb

The results obtained for a 6-bit converter considering that one wants to comply with a DNL and an INL specification of 0.5 LSB and 1.0 LSB, respectively , are presented in Figure 4. So in order to design a

6-bit converter with an acceptable yield, the comparator offset standard deviation should not exceed the value of 0.11 times the least significant bit. Next section will introduce models to calculate the offset standard deviation of the comparator. Together with this information the trade-off between speed, power and accuracy will be shown in section 3.

Figure 5 shows an overview of some very high-speed ADC's (references at the end of the article). The static resolution of all these converters is 6 bit and the used technology is indicated on the figure. Sampling speeds are increasing due to the technology scaling, without consuming more power. This is one of the implications of the trade-off between speed-power-accuracy that will be discussed later on.

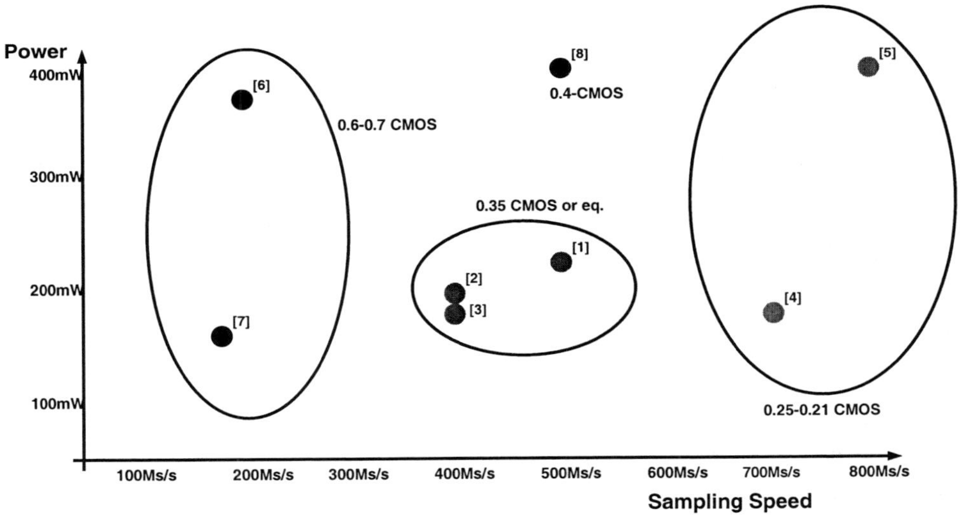

Figure 5: Overview high-speed 6 bit converters (refer. At the end of this article)

Models for matching in deep sub-micron technologies

1. What is transistor mismatch ?

Two identical designed devices on an integrated circuit have random differences in their behaviour and show a certain level of random mismatch in the parameters which model their behaviour. This mismatch is due to the stochastic nature of physical processes that are used to fabricate the device. In [10] the following definition for mismatch is given: mismatch is the process that causes time-independent random variations in physical quantities of identically designed devices.

2. Modelling of CMOS transistor mismatch

The mismatch of two CMOS identical transistors is characterized by the random variation of the difference in their threshold voltages V_T, their body factor γ and their current factor β and their distance from each other. For technologies with a minimal device size larger than typically 2µm, a widely accepted and experimentally verified model for these random variations is a normal distribution with mean equal to zero and a variance dependent on the gate-width W, the gate-length L and the mutual distance D between the devices [10]:

$$\sigma(\Delta V_T) = \frac{A_{V_T}}{\sqrt{W.L}} + S_{V_T}.D \tag{1}$$

$$\sigma\left(\frac{\Delta\beta}{\beta}\right) = \frac{A_\beta}{\sqrt{W.L}} + S_\beta.D \tag{2}$$

A_{Vt}, A_β, S_{Vt} and S_β are process-dependent parameters. In table 1 the proportionality constants for several processes are summarized. Experimental data show that the correlation between the V_T and β mismatch is very low although both parameters depend on the oxide thickness [11].

A corner distance D_m can be defined as the distance D for which the mismatch due to the distance effect is equal to the mismatch due to the size dependence for a minimal size device. The obtained critical distances D_m are very large compared to the typical size of an analog circuit. Therefore, the distance dependence of the parameter mismatch will be neglected in the discussion of the impact of transistor mismatch on analog circuit and system performance. In equations (1) and (2) the standard deviation of the difference of the parameters of two transistors is given. The following relations are valid for the variance of the absolute parameters of a single transistor:

$$\sigma(V_T) = \sqrt{2}\sigma(\Delta V_T) \tag{3}$$

$$\left(\frac{\sigma(\beta)}{\beta}\right) = \sqrt{2}\left(\frac{\sigma(\Delta\beta)}{\beta}\right) \tag{4}$$

Table 1 shows clearly the dependence of A_{Vt} on the technology whereas A_β remains almost constant.

Technology	Type	A_{Vt}	A_β	$(V_{GS}-V_T)_m$
		[mV.μm]	[%.μm]	[V]
2.5 μm	nMOS pMOS	30 35	2.3 3.2	2.6 2.2
1.2 μm	nMOS pMOS	21 25	1.8 4.2	2.3 1.2
0.7 μm	nMOS pMOS	13 22	1.9 2.8	1.4 1.6
0.5 μm	nMOS pMOS	11 13	1.8 2.3	1.3 1.2
0.35 μm	nMOS pMOS	9 9	1.9 2.25	1 0.8
0.25 μm	nMOS pMOS	6 6	1.85	+/- 0.8

Table 1: Mismatch parameters of different CMOS technologies

It can be shown that the threshold mismatch parameter A_{Vt} is proportional to the oxide thickness of the process. The oxide thickness scales with the minimum technology length to reduce the short channel effects [21]. Consequently, the use of a deep sub-micron technology improves the threshold matching.

Figure 6 shows the decrease of the threshold mismatch parameter A_{Vt} as a function of the technology (or the minimum transistor length). The same is done for the current mismatch parameter A_β as shown in Figure 6. As long as the threshold mismatch dominates over the β mismatch, scaling of technology improves the overall mismatch. This will be discussed in detail in section 3.

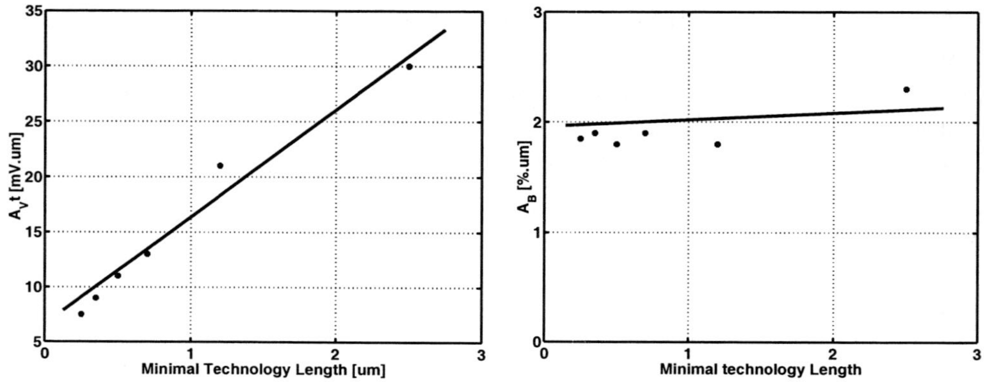

Figure 6: Threshold mismatch and β-mismatch as a function of minimal transistor length

In high-speed analog designs, the designer prefers to use small gate-lengths so that the highest intrinsic speed f_t for the transistor is obtained; accurate models for minimum sized transistors are thus necessary. For the accurate modelling of the threshold mismatch in sub-micron technologies, the simple linear model has to be extended for short and narrow channel effects. The threshold voltage is dependent on the flat-band voltage, the surface potential, the depletion charge and the gate capacitance. It has been experimentally verified that the mismatch of the threshold voltage is mainly determined by the mismatch of the bulk depletion charges in the two devices. In sub-micron technologies, two effects introduce errors in the model. Due to the presence of the source and the drain diffusion areas and the charge sharing effect, part of the channel depletion charge is not controlled by the gate voltage anymore. For devices with a small gate-length, this charge is a relatively large part of the depletion charge therefore the threshold voltage lowers for small gate-lengths whereas the variance of the V_T mismatch increases. A similar explanation can be given for small gate-widths. These effects can be taken into account if following formulas are used [11]:

$$\sigma^2(\Delta V_T) = \frac{A_{1Vt}^2}{W.L} + \frac{A_{2Vt}^2}{W.L^2} - \frac{A_{3Vt}^2}{W^2.L} + S_{Vt}^2.D^2 \tag{5}$$

$$\sigma^2\left(\frac{\Delta\beta}{\beta}\right) = \frac{A_{1\beta}^2}{W.L} + \frac{A_{2\beta}^2}{W.L^2} - \frac{A_{3\beta}^2}{W^2.L} + S_{\beta}^2.D^2 \tag{6}$$

As shown in the first section, a flash ADC consists of an array of pre-amplifiers (differential pair structure) followed by comparators. To end this section and introduce the next section, which deals with the trade-off between speed, accuracy and power, mismatch formulas for a differential pair configuration, shown in Figure 7, are deduced.

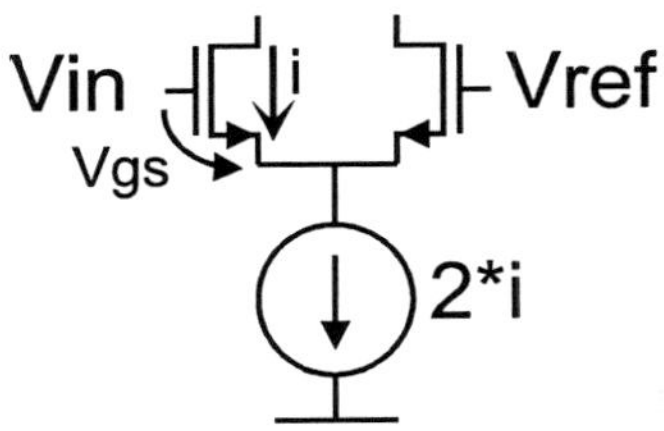

Figure 7: Differential Pair Configuration

Next formula shows the current of a transistor as a function of the β-factor and the threshold voltage V_T [20]:

$$I = \frac{\beta}{2}(V_{GS} - V_T)^2 \text{ with } \beta = \frac{KP \times W}{L} \tag{7}$$

The input-referred offset of a differential pair can be derived from this formula and is given by :

$$\sigma^2(\Delta V_{GS}) = \sigma^2(\Delta V_T) + \left[\frac{(V_{GS} - V_T)}{2}\sigma^2\left(\frac{\Delta\beta}{\beta}\right)\right]^2 \tag{8}$$

After substituting the formulas for the mismatch (1) and (2) into (8) the offset voltage can be written in terms of the mismatch parameters A_{Vt} and A_β of the used technology.

$$\sigma^2(\Delta V_{GS}) = \frac{1}{W.L}\left[A_{V_T}{}^2 + \frac{A_\beta^2}{4}(V_{GS} - V_T)^2\right] \tag{9}$$

From (9) we can conclude that the current and threshold matching depends on A_{Vt} and A_β and that the relative importance of threshold mismatch and current mismatch depends on the gate overdrive voltage. A corner gate overdrive voltage $(V_{GS} - V_T)_m$ is defined for which the effect of the V_T and β mismatch on the gate-voltage or drain current is of equal size (see table 1 for values):

$$(V_{GS} - V_T)_m = 2\,{}^{A_{Vt}}\!/_{A_\beta} \tag{10}$$

For circuits with a bias point with a $(V_{GS} - V_T)$ smaller than $(V_{GS} - V_T)_m$ the effect of V_T mismatch is dominant, whereas for a $(V_{GS} - V_T)$ larger than $(V_{GS} - V_T)_m$ the effect of β mismatch dominates. It is clear that in practical circuits the $(V_{GS} - V_T)$ will be smaller than the corner gate overdrive voltage so that only the V_T mismatch is dominant. In practice, the offset voltage can be approximated by:

$$\sigma^2(\Delta V_{GS}) = \frac{1}{W.L}\left[A_{V_T}{}^2\right] \qquad (11)$$

The approximation error is equal to $((V_{GS} - V_T)/(V_{GS} - V_T)_m)^2/2$ and is small for typical bias conditions.

Speed-Power-Accuracy Trade-off

The trend in analog circuit design has always been towards higher speed, higher accuracy and lower power drain. However, it will be shown that the speed-accuracy-power trade-off is simply limited by technology parameters only, and more especially the mismatch parameters of the technology.

One possible way to overcome this problem is by using offset compensation or auto-zero techniques (analog or digital, background or foreground). However, those compensation techniques require calibration phases during which the normal system operation is interrupted and the offset voltages of the building blocks are sampled and dynamically stored in a memory. This reduces the maximum processing speed and requires a lot of extra chip overhead to provide calibration and replica circuits. In many high-speed, low-power circuits, the interruption of the system cannot be tolerated or the required continuous operation is too long to ensure the offset correction. Therefore, the accuracy completely depends on the matching performances of the technology. The bit accuracy that can be achieved is proportional to the matching of the transistor. To improve the system accuracy, larger devices are required, but at the same time the capacitive loading of the circuit nodes increase and more power is required to attain a certain speed performance. This can very easily be verified for a typical ADC system, which consists of a cascade of different stages each stage with their own gain (as seen in figure 8).

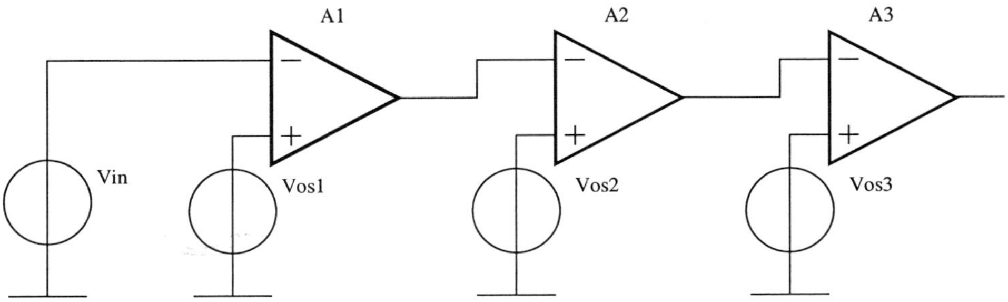

Figure 8: Schematic Representation of a multi-stage voltage processing system

The accuracy of the system will be determined by the equivalent input offset voltage, which is given by:

$$\sigma(V_{OS,eq}) = \sqrt{(\sigma(V_{OS1}))^2 + \left(\frac{\sigma(V_{OS2})}{A_1}\right)^2 + \left(\frac{\sigma(V_{OS3})}{A_2}\right)^2 + \ldots} \approx \sigma(V_{OS1}) \quad (12)$$

The simplification is only valid if the interstage gains are high enough. On the other hand the accuracy that can be achieved in a system is proportional to the matching accuracy of the components. In section 2, a formula was derived for the offset of a differential amplifier with a transistor area of W.L, so the equivalent input referred offset is given by ($C_{in} = 2/3.C_{OX}.W.L$):

$$\sigma(V_{OS1}) = \sqrt{\frac{2/3.C_{OX}.A_{Vt}^2}{Cin}} \quad (13)$$

The accuracy of the ADC system is given by the ratio of the maximal input signal and the offset voltage[1]:

$$Accuracy = \frac{V_{IN,rms}}{3.\sigma(V_{OS1})} \quad (14)$$

The power to drive the input capacitance C_{in} is delivered by the system itself to reduce the loading of the signal source. The best power efficiency is obtained for class B systems and the power required to drive a signal Vin,rms with a frequency f across a capacitor C_{in} is given by

[1] The 3 sigma value ensures that the accuracy specification is met with a probability of 99.7 %

$$P = 8 \cdot f \cdot C_{IN} \cdot V_{IN,rms}^{2} \tag{15}$$

If the equations 13, 14 and 15 are combined into the Speed.Accuracy^2/power product the result is

$$\frac{Speed \times Accuracy^2}{Power} \approx \frac{1}{C_{OX}.A_{Vt}^{2}}$$

On the other hand, the fundamental limit in the speed-power-accuracy trade-off is imposed by thermal noise [17]:

$$\frac{f.DR^2}{P} \approx \frac{1}{k_B.T} \tag{16}$$

where k_B is the Boltzmann constant and T is the absolute temperature. This fundamental limit is independent of technology. For modern technologies this fundamental limit is orders of magnitude lower than the technological limit derived before. In other words, for present-day CMOS technologies, the performance of precision analog circuits is limited by transistor mismatch and not by noise.

It has been shown that the trade-off between speed, accuracy and power still holds for more complex circuits [9], such as current processing circuits (current mirrors), voltage processing circuits (differential pairs and operational amplifiers) and even multi-stage circuit designs. The impact of the relationship above is that for the circuits of today which are after high speed, high performance or accuracy, and low power drain, a technological limit is encountered, namely the mismatch of the devices. This means that for a given technology, if high speed and high accuracy is required, this can only be achieved by consuming power. For example, if one bit extra accuracy is required in the design of AD and DA converters the power drain for the same speed performance will increase with a factor 4 !

This trade-off has also been shown in a fitting model for high-speed ADC in [14]. In this article following formula is derived:

$$P \approx \frac{L_{MIN} \cdot \left(F_{sample} + F_{signal}\right)}{10^{\left(-0.1525 ENOB + 4.838\right)}} \tag{17}$$

ENOB can be taken as an accuracy measurement, F_{sample} as a speed factor and L_{MIN} as a technological constant.

The derived performance limit caused by mismatch is of course only valid for converter architectures for which the accuracy relies on component matching (not like ΣΔ architectures for which the accuracy is limited by noise).

Impact of voltage scaling on Trade-off

In the previous section, a fundamental trade-off between speed, accuracy and power has been deduced:

$$\frac{Speed \times Accuracy^2}{Power} \approx \frac{1}{C_{OX} \cdot A_{Vt}^2} \tag{18}$$

Wat happens with this trade-off when technology scales down ?
To reduce the short channel effects in deep-sub micron transistors, the oxide thickness is scaled down together with the minimum transistor length. As shown in the second section, the threshold mismatch parameter is proportional with the oxide thickness. Consequently, the threshold mismatch parameter A_{Vt} decreases as technology scales down. The gate-oxide capacitance on the other hand increases when technology scales down (inverse proportional with the oxide thickness). Therefore, $C_{OX} \cdot A_{Vt}^2$ increases as technology scales down and as a result, the trade-off becomes better. This means that e.g. for the same speed and accuracy, less power is needed when technology is scaled down.
However, the maximal supply voltage also reduces for smaller oxide thickness (see Figure 9), so that smaller signal levels have to be used (0.25 μm technology uses a 2.5 V power supply, 0.18 μm uses a 1.8 V power supply). When the supply voltage becomes smaller, the input swing of the differential pair decreases leading to smaller values for the least significant bit. As a result, the maximum allowable offset also decreases. Consequently, the scaling advantage for the trade-off with smaller technology line-widths is reduced. Moreover, the increasing substrate doping levels in deeper sub-micron technologies make the parasitic drain to bulk and source to bulk capacitors relatively more and more important compared to the gate-oxide capacitance. This effect is clearly seen in Figure 10 where the f_t and the f_{-3dB} is plotted as a function of minimum technology length:

$$F_T = \frac{g_m}{2\pi . C_{GS}} \quad \text{and} \quad F_{-3dB} = \frac{g_m}{2\pi(C_{GS} + C_{DB})} \tag{19}$$

In fact, the f_{-3dB} is a staircase function (now and then technology modifications reduce the drain-bulk capacitance [16]). This results in extra capacitive loading of the signal nodes and requires extra power to attain high-speed operation.

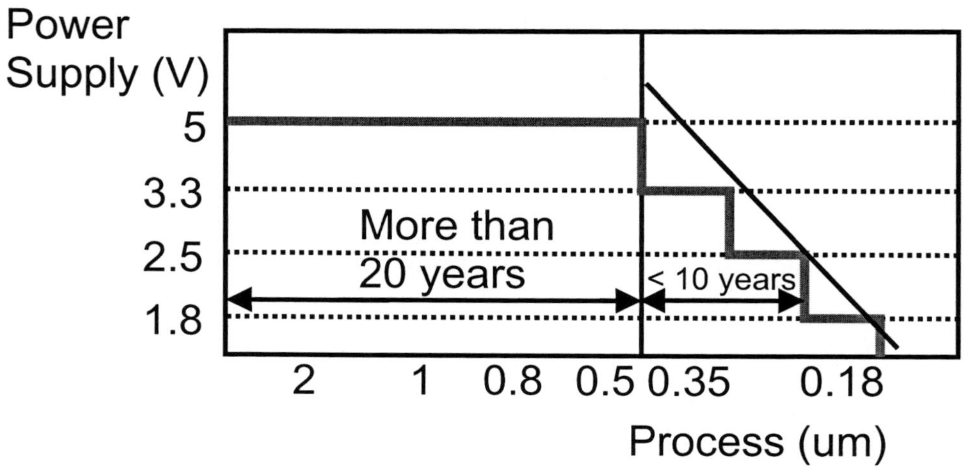

Figure 9: Power Supply scaling as a function of process scaling

Therefore, although the intrinsic matching quality of the technology improves for sub-micron and deep-sub-micron technologies, practical limitations make the theoretical boundary harder to achieve.

At the start of the mismatch analysis, we compared the relative importance of threshold voltage and current factor mismatch. For present-day processes, the impact of the V_T mismatch is clearly dominant. When the scaling trends for A_{Vt} and A_β are compared, it is evident that the β mismatch gains in importance for deeper sub-micron technologies. This trend is confirmed by the decreasing values of the corner gate-overdrive voltage in table 1 for different technologies.

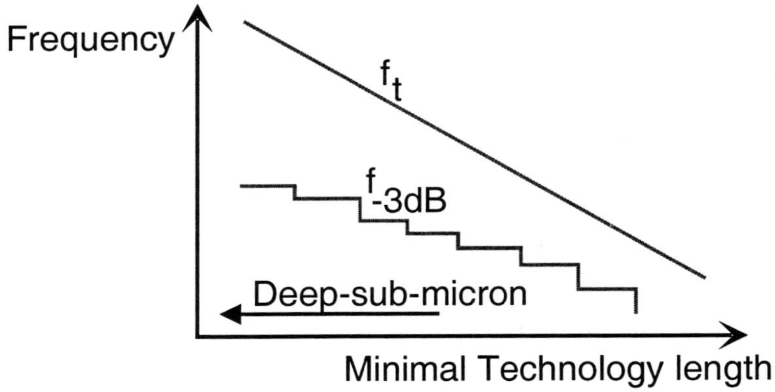

Figure 10: F_t and F_{-3dB} as a function of technology

For some technology in the future the β mismatch will be at least as important and even more important as the V_T mismatch for the calculation of the accuracy of circuits in the whole strong inversion region. At that point, the minimal power consumption for a given speed and accuracy is proportional to $C_{OX}.A_{Vt}.A_{\beta}$; this indicates that a further scaling of the technology would not further improve the performance.

An example will further illustrate the scaling issues which degrade the speed-power-accuracy trade-off in high-speed ADC.

Consider a 6-bit, 500 MSample/s CMOS ADC in two technologies, e.g. 0.5 μm and 0.25 μm CMOS. First, the supply voltages are supposed to be equal, second the mismatch is expected to be dominated by the threshold mismatch and the drain-bulk capacitance is neglected.

Because the two ADC's have the same resolution, following formula can be proven[2]:

$$\frac{A_{Vt1}^2}{W_1L_1} = \frac{A_{Vt2}^2}{W_2L_2} \tag{20}$$

To achieve the same acquisition speed, the regenerative time constant should be the same leading to the next formula:

$$\frac{g_{m1}}{C_{gs1}} = \frac{g_{m2}}{C_{gs2}} \Rightarrow \frac{2I_1}{W_1L_1C_{ox1}.(V_{gst1})} = \frac{2I_2}{W_2L_2C_{ox2}.(V_{gst2})} \tag{21}$$

[2] Index 1 is used for the 0.5 μm technology, index 2 for the 0.25μm technology.

Assuming equal gate-overdrive voltages, the power drain can be compared (A_{Vt} is proportional to the oxide thickness) :

$$\frac{P_1}{P_2}=\frac{I_1}{I_2}=\frac{W_1L_1C_{ox1}}{W_2L_2C_{ox2}}=\frac{A_{Vt1}^2}{A_{Vt2}^2}.\frac{t_{ox2}}{t_{ox1}}=\frac{t_{ox1}}{t_{ox2}} \tag{22}$$

Therefore, to achieve the same speed and accuracy, the power in the down-scaled technology is smaller, because of the improved matching of this technology.

Now, some modifications will be done on these formulas to include the supply-voltage scaling and the relatively increasing importance of the drain-bulk capacitance compared to the gate-oxide capacitance.

Normally the input range of the ADC is made as high as possible. One assumption made then, is that the least significant bit of the converter scales down together with the supply voltage, leading to a smaller allowable mismatch:

$$\frac{A_{Vt1}^2}{W_1L_1}=m^2\frac{A_{Vt2}^2}{W_2L_2}\ \text{with}\ m=\frac{V_{DD1}}{V_{DD2}}>1 \tag{23}$$

The speed formula can be rewritten, now including the drain bulk capacitance.

$$\frac{g_{m1}}{C_{gs1}+C_{gd1}}=\frac{g_{m2}}{C_{gs2}+C_{gd2}}\ \text{or}$$
$$\frac{2I_1}{(W_1L_1C_{ox1}+W_1C_{db1}).(V_{gst1})}=\frac{2I_2}{(W_2L_2C_{ox2}+W_2C_{db2}).(V_{gst2})} \tag{24}$$

Again assuming equal gate-overdrive voltages, the power drain can be compared[3]:

$$\frac{P_1}{P_2}=\frac{I_1.V_{DD1}}{I_2.V_{DD2}}=\frac{(W_1L_1C_{ox1}+W_1C_{db1})}{(W_2L_2C_{ox2}+W_1C_{db2})}.\frac{V_{DD1}}{V_{DD2}}=\frac{W_1L_1C_{ox1}.(1+(C_{db1}/L_1.C_{ox1}))}{W_2L_2C_{ox2}.(1+(C_{db2}/L_2.C_{ox2}))}.\frac{V_{DD1}}{V_{DD2}}$$
$$=\frac{1}{m^2}.\frac{A_{Vt1}^2}{A_{Vt2}^2}.\frac{t_{ox2}}{t_{ox1}}.m.\frac{(1+(C_{db1}/50.\varepsilon_{ox}))}{(1+(C_{db2}/50.\varepsilon_{ox}))}=\frac{1}{m}.\frac{t_{ox1}}{t_{ox2}}.\frac{(1+(C_{db1}/50.\varepsilon_{ox}))}{(1+(C_{db2}/50.\varepsilon_{ox}))}$$

Because ground rules don't scale at the same rate as technology minimal length, the last factor $\frac{(1+(C_{db1}/50.\varepsilon_{ox}))}{(1+(C_{db2}/50.\varepsilon_{ox}))}$ is smaller than 1. This formula shows the relatively increasing power consumption when down-scaling the technology (m > 1).

[3] The typical assumption of [tox=L/50] has been used in this formula.

This trend towards relative increasing power consumption is also shown in next figure.

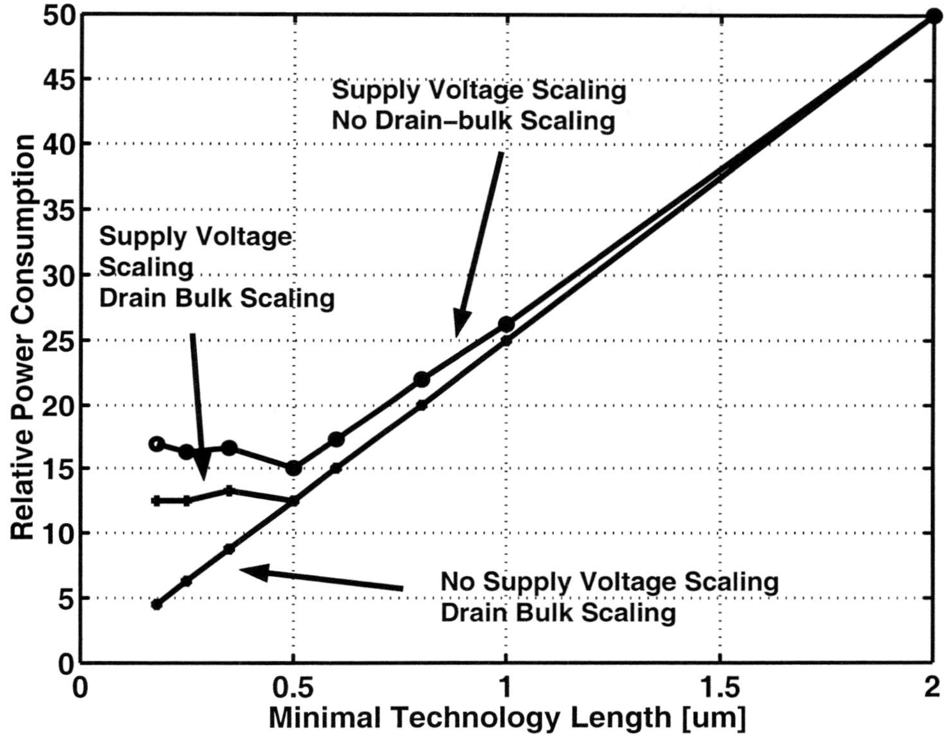

Figure 11: Influence of Voltage scaling and Drain/Bulk Capacitance Scaling on Power consumption ADC

To conclude, the expexted power-decrease is counteracted by the more stringent mismatch demand and the relatively increasing drain-bulk capacitance.

When technology scales further, the β-mismatch becomes dominant leading to following formula

$$\frac{P_1}{P_2} = \frac{I_1.V_{DD1}}{I_2.V_{DD2}} = \frac{1}{m^2}.\frac{A^2_{V\beta 1}}{A^2_{\beta 2}}.\frac{t_{ox2}}{t_{ox1}}.m.\frac{(1+(C_{db1}/50.\varepsilon_{ox}))}{(1+(C_{db2}/50.\varepsilon_{ox}))} \quad (25)$$

$$= \frac{1}{m}.\frac{t_{ox2}}{t_{ox1}}.\frac{(1+(C_{db1}/50.\varepsilon_{ox}))}{(1+(C_{db2}/50.\varepsilon_{ox}))}$$

which makes the case even worse (power goes up!). In this analysis, nothing has been said about the susceptibility of the high-speed A/D converter to substrate noise, power supply and ground noise. These noise sources become relatively more important if the supply voltage scales down.

Averaging is unavoidable

In the previous section, the fundamental trade-off between speed, power and accuracy has been discussed. It has been shown that without other precautions, technology scaling will increase power consumption of high-speed A/D converters in the future. To circumvent this power increase, modifications have to be found.
From a general viewpoint, modifications can be done on three levels: system level, architectural level and technology level.

1. Technological Modifications

Not only analog circuits have problems with the decreasing power supply voltage and mismatch, also digital circuits suffer from the mismatch between identical devices, e.g. offsets in a SRAM cell. Because of the enormous economical impact of digital circuits, maybe more effort will be spent at extensive research to achieve much better mismatch parameters in future technologies. Here, for once digital demands go hand in hand with analog demands. Another technological adaptation is the use of dual oxide processes which can handle the higher supply voltages necessary to achieve the required dynamic range in data converters.

2. System level

Good system level design can substantially decrease the needed performance of the data converter in the system. High level design decisions can have a huge impact on the speed-power-accuracy of the ADC. This high level design needs behavioural models, including power estimators [14].

3. Architectural Level

In this section some possible architecture modifications are presented to break through this trade-off. Two possibilities will be discussed: analog preprocessing techniques and averaging techniques.

Analog pre-processing techniques reduce the input-capacitance of the flash A/D converter and the number of preamplifiers. Examples are interpolating (voltage/current), folding. These techniques do not really improve the speed-power-accuracy trade-off, they only decrease the input capacitance (limiting the highest input frequency) and the number of pre-amplifiers or comparators.
Averaging is a technique which reduces the offset specification for high-speed A/D converters without requiring larger transistors areas. Averaging was first presented in 1990 by [12], where the outputs of the differential bipolar preamplifiers were combined by a resistive network (shown in figure 12). This technique makes a trade-off between the improvement in DNL/INL and the gain of the preamplifier. An improved version of this technique is presented in [13] where the improvement in DNL/INL only depends on the number of stages which contribute the the averaging.

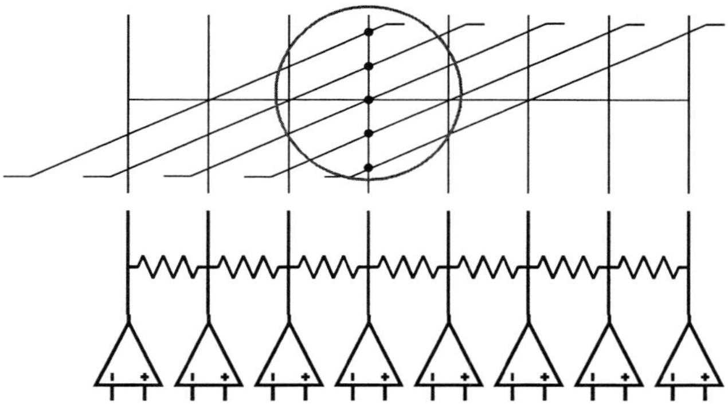

Figure 12: Averaging of preamp-outputs

Averaging can be seen as taking the average value of neighbouring node-voltages and thereby reducing the offset demand. The offset of the averaged value is equal to the original offset devided by the square root of the number of values one has averaged:

$$B_1 = \frac{A_1 + A_2}{2} \Rightarrow \sigma_{B_1} = \frac{\sqrt{\sigma_{A_1}^2 + \sigma_{A_2}^2}}{2} \Rightarrow \sigma_{B_1} = \frac{\sigma_A}{\sqrt{2}} \qquad (26)$$

A modification to this technique, called shifted averaging, was first presented in [15]. This technique eliminates the need for averaging resistors connecting neighbouring stages, but the overall reduction in DNL/INL is fixed.

The same principle as in shifted averaging has been used in [18] where "re-interpolation" is done to reduce the input-referred offset, shown in figure 13.

In pipelined structures, error correction (digital or analog) is

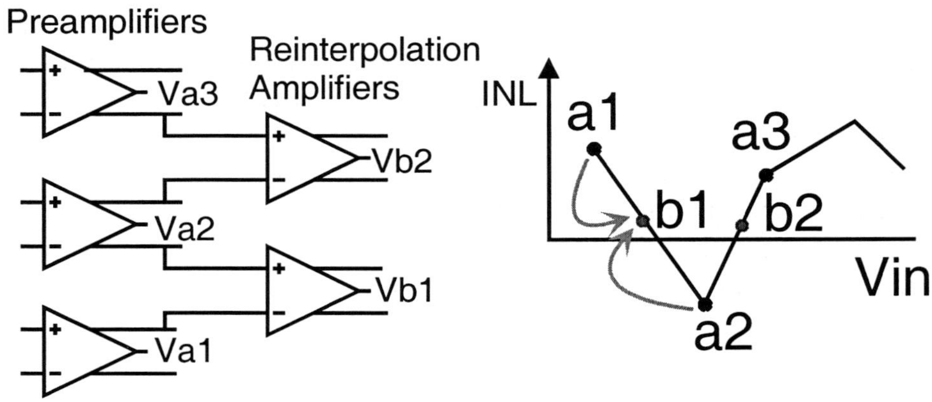

Figure 13: Re-interpolation Architecture and effect on INL

performed to reduce the offset demands on the comparators.

CONCLUSIONS

In this paper an overview of the state-of-the-art high-speed A/D converters has been given. Mismatch models for deep sub-micron technologies have been discussed followed by an analysis of the speed, power and accuracy trade-off in these A/D converters. This speed, power and accuracy trade-off is only dependent on the mismatch specifications of the technology used for the design of the A/D converter. An in-depth analysis on the influence of technology scaling (together with supply voltage scaling) on this trade-off has been made. It is shown that without extra modifications to the design or technology, power consumption will become a problem for future high-speed A/D converters. Some solutions to circumvent this trade-off (and thus lower the power consumption) are discussed and averaging is seen as the only way out of the fundamental trade-off.

Scaling down technology has a good impact on the raw matching properties (e.g. A_{Vt} in Figure 6). In addition, the use of better technology steps, e.g. silicide and multi-metal layers, decrease the parasitics and so increase the achievable speed. On the other hand, devices are getting smaller (so mismatch increases for the same transistor area) and there is only a moderate increase in speed because of the drain-bulk capacitances (Figure 10). So in fact, we do not really need sub-micron *transistors* but we do need sub-micron *technologies*. The better the mismatch of devices is modelled and characterized, the smaller area's the designer can safely use while keeping a high circuit yield; consequently the circuits will consume less power for the specified accuracy and speed. Technology scales so fast that mismatch parameter extraction and mismatch model generation must be generated in much less time. Extrapolating mismatch data from previous processes can substantially differ from the exact data, so non-optimal data converter design is done.

References

[1] Iuri Mehr and Declan Dalton, "A 500 MSample/s 6-Bit Nyquist Rate ADC for Disk Drive Read Channel Applications" , IEEE Journal of Solid-State Circuits , Sept. '99.

[2] M. Flynn and B. Sheahan, "A 400 MSample/s 6b CMOS Folding and Interpolating ADC", ISSCC '98, Feb. 1998.

[3] Sanruko Tsukamoto et al., "A CMOS 6b 400 MSample/s ADC with Error Correction", ISSCC '98, Feb. 1998.

[4] K. Nagaraj et al., "A 700 MSample/s 6b Read Channel A/D converter with 7b Servo Mode", ISSCC '00, Feb. 2000.

[5] K. Sushihara, " A 6b 800 MSample/s CMOS A/D Converter", ISSCC '00, Feb. 2000.

[6] Declan Dalton et al., "A 200-MSPS 6-Bit Flash ADC in 0.6-μm CMOS" , IEEE Journal of Solid State Circuits, Nov. 1998.

[7] R. Roovers and M. Steyaert, "A 6bit, 160mW, 175 MS/s A/D Converter", IEEE Journal of Solid-State Circuits, July '96.

[8] Yuko Tamba, Kazuo Yamakido, "A CMOS 6b 500Msample/s ADC for a Hard Disk Read Channel", ISSCC '99, Feb. 1999.

[9] P. Kinget and M. Steyaert, "Impact of transistor mismatch on the speed-accuracy-power trade-off of analog CMOS circuits", Proceedings CICC, May 1996.
[10] M. Pelgrom et al., "Matching properties of MOS Transistors", IEEE Journal of Solid-State Circuits, vol. 24, no. 5, pp. 1433-1439, Oct. 1989.
[11] J. Bastos et al., "Mismatch characterization of small size MOS Transistors", Proc. IEEE Int. Conf. On Microelectronic Test Structures, vol. 8, pp. 271-276, 1995.
[12] K. Kattmann and J. Barrow, "A Technique for reducing differential non-linearity errors in flash A/D converters", 1991 IEEE ISSCC Dig. Of Tech. Papers, pp. 170-171, Feb. 1991.
[13] K. Bult and A. Buchwald, "An embedded 240mW 10b 50Ms/s CMOS ADC in 1mm^2 ", IEEE Journal of Solid-State Circuits, Vol. 32, pp. 1887-1895, Dec. 1997.
[14] E. Lauwers and G. Gielen, "A power estimation model for high-speed CMOS A/D Converters", Proc. DATE , March 1999.
[15] G. Hoogzaad and R. Roovers, "A 65-mW, 10-bit, 40-Ms/s BICMOS Nyquist ADC in 0.8 mm^2", IEEE Journal of Solid-State Circuits, Dec. 1999.
[16] Q. Huang et al., "The Impact of Scaling Down to Deep Submicron on CMOS RF Circuits", IEEE JSSC, Vol. 33, no. 7, July 1998.
[17] E.A. Vittoz, "Future of Analog in the VLSI Environment", ISCAS 1990, pp. 1372-1375, May 1990.
[18] Yun-Ti Wang and B. Razavi, "An 8-bit, 150-MHz CMOS A/D Converter", Proceedings Custom Integrated Circuits Conference, pp. 117-120, May 1999.
[19] M.J.M. Pelgrom, A.C.J. v. Rens, M. Vertregt and M. Dijkstra, "A 25-Ms/s 8-bit CMOS A/D Converter for Embedded Application", IEEE Journal of Solid-State Circuits, vol. 29, no. 8, Aug. 1994.
[20] W.M.C. Sansen and K.R. Laker, "Design of analog integrated circuits and systems", McGraw-Hill International Editions, 1994.
[21] R. K. Watts, "Sub-micron Integrated Circuits", Wiley-Interscience Pub. – John Wiley & Sons, 1989.

A DUAL MODE 700 MSAMPLES/s 6-BIT, 200MSAMPLES/s 7-BIT A/D CONVERTER IN 0.25 MICRON DIGITAL CMOS

K. Nagaraj, David Martin, Mark Wolfe[1], Ranjan Chattopadyay
Shanthi Pavan, Jason Cancio, and T.R. Viswanathan[1]

Texas Instruments
Warren, NJ and Dallas, TX[1], U.S.A

Abstract

The design of a high speed A/D converter for hard disk drive read channels is described. It has 6-bits of resolution at full speed, as well as a 7-bit mode operating at a lower speed. The 7-bit mode is useful for servo signal processing. This A/D converter has been implemented in a four level metal, single poly 0.25μm CMOS technology. The chip operates at a speed of up to 700 MSamples/s in the 6-bit mode while maintaining an SNDR of greater than 35 dB at input frequencies of up to one fourth the sampling rate. In the 7-bit mode, the device operates at up to 200 MSamples/s with a SNDR greater than 40 dB. It occupies an active area of 0.45 sq. mm and consumes less than 187 mW of power.

1 Introduction

Very high speed, medium resolution A/D converters are an essential part of modern data communication receivers and hard disk drive read channels. With the trend towards the integration of larger systems, it is important to realize such A/D converters in CMOS technologies. Area and power consumption are also important considerations in these applications.

This paper describes a 6-bit CMOS A/D converter that has been

R. J. van de Plassche et al. (eds.), Analog Circuit Design, 25–45.

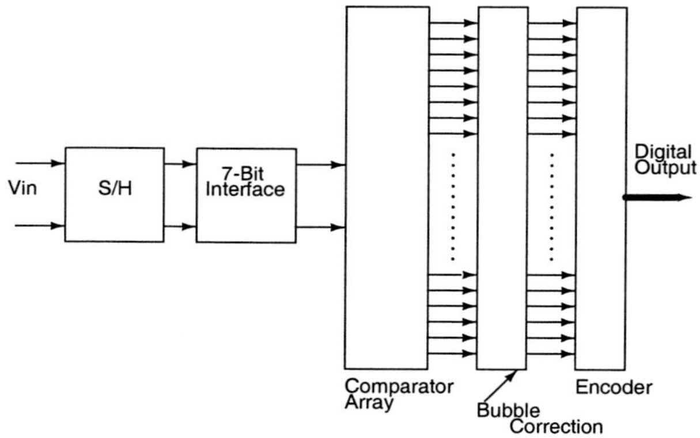

Figure 1: Block schematic of the A/D converter

designed for hard disk drive applications. The design has been very carefully optimized by taking into account system considerations. The prototype exhibits 6-bit performance at a sampling frequency of up to 700 MSamples/s. The converter also has a 7-bit mode working at up to 200 MSamples/s. This mode is useful for servo signal processing.

The top level block schematic of the converter is shown in Figure 1. The input is sampled and held by the sample and hold (S/H) circuit. The output of the S/H is processed by a circuit block called the 7-bit interface which facilitates the operation of the 7-bit mode. The operation of this circuit will be described later. In the 6-bit mode the 7-bit interface behaves like a short circuit. The output from this circuit is fed into the comparator array that converts the input signal into a digital thermometer code. This digital output is connected to a bubble correction logic that converts the thermometer code into a 1 of 64 code. This in turn is fed into a ROM type encoder that generates the final 6-bit digital output.

2 Sample and Hold Circuit

The S/H circuit employs a pseudo-differential architecture made up of two single ended S/H circuits, as shown in Figure 2. The block schematic of each single ended S/H circuit is shown in Figure 3. An

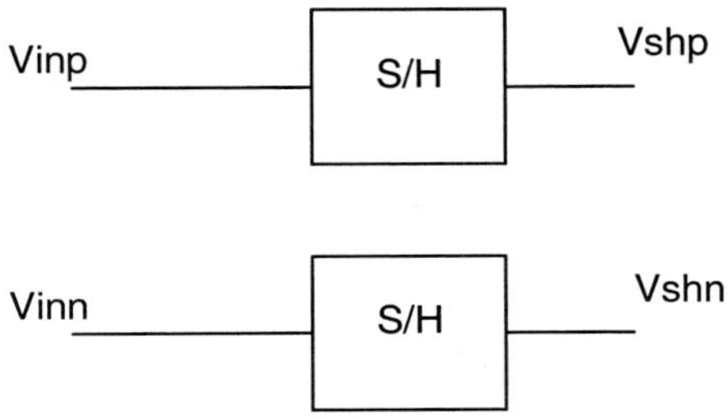

Figure 2: Block schematic of the S/H circuit

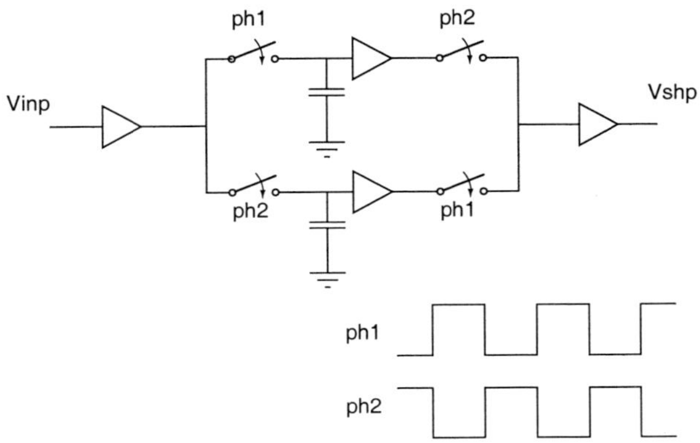

Figure 3: Block schematic of one single-ended S/H circuit path

important feature of this architecture is that it uses two inter-leaved S/H circuits operating at half the sampling frequency. The input signal is first buffered by an input buffer before being fed into the two inter-leaved paths. Each of these paths consists of a sampling switch which is followed by another buffer. The two inter-leaved outputs are re-combined using a set of pass gates. The re-combined signal is fed into a common output buffer that drives the comparator array. The inter-leaving has two advantages. First, the acquisition time available for each S/H is twice that which would be available if a single S/H circuit was used. This makes the design of the S/H circuit more manageable. A second important advantage of inter-leaving is that the final output of the S/H is a 'held' signal for an entire clock interval. This dramatically eases the design of the output buffer that drives the comparator array.

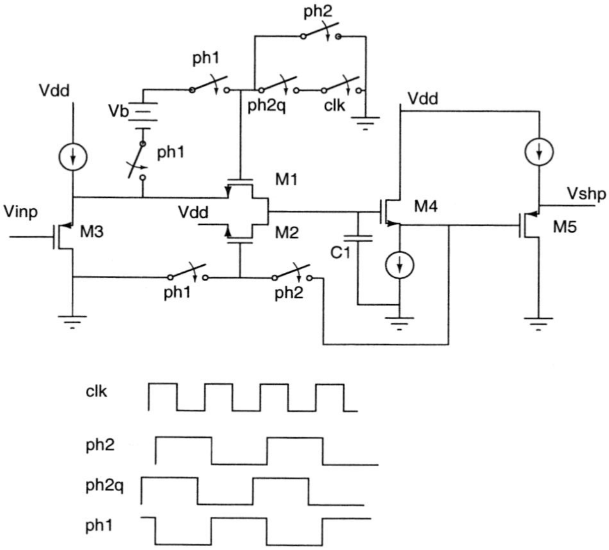

Figure 4: Circuit schematic of an inter-leaved path in the S/H

The details of one inter-leaved path are shown in Figure 4. The source follower M3 constitutes the input buffer and the source follower M5 constitutes the final output buffer. These are common to the two inter-leaved paths. The core S/H circuit consists of the sampling switch M1, the hold capacitor C1 and the source follower M4 which constitutes the internal buffer for the core S/H. Several measures have been taken in this circuit to achieve the required level of performance. First, a constant voltage is applied between the gate and source of M1 during the tracking mode. This ensures that the gate overdrive is independent of the input level, thus eliminating distortion due to signal dependent switch-feedthrough. Another source of signal dependent feed through is the gate-drain capacitance of M1. This is minimized by using the dummy transistor M2 which is identical to M1. When the S/H goes from the tracking mode to the hold mode, the gate of M2 is switched from ground to the output of the S/H. Thus the gate drain capacitances of M1 and M2 experience equal and opposite transitions, canceling their feed-through. Note that M2 is always turned off.

A potential problem with inter-leaving is the mismatch between the

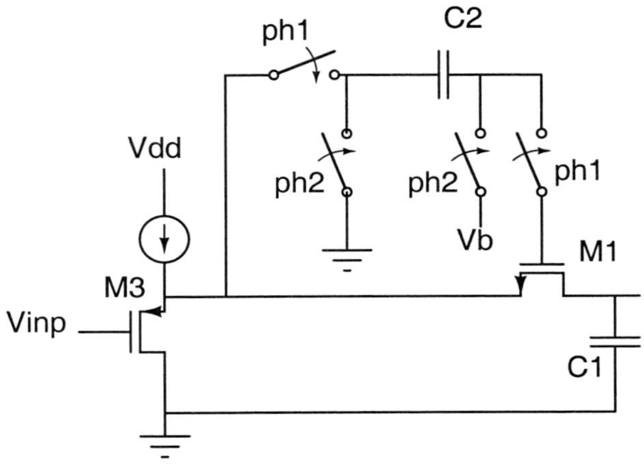

Figure 5: Details of the sampling switch showing the generation of the constant gate over drive

two channels. There are three possible sources of mismatch. Any timing mismatch or gain mismatch results in an intermodulation between the input frequency and half the sampling frequency. Any offset mismatch results in a tone at the half the sampling frequency. Timing mismatch is the most serious among these sources of mismatch. To minimize the effect of this, the circuit of Figure 4 synchronizes the two inter-leaved paths with the master clock. This is achieved by means of the switches connected to the gate of M1. The clock signal ph2q goes high a little before clk goes high, whereas ph2 goes high a little after clk goes high. Thus, as soon as clk goes high the gate of M1 is pulled low, causing the S/H to go into the hold mode. When ph2 goes high it shorts the gate of M1 to ground through a parallel switch. This ensures that the S/H continues to be in the hold mode until ph2 goes low. The other inter-leaved path has the same arrangement except that ph2 and ph2q are replaced by ph1 and ph1q. The instant at which either of the inter-leaved paths goes into the hold mode is synchronized to the low to high transition of clk, eliminating the error due to any mismatches between ph1 and ph2.

The constant gate over drive for M1 is achieved by using a switched-capacitor arrangement, as shown in Figure 5. Here the capacitor C2 is charged to the voltage Vb during the hold phase (ph2 in Figure 3). During the tracking phase ph1, the bottom plate of C2 is connected to the input voltage whereas the top plate is connected to the gate of M1. Thus, the gate voltage of M1 during the tracking mode is equal to Vinp + Vb, making its gate source voltage equal to Vb - Vt where Vt is the threshold voltage of M1.

3 Comparator Design

The overall structure of the comparator array is shown in Figure 6. The comparator array refers to all of the circuitry in Figure 6 except the S/H circuit and encoder.

The output of the S/H circuit is compared against 2^N references, the differences amplified by the preamps and then latched by the latches. The latches take the analog input (now amplified by the preamps) and convert them to ones or zeros. The bubble correction logic eliminates some types of bubbles and converts the thermometer code into a '1 of 64' code. The encoder then encodes this to a 6 bit output.

To reduce the input capacitance of the comparators and to save area and power, the preamps use interpolation to eliminate half of the first stage of preamps.

Although not shown in the figure, the analog signal is differential.

The offset voltages of the preamps are cancelled during a special autozero period. This period lasts approximately 50ns and takes place during idle times that periodically occur (approximately every 100μs) in a hard disk drive read channel.

3.1 First Stage Preamp

The operation of the first stage preamp (P1) is shown in Figure 7. During an autozero period, the reference voltages, produced by a resistor ladder, are connected to one side of the input capacitors while feedback loops are connected around the preamp. For conversion cycles, the feedback around the preamp is opened, and the capacitors are connected to the S/H output instead of the resistor ladder. The voltage stored

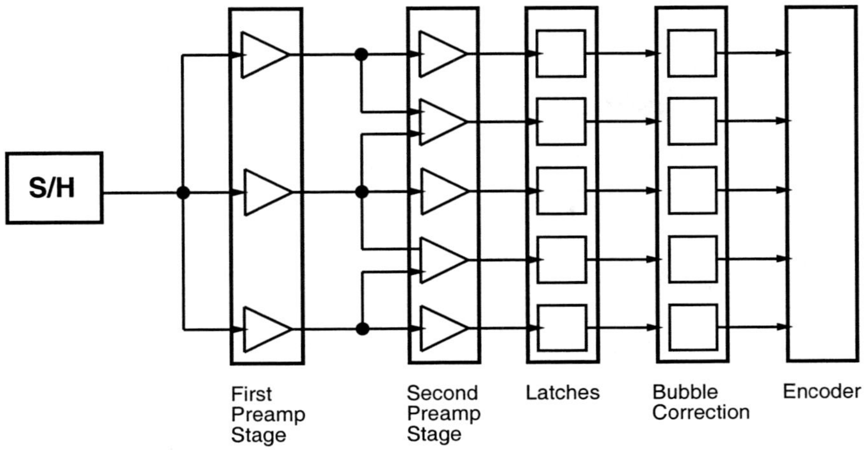

Figure 6: Overall Comparator Structure

on the capacitors is equal to the reference voltage minus the preamp's common mode voltage. This is how the reference voltage is subtracted from the analog input for each comparator. Note that this scheme also cancels the offets of the first stage preamp since the offset voltage is also stored on the capacitors.

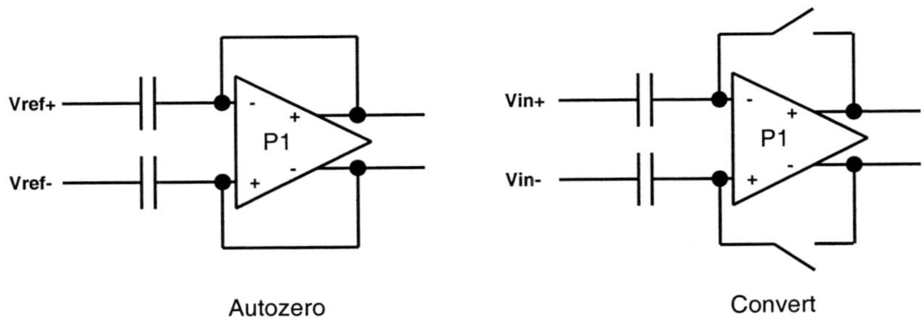

Figure 7: First Stage Preamp Operation

The schematic for P1 is shown in Figure 8. The reset switch is an NMOS transistor. M3 and M4 constitute the input differential pair, while M5 and M6 serve as constant current loads. M1 and M2 serve two purposes. First, they form the tail current source for M3 and M4.

Second, since the gates of M1 and M2 are tied to the outputs, they provide the common mode feedback. If the output common mode goes up, this pulls the common mode of the gates of M1 and M2 up, which causes more current to flow, which tends to bring the output common mode down. The preamp is reset during one half of every clock cycle by turning on the reset switch. As soon as the reset goes low, the amplifier simply 'integrates' the input and the output grows in a linear fashion. This is in contrast to amplifying type preamps that have been used in other high speed A/D converters [1]. The integrating type preamp used above requires a smaller ampunt of power to achieve a desired dynamic gain. It also has a larger DC gain that helps in better offset cancellation.

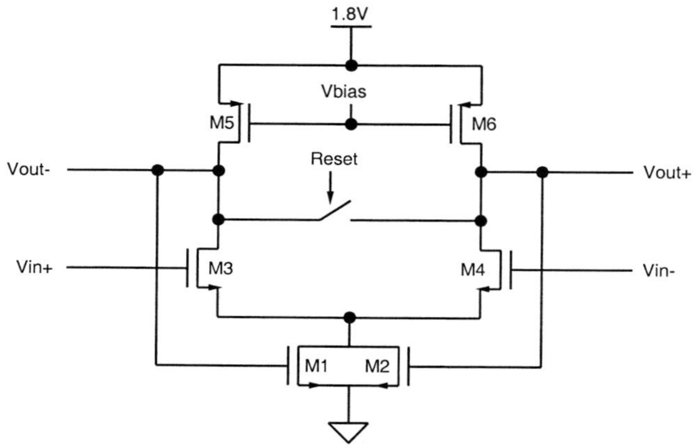

Figure 8: First Stage Preamp Schematic

3.2 Second Stage Preamp

The operation of the second stage preamp (P2) is shown in Figure 9. P2 is autozeroed at the same time as P1. To make sure that the differential input of P2 is zero during autozero, P1 is reset during that period. During this autozero the outputs of P2 are looped back to auxiliary inputs so that the differential output is approximately equal to the offset of P2 referred to the auxiliary input. This voltage is stored on the capacitors C_{hold+} and C_{hold-}.

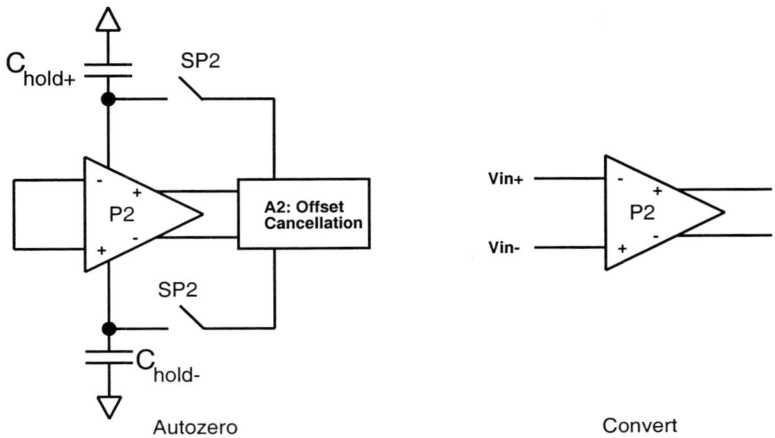

Figure 9: Second Stage Preamp Operation

The schematic for P2 is shown in Figure 10. M1-M2 and M3-M4 serve as the two input differential pairs while M5 and M6 serve as current source loads. M11 and M12 constitute the auxiliary input pair for offset cancellation. M7, M8, M9, and M10 serve the same purpose as M1 and M2 in the first stage preamp.

M14 and M15 serve as hold capacitors for the gate voltages of M12 and M11. This is the C_{hold} of Figure 9 where the offset of the preamp is stored. The switches S1 and S2 connect the outputs to the gates of M12 and M11. This feedback around the differential pair then acts to cancel the preamp's offset. S1 and S2 are formed using NMOS transistors. They are the $SP2$ of Figure 9.

Since P2 may have inputs from two different first stage preamps, it has two sets of differential inputs and thus four input transistors. The drains of the input transistors are connected so that the positive inputs steer current into one leg of the preamp, and the negative inputs into the other leg. Note that the input transistors have two different current sources rather than one. The reason for this is that it is possible that two first stage preamps might have slightly different common mode output voltages due to mismatches. If the second stage preamp's input transistors all shared the same current source, then the two input transistors connected to the first stage preamp with the higher common mode output voltage would tend to use more current than the other two input transistors. In that case, the second stage preamp would no longer be performing an exact interpolation between the two first stage preamps. Rather, one first stage preamp would dominate the result. This problem is avoided by using two current sources for the input transistors of the second stage preamp.

The timing diagram for the operation of the pre-amplifiers during conversion is shown in Figure 11.

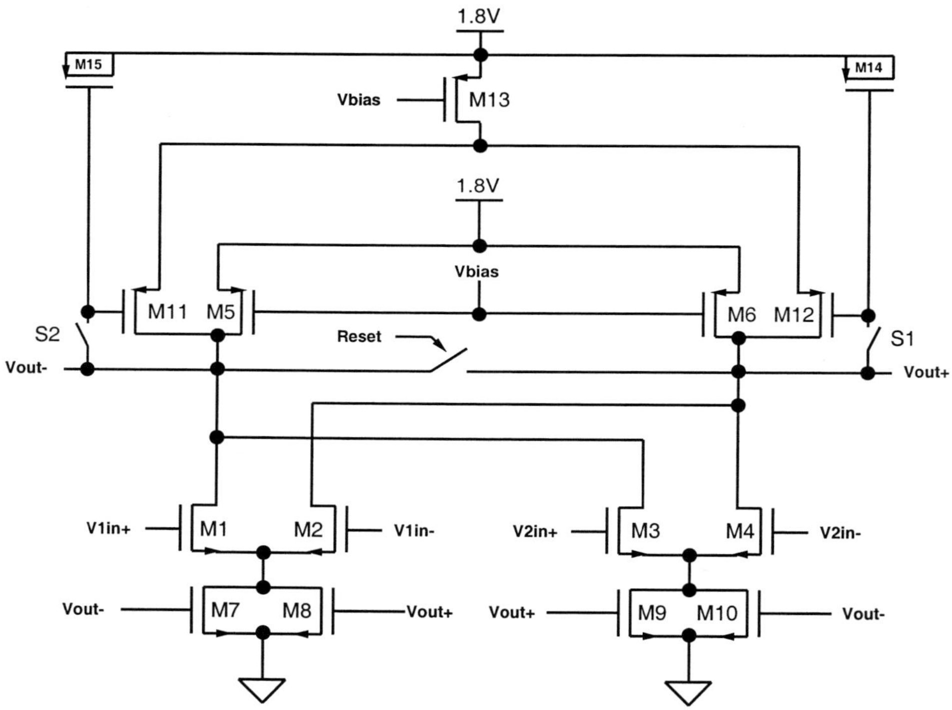

Figure 10: Second Stage Preamp Schematic

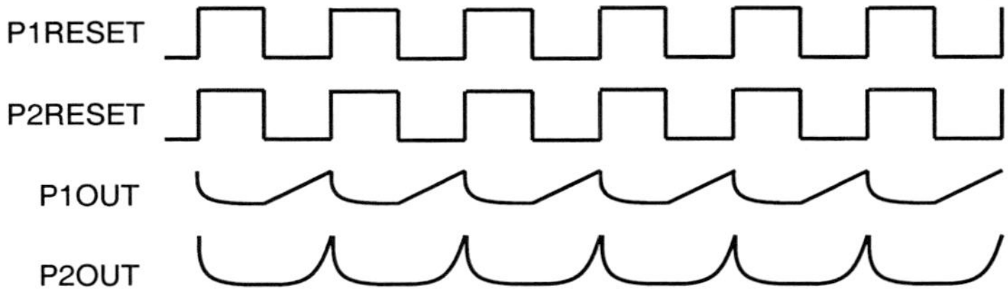

Figure 11: Timing diagram of the operation of the pre-amplifiers

3.3 Autozero Timing

As seen above, for proper offset cancellation the autozero controls to the two preamplifiers need to follow a particular sequence. The clock signals for an autozero are shown in Figure 12. First, AZ, $CANCEL1$, and $CANCEL2$ all go high. During the first part of the autozero cycle (when $CANCEL2$ is high) the second stage preamps cancel their offset. Then, halfway through an autozero cycle, $CANCEL2$ goes low, and then, $P1RESET$ goes low. This removes the reset on the first stage preamps and allows them to autozero themselves. When this is completed, $CANCEL1$ goes low. Finally, AZ goes low, and conversion cycles can begin.

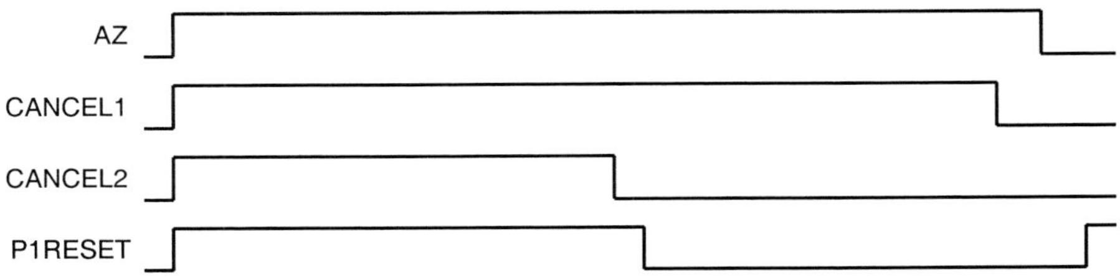

Figure 12: Clock signals during an autozero

3.4 Latch

The schematic of the latch is shown in Figure 13. It consists of two cross coupled inverters whose power supplies are turned on by a $STROBE$ signal, shown as S in Figure 13. It has pass gates so that it can disconnect itself from the preamp while latching, and it has reset switches so that it can reset itself before being connected back to the preamp.

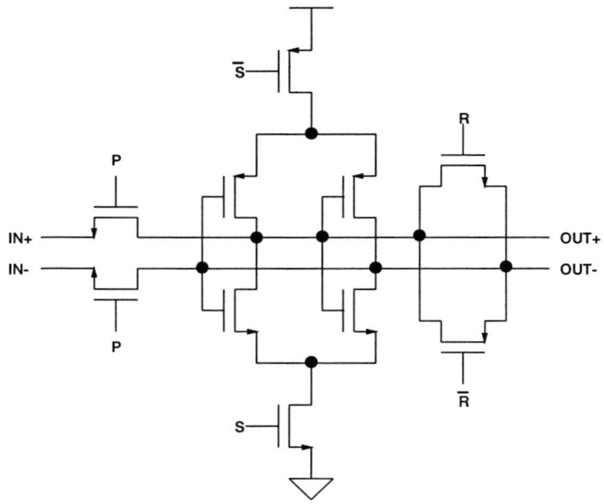

Figure 13: Latch Schematic

3.5 Bubble Logic

The bubble logic uses a majority logic circuit, as shown in Figure 14. $D(n)$ is the digital output of comparator n. $Dout(n)$ goes to the encoder.

$CKDIG$ is high during the first part of a clock cycle. While $CKDIG$ is high, $Dout(n)$ goes low. When $CKDIG$ goes low, $Dout(n)$ will go high only if two of $D(n-1)$, $D(n)$, and $D(n+1)$ are high, and two of $D(n)$, $D(n+1)$, and $D(n+2)$ are low. This will normally happen if $D(n)$ is the top of the comparators whose output is one (called the thermometer code). The logic also eliminates some types of bubbles in the thermometer code.

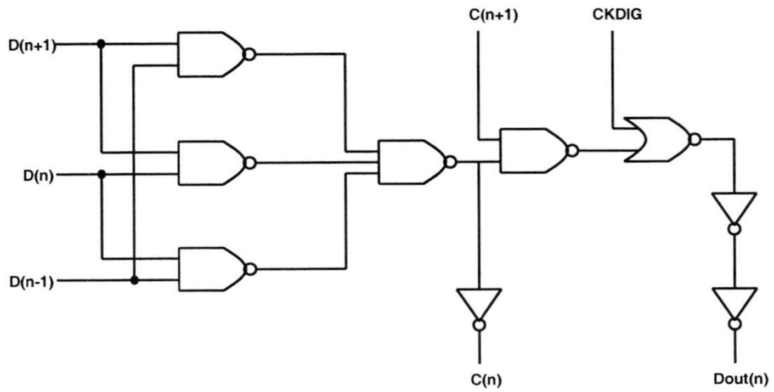

Figure 14: Bubble Logic

4 Reference Generator

A block schematic for the reference voltage generator for the comparator array is shown in Figure 15. It receives two inputs; a band gap referenced voltage V_{BG} and an input common mode reference voltage VCMI. The function of the reference generator is to impose the reference voltages Vrefp and Vrefn across the two resistor ladders such that the difference between Vrefp and Vrefn equals half the required full scale reference voltage, and their common mode level is equal to the common mode output level of the S/H circuit. This common mode level is dependent on the gate-source voltage drops of the source followers in the S/H, and is thus process and temperature dependent. To overcome this problem the reference generator uses the replica circuit SHREF that accepts VCMI as an input and generates a signal Vcmr which is equal to the common mode level of the S/H output. The circuit SHREF consists of three sources followers that are scaled versions of those in the main S/H circuit. From Vcmr, the voltage Vrefp is derived by means of the amplifier A2 and resistor R1. By setting R1 equal to the total resistance of the ladder, we can ensure that Vrep = Vcmr + (Iref)(R1), which is the correct value required at the top of the resistor ladder. This voltage is applied to the top of the ladder. Simultaneously a current sink equal to Iref is attached to the bottom of

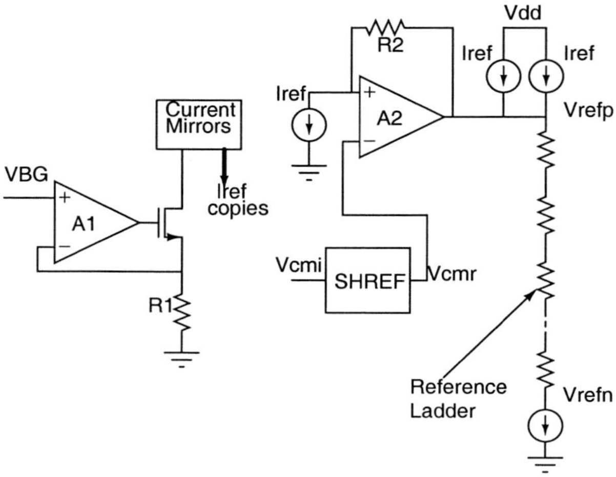

Figure 15: Block schematic of the reference generator

the ladder. This ensures that voltage across the ladder has the correct value. Note that two current sources equal to Iref are connected to the top of the resistor ladder. Thus, the amplifier A2 is not required to sink or source any significant amount of current. The current Iref is derived from Vref using a V to I converter made up of A1, M1 and R1. The current output from this circuit is fed into a current mirror that generates the current sources and sinks that are required by the other parts of the reference generator.

5 Operation of the 7-bit mode

7-bit resolution is required in a read channel for the processing of servo data. Because servo data has a significantly lower data rate than normal data, the 7-bit mode can operate at a slower rate. Taking advantage of this, 7-bit operation is achieved here using a two step technique [2]. This is illustrated in Figure 16. During the first step, a 6-bit A/D conversion is performed with an analog voltage equal to 1/2 LSB at the 6-bit level added to the output of the S/H. The LSB (b1 in Figure 7) from this operation is stored in a 1-bit memory element D. During the second step a 6-bit A/D conversion is performed without the 1/2

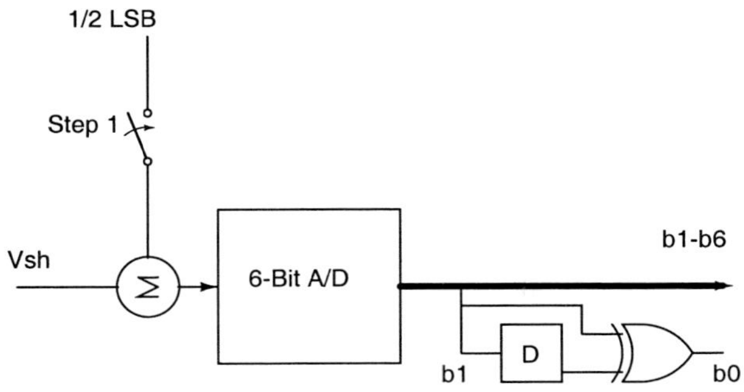

Figure 16: Principle of operation of the 7-bit mode

LSB added to the input. If the input were to lie in the upper half of a 6-bit LSB interval, adding the 1/2 LSB would push the result of the conversion into the next higher digital code. Thus, the results from the first and second steps would be different. This condition is detected by means of an exclusive-OR operation between the 6th bits from the two steps. The output of the exclusive-OR gate gives the 7th bit. An important advantage of the above technique is that the digital bits from the first step are not required for the second step. This is in contrast to conventional two step architectures where the digital bits from the first step are required for the second step, thus limiting their speed of operation.

The addition of the 1/2 LSB voltage is achieved by the 7-bit interface circuit shown in Figure 8 (a single ended equivalent is shown; the actual circuit is differential). A resistor equal to one half of each element in the ladder is inserted in series with the S/H output. A current Iref equal to the current in the reference ladder is applied to this resistor during the first step, resulting in the required 1/2 LSB offset. Note that during both the steps the current outputs ultimately flow into the sources of the buffer in the S/H circuit ensuring that the output of the S/H itself does not change from the first step to the second.

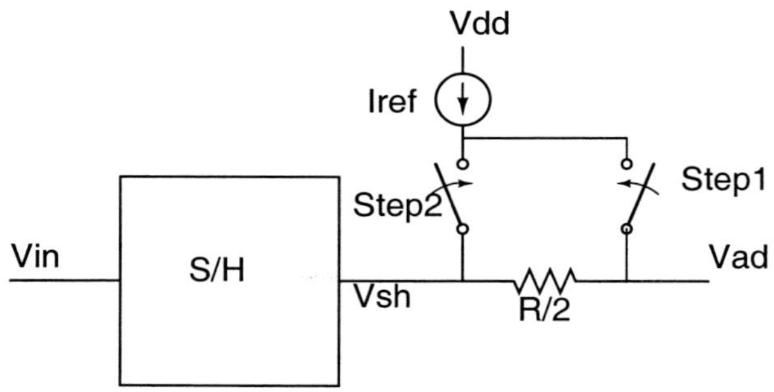

Figure 17: Circuit schematic of the analog 1/2 LSB adder

6 Output Interface

Driving the digital outputs is a problem in very high speed A/D converters because of the large currents required to charge and discharge the load capacitances. This can lead to a large bounce on the supply and ground leads. To minimize this, current steering type output buffers have been used in this device. This ensures that the total current drawn by the output buffers is constant. Further, the updating of the outputs is controlled by an external clock that can be a sub multiple of the main clock. By a proper choice of the relationship between the input frequency and the main clock frequency, we can obtain the necessary spectral information even from such undersampled outputs.

7 Results

The A/D converter was fabricated in a four level metal, single poly, 0.25μm, digital CMOS process. A photomicrograph of the chip is shown in Figure 23. Measured performance is summarized in Table 1.

In the 6 bit mode, at 700MSamples/s, with 3.3V and 1.8V supplies, the A/D converter consumes 187mW of power. Figure 18 shows the output spectrum of the A/D converter output in the 6 bit mode; the SNDR is 35.2dB for Fin=136MHz, Fs=700MSamples/s. At Fin=247MHz, Fs=500MSamples/s, and Vin=0.6 of fullscale, the measured SNDR is 31.8dB. Measured differential non-linearity (DNL) and

CMOS Technology	1-poly, 4-metal, .25μm	
Supply Voltages	3.3V, 1.8V	
Input Range	1.0V p-p	
A/D converter Area	.45 mm^2	
Resolution	6 bit	7 bit
Conversion Rate	700MS/s	200MS/s
Power Consumption	187mW	143mW
DNL	$<$ 0.4 LSB	$<$ 0.4 LSB
INL	$<$ 0.4 LSB	$<$ 1.0 LSB
SNDR	35.2dB	41dB
Fin for SNDR measurement	136MHz	53MHz
Tone at Fin +/− Fs/2	-55 dB at Fin=136 MHz, Fs=600MHz	

Table 1: A/D converter Performance Summary

integral non-linearity (INL) for the 6 bit mode are shown in Figure 19. In the 7 bit mode, at 200MSamples/s, with 3.3V and 1.8V supplies, the A/D converter consumes 143mW of power. Figure 18 shows the output spectrum of the A/D converter output in the 7 bit mode; the SNDR is 40.66dB for Fin=53MHz, Fs=200Msamples/s. Measured DNL and INL for the 7 bit mode are shown in Figure 21.

To measure the intermodulation distortion due to the inter-leaving in the S/H, the S/H output was measured directly through a special test port that uses source follower buffers. The distortion due to interleaving is 56dB below the fundamental, as shown in the measured spectrum in Figure 22.

Acknowledgements :

The authors would like to thank Mark Chambers, Anthony Brewster, Vic Pierotti, Mark Spaeth, Mark Peng and Brian Liebowitz for technical contributions, Mark Barnett for assistance with the layout and Kris Kistner for assistance with the test board.

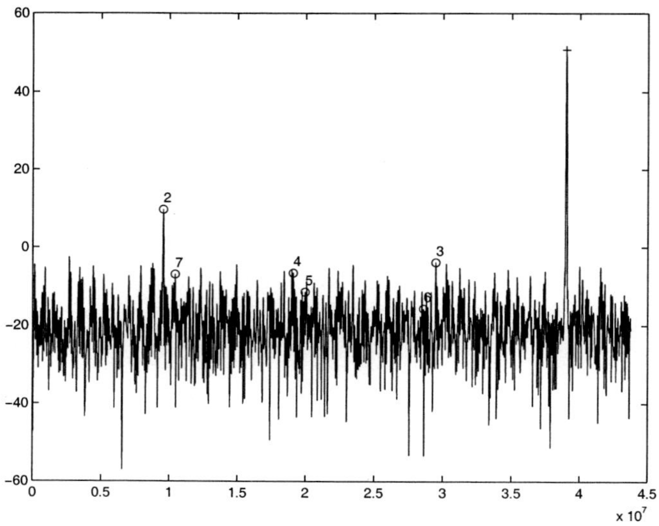

Figure 18: Output spectrum (in dB) for the A/D converter, 6 bit mode. Fin=136MHz, Fs=700MS/s, undersample ratio=8

References

[1] I. Mehr and D. Dalton, "A 500-MSamples/s, 6-bit Nyquitst rate ADC for disk drive read-channel applications", IEEE Journal of Solid State Circuits, Vol.34, pp 912-919, July 1999

[2] K. Nagaraj, "2 1/2 step flash A/D converter", Electronics Letters, Vol.28, pp 1975-1976, October 1992.

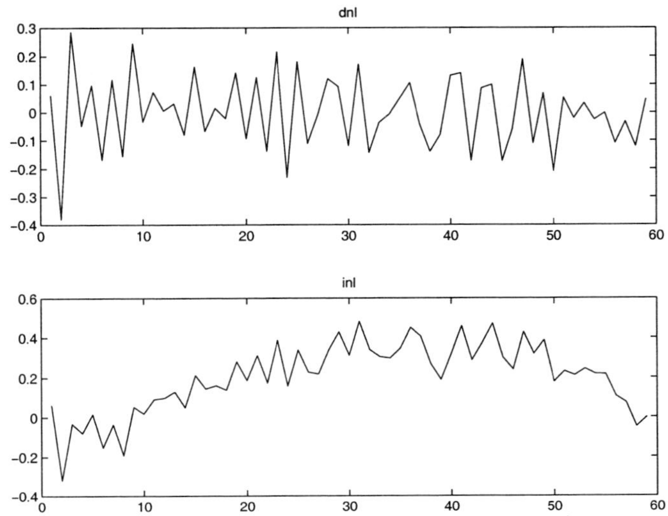

Figure 19: DNL and INL (in LSB) for the A/D converter, 6 bit mode. Fs=700MS/s. x-axis shows the output code

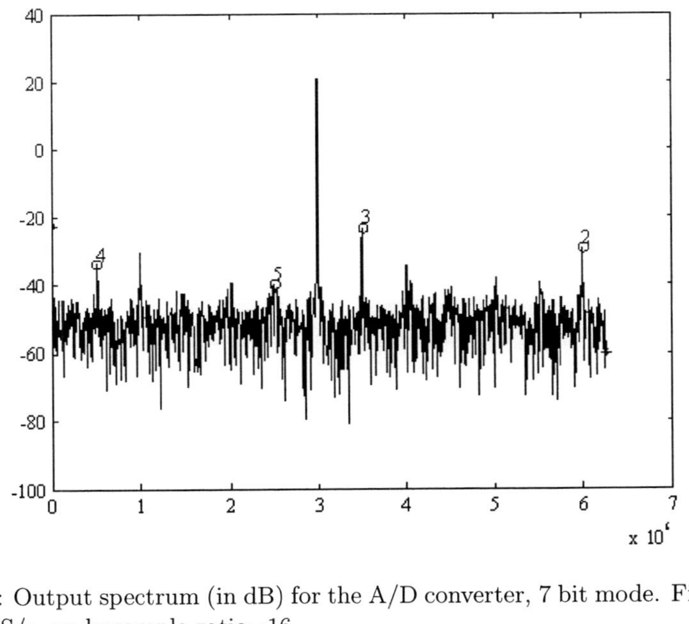

Figure 20: Output spectrum (in dB) for the A/D converter, 7 bit mode. Fin=53MHz, Fs=200MS/s, undersample ratio=16

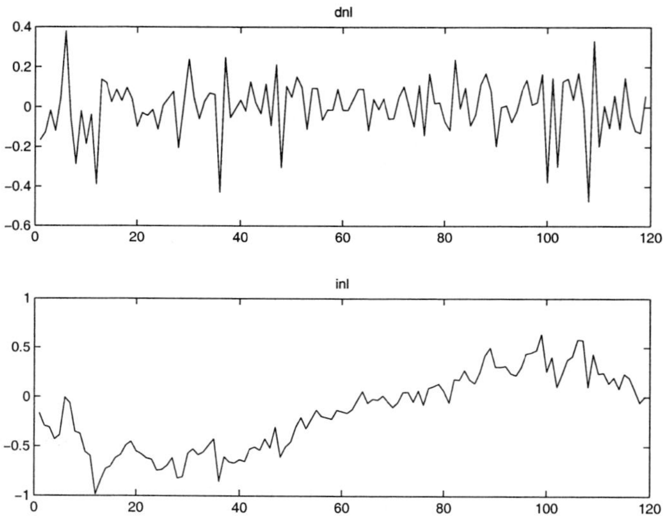

Figure 21: DNL and INL (in LSB) for the A/D converter, 7 bit mode. Fs=200MS/s. x-axis shows the output code

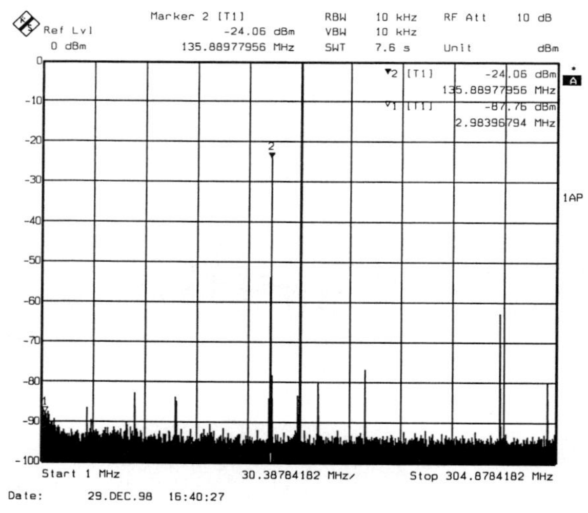

Figure 22: Intermodulation Distortion (in dB). Fs=600MS/s, Fin=136MHz, Intermodulation Tone is at 164MHz

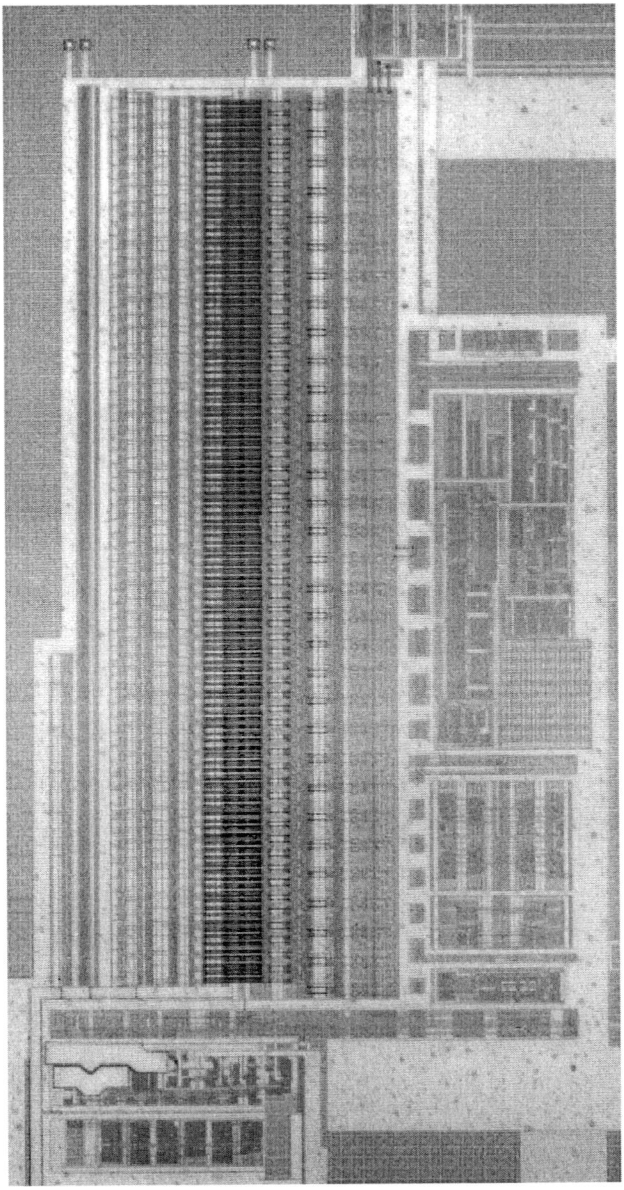

Figure 23: Photomicrograph of the A/D converter

A 3.3-V 12b 50-MS/s A/D Converter in 0.6-μm CMOS with over 80-dB SFDR

Hui Pan, Masahiro Segami, Michael Choi, Jing Cao, Fumitoshi Hatori, and Asad A. Abidi

Integrated Circuits & Systems Laboratory
Electrical Engineering Department
University of California
Los Angeles, CA 90095-1594

ABSTRACT

This paper discusses the impact of SFDR specification on the design of A/D converter (ADC) in CMOS technology and describes the implementation of a prototype optimized for wideband SFDR performance for use in modern wireless base stations. The 6b-7b two-stage pipelined ADC using bootstrapping to linearize the sampling switch of on-chip track-hold achieves over 80 dB SFDR for signal frequencies up to 75 MHz at 50 MS/s without the need for trimming, calibration and dithering. INL is 1.3LSB, DNL is 0.8LSB. The 6b and 7b sub-ADC's are made efficient with averaging and folding. In 0.6μm CMOS, the 16mm^{2}ADC dissipates 850mW from 3.3V supply.

I. INTRODUCTION

Modern wireless base stations digitize the entire received band, and then separate individual channels with digital filters [1] (Figure 1). Digitizing at IF poses a challenge on the design of the A/D converter (ADC) for two reasons. First, the spurious free dynamic range (SFDR) specification becomes paramount; it must be over 80 dB. For this application, SFDR is defined as the difference between the full scale (FS) fundamental and the maximum spurious tone in decibel (dB) in an ADC output spectrum. It is the spur, not the noise, that limits the system sensitivity. The signal-to-noise ratio (SNR) requirement is much relaxed because the noise is divided into many channels bundled in the IF band. For example, the SNR can be 20 dB lower than the SFDR, given 100 channels. Second, the conversion rate must be on the order of 50 MS/s to accommodate the wideband IF, which is usually over 20 MHz. It is also desirable to extend the SFDR performance beyond Nyquist input frequency to give more freedom in the frequency planning of IF. This requires a very good track-and-hold (T/H) on chip. So far only bipolar and BiCMOS ADC's have barely met these specifications [3] [4] [5].

This paper addresses the impact of the SFDR specification to both architecture and circuit design, and describes a low-Vdd CMOS ADC capable of over 80 dB SFDR for input frequencies up to 75 MHz at 50 MS/s without the need for trimming, calibration and dithering. The prototype is implemented in a 0.6μm 3M1P p-epi on p$^+$ process and operates from a 3.3V supply. The paper starts with the analysis on the SFDR of ideal quantizers and finds out the key to high SFDR in Section II, The key is then applied to the architecture design in Section III. Section IV and Section V discuss how to make the ADC efficient (compact and low power). Section VI is

R. J. van de Plassche et al. (eds.), Analog Circuit Design, 47–73.

dedicated to the T/H design and Section VII describes each building blocks at the circuit level. Design methodology played an important role in the success on first silicon and is briefly covered in Section VIII. The experimental results are presented in Section VIII, followed by conclusions in Section IX.

II. FUNDAMENTALS ON SFDR

A. Ideal quantizer

It is well known that the SNDR expression for a full-scale (FS) sinewave quantized by an ideal *n*-bit uniform midriser quantizer is 6*n* +1.76 (dB) [6] [7]. In contrast, the SFDR expression is not quite popular. It can be shown, by applying Fourier series expansion on the quantized sinewave or on the sawtooth quantization error characteristics [8] [9], that the harmonics peaks at $2^n\pi$ fin with value about -(9*n* - 6) dBc, that is [10] [11],

$$\mathbf{SFDR} \approx 9\mathbf{n} - 6 \quad \text{(dB)}, \tag{1}$$

and the low-order harmonics are 9*n* dB below the fundamental. The 9-dB-per-bit improvement can be easily understood from energy conservation. As *n* increases by one, the total quantization error energy $LSB^2/12$ is reduced by 6 dB, asymptotically independent of signal distribution. In addition, the number of harmonics doubles due to the doubled segmentation of the sawtooth error characteristic, and the spur level must be down by additional 3 dB to keep the total harmonic energy unchanged.

A/D conversion consists of sampling and quantization. For ideal ADC, sampling can be considered after quantization. Due to aliasing all the high-order harmonics from quantization appear as spur in the Nyquist band. Based on the intuitive insight into the 9dB/bit of SFDR, the key to high SFDR is to spread a given error energy over harmonics as many as possible so that the overall harmonic level can be reduced. In terms of the error transfer characteristic, this corresponds to randomizing it by either increasing the segmentation or dynamically perturbing the segmentation or anything to this effect such as dithering [12], [13]and dynamic element matching (DEM) [14], [15]. Clearly, 9n - 6 dB is not the fundamental limitation of SFDR for an *n*-bit quantizer due to the possible enhancement by dithering from the inherent thermal noise of the system. Note noise shaping technique does not apply.

Deviation from the ideal case by a fraction of LSB can cause drastic change in the amplitude of individual harmonics of the quantized sinewave. Ideally, the spectrum contains no even harmonics. However, if the quantizer input offsets by as little as one quarter LSB (assume the amplitude is reduced enough to avoid overloading the quantizer), even harmonics become comparable to the odd ones, because the offset equivalently breaks the odd-symmetry in the error characteristic. Since the emerging even harmonics share at most half energy with the odd harmonics, overall harmonics level is expected to drop by at most 3 dB. The low-order harmonics correspond to the slowly varying portion of the error waveform arising from quantizing the region around the zero-slope maxima. As the input amplitude starts reducing from full scale, this region and error shrink, and at some point, certain low-order harmonics disappear. Therefore, low-order harmonics are very sensitive to tiny amplitude variation. The peak harmonic corresponds to the fundamental of the sawtooth-like portion of the error waveform arising from quantizing the zero-crossing region within which sin(*x*) deviates *x* by less than quarter LSB, i.e., $|x - \sin(x)| \approx |x^3/6| < LSB/4$. Beyond this region the sawtooth is stretched out-of-phase and does not contribute to the peak harmonic. Indeed, the asymptotic frequency of the sawtooth is $2^n\pi$ fin and this portion, which occupies $1.44/(2^{n/3})$ FS, dominates other portions at lower frequencies. Obviously, tiny amplitude change has little impact on the maximum tone. Variation in the phase of input sinewave has no effect on the harmonic amplitude. In conclusion, in spite of the sensitivity of individual tones to input variation, the SFDR performance of an ideal quantizer is still robust and Eq. (1) is a very good approximation.

It should be noted that SFDR of ideal quantizer is extremely sensitive to noise or dithering at the input. When the noise (rms) exceeds quarter LSB, the error waveform becomes almost random and the spurs virtually disappear.

B. Random threshold offsets

If the quantization thresholds within the zero-crossing region, $|x| < 1.44/(2^{n/3})$ FS, have uncorrelated offsets on the order of a quarter LSB, the periodicity of this sawtooth portion significantly diminishes, and so does the corresponding peak harmonic at $2^n\pi$ fin. With this peak removed, SFDR is limited by the low-order -9n dB harmonics, 6-dB improvement on SFDR is possible.

It is interesting to notice that the effect of random offsets on the lower-order harmonics is opposite to that on the high-order peak. If the thresholds beyond the zero-crossing region have random offsets of quarter LSB, some lower-order harmonics may stick out above -9n dB, and SFDR is degraded. The low-frequency portion of the error waveform exhibits little local periodicity within one period of the sinewave, though the periodicity appears globally across many periods of input sinwave. The random offsets have much less periodicity to break of the low-frequency portions but alter the size more than that of the high-frequency portion. Some portions expand, while others shrink. The enlarged portions correspond to higher energy, and therefore, higher harmonics.

Even though the random offsets boost some low-order harmonics, their effect is very limited. In reality, it is always the systematic INL that determines the SFDR performance. The basic reason for this phenomenon is that the concentration of error energy on certain harmonics in the frequency domain arises from waveform regularity in time domain. (The extreme case is a single tone corresponding to a perfect sinewave.) Regularity in systematic INL causes regularity in the error waveform. Systematic INL usually has to do with quantizer architecture which transforms the inaccuracy of each device into certain structured INL. Therefore, architecture must be designed carefully if SFDR performance is important. As for the effect of the architectural INL on SFDR, it turned out that the basic principles derived for ideal quantizer still hold and even Eq. (1) applies directly, as is to be shown in the following three subsections.

C. Interstage gain error

From Eq. (1), 85 dB SFDR requires about 11 ~ 12b quantizer. Given the CMOS comparator (with preamp) offset on the order of a few mV [19], [20] and available input full scale on the order of 2V, 12b resolution can not be realized without some technique either to effectively reduce the offset or effectively amplify the input full scale. Residue amplification [21], which effectively amplify the FS without running into headroom problem, is a common choice for CMOS ADC to overcome the excessive offsets. It is therefore of interest to study the SFDR performance of residue amplification architectures.

Residue gain error of a two-stage sub-range architecture contributes a sawtooth error transfer characteristic, with the number of segments equal to 2^{n1}, where n1 is the bit number of the first stage. For each bit increment in the first stage, a 9 dB improvement can be expected in SFDR, if the residue gain error dominates SFDR, and the max spur locates at $2^{n1}\pi$. Simulations with an ADC programmed in Matlab show 2~3dB-per-1st-stage-bit improvement for $n1 > 2$, given the same SNDR (Table 1). Apparently, by increasing n1, the accuracy bottleneck in residue gain can be removed.

Dithering suppresses spurs from the gain error, but suffers SNR degradation. Dithering is realized by intentionally varying the quantization thresholds, or effectively by injecting noise at the input. To remove the regularity in INL, the rms of dithering noise must be comparable to the period of the error characteristic, i.e., the LSB of the first stage. Obviously, such high level of input dither has to be applied out-of-band [2]. If the thresholds of the first-stage sub-ADC are directly dithered by about one LSB [13], the maximum output error contributed by the gain error is doubled and hence increase the error energy by 3 ~ 6 dB. It is noted that the so called "digital correction" corrects the threshold error, but not the gain error. Threshold dithering amplifies the effect of gain error.

D. DAC nonlinearity

Reconstruction DAC is an integral part of residue amplification architectures. The error characteristic contributed by reconstruction DAC comprises 2^{n1} flat segments between the reference levels (or the taps). The segments shift up and down around zero as a result of the nonlinearity. For each additional bit in the first stage, the spurs from the random nonlinearity are spread out one more time due to doubled segmentation, and SFDR improves by 3 dB, which is verified by Monte-Carlo simulations with the programmed ADC. The correlated error from tap to tap is very harmful, since it gives rise to strong spurs that can not be spread out by more segmentation. Therefore, the DAC INL is usually the bottleneck in SFDR, and careful layout is essential in avoiding systematic mismatch in DAC elements.

Dithering is not effective in suppressing the spurs dominated by DAC nonlinearity. It causes the output error to toggle between the errors in the taps adjacent to the input sample. This effectively smooth out the DAC INL, but the low frequency components remain which are the major contributor to the dominant spurs. It is much more effective to dynamically perturb the level of each error segment using DEM, which centers the average level of each error segment at zero [15], [16], [17], [18].

E. Sub-ADC INL

The INL of the first sub-ADC does not cause error at the ADC output so long it is kept within +/- 0.5 LSB of the first stage for 1b over-range in the second sub-ADC. The INL of the last stage sub-ADC directly contributes to the output error. The overall error characteristic consists of 2^{n1} segments of variable portion of the last stage INL profile. The segmentation and the variation in the segmentation caused by the first stage INL tend to improve the SFDR. The 9-dB-per-bit improvement on SFDR also holds.

F. Some subtleties

Aliasing drastically affects the maximum spur level at coherent sampling (i.e. sampling frequency fs = multiple of input frequency fin) when all the harmonics are folded back and concentrate on a few low-order harmonics. This situation is avoided in practice by careful frequency planning. Aliasing and windowing effect are not taken into account in the definition of ADC SFDR. In stead, they are up to the system designers to consider. Due to the extra 3-dB-per-bit improvement, the sub-ADC should not be the bottleneck in SFDR with 4 or 5 bits in the first stage.

Unlike the error energy, the maximum spur is sensitive to the input waveform. Full-scale sinewave input is chosen for definition of SFDR because it corresponds to the worst case in-band interference. Multi-tone interferers cause less maximum spur from ideal uniform quantization. This is true because the former exhibits more regularity than the later which corresponds to the former modulated in amplitude. The regularity in the input waveform causes more regularity in the quantization waveform. For example, if the input waveform is full-scale sawtooth, the error waveform is also a perfect sawtooth, and all the error energy are concentrated on the 2^n-th and its multiple harmonics. The maximum spur is much higher than -(9n - 6) dBFS, even the quantizer is ideal. In real quantizer, the INL may happen to boost the maximum spur from certain multi-tone interference, and this necessitates multi-tone test. Those uncertainties must be considered carefully at the system level where the ADC is used.

III. ARCHITECTURE DESIGN

A. *1.5b/stage*

The most popular architecture for high speed and high resolution (10b and above) ADC's in CMOS is pipelined 1.5b/stage [21], [22]. Only one additional scaled stage is needed to resolve each additional bit. This makes the architecture very efficient at high resolution where the overhead becomes much less important. Since the interstage switch-cap (SC) amplifiers are built-in with the required pipelining track-and-hold (T/H), 1.5b/stage is most suitable for implementation in CMOS. The other advantage is that the reconstruction DAC can be made perfectly linear because only three reference levels are needed. The differential zero level is obtained by shorting to common mode (CM); differential positive and negative levels are realized by flipping the differential connection. However, the interstage gain accuracy of the front-end stages are the bottleneck to high resolution. For less than 3dB degradation in SNDR, the gain accuracy of each stage must be comparable to the remaining resolution to be resolved in the following stages; this sets the nominal gain accuracy requirement. For example, the first stage gain must be 11b accurate for a 12b ADC - almost as accurate as the overall resolution. This sets the nominally required accuracy in the interstage gain. Ideally, the SC amplifier gain is determined by the capacitor ratio Cs to Cf, where Cs is the sampling capacitor and Cf is the feedback capacitor. The gain error comes from capacitor mismatch, finite DC gain of the operational transconductance amplifier (OTA), the parasitic capacitor between the input and output nodes, and the incomplete settling. As a result, calibration [23], [24], [25], [26], [27], trimming, and error averaging [28], are necessary for resolution of 12b and above.

Unfortunately, even with the nominal gain accuracy, a 12b ADC in 1.5b/stage architecture still can not comfortably meet the SFDR specification of over 80 dB (Table 1). Gain accuracy over the nominal value is necessary for the required SFDR.

B. *multibit/stage*

This research investigates the fundamental limitations to SFDR with pure analog solution. Without calibration and trimming, the accuracy bottleneck of 1.5b/stage has to be removed by increasing the first stage resolution [29]. Issues arise from the multibit/stage implementation. First, the increased interstage gain must be pipelined to avoid the possible bandwidth (BW) bottleneck due to increased interstage gain [31]. Fortunately, pipelining in CMOS is automatic. Second, the multibit sub-ADC's must be made efficient to avoid complexity explosion. Efficient sub-ADC is possible with flash two-step, or folding, architecture [32], [33] combined with offset averaging [34]. Third, very linear DAC is required. The accuracy requirement is somehow transformed into the linearity requirement. However, the transformation makes sense, since the required linearity can be relaxed by the increased first stage resolution and is attainable by capacitor matching [29].

C. *The top-level architecture*

A pipelined two-stage architecture is chosen with 6b-7b partition (Figure 2). With the 6b first stage, SFDR of 88 dB requires interstage gain accurate to only 1.6% which is easily attainable without the need for calibration and trimming. The 7b sub-ADC can be implemented with 1.5b/stage, which may have the advantage of lower power. To reuse the design of the 6b sub-ADC, the 7b is implemented with two 6b in parallel. The interstage gain of 2^5 is implemented with a cascade of pipelined five 2x SC amplifiers. The first T/H ensures above Nyquist operation. The second T/H pipelines the regeneration of the comparator latch in the 6b sub-ADC. The regeneration takes half clock cycle. The throughput is doubled at the expense of one more stage of KT/C noise and power.

IV. EFFICIENT SUB-ADC'S

Generally, A/D converter is made efficient using multi-step conversion. 1b/stage is the extreme multi-step. The multi-step sub-ADC must also be flash, since it is in the critical path. Folding is the only way to realize multi-step flash ADC.

In the 6b sub-ADC design, two-step folding is employed with 3b-3b partition (Figure 3). The coarse quantizer or cycle pointer is implemented in simple flash topology with a single preamp stage preceding the latches. The 3b fine quantizer consists of three cascaded gain stages before the latches with an overall gain of over 15 to overcome the latch dynamic offset. Signal folding is merged with the second and third amplifier stages. The two cascaded folding-by-3 stages generate 9 folds with 8 folds effective in the input FS. The extra fold is used as dummy with half fold at each end. The 8x folding reduces the complexity of the latches and digital encoders by 8x, but a total of 2^6-1 = 63 zero-crossing (ZX) must be generated. Offset averaging and interpolating are employed to reduce the size and power dissipation of the ZX generators (i.e. the preamps) and the loading to the T/H.

This sub-ADC architecture is similar to the previous work [34], but is different in that a preamp stage is inserted before the folding stages. The preamp array allows for optimum averaging and unifies the input CM voltage of the folding amplifiers.

A. *optimum preamp BW.*

For a certain required overall gain G_T, the overall BW is optimized with respect to the number of cascading gain stages and the gain of each stage. Under the assumption that $BW_i = BW_u/G_i$, where G_i, i = 1, 2, ..., m, is the i-th stage DC gain, and BW_u is the unit-gain BW of each stage, the optimal condition is $m = \ln A_T$ and $A_i = A_T^{1/m}$. For example, in CMOS, we can reasonably assume latch offset (including dynamic offset) = 30 mV. $A_T = 15$ is required for 2 mV referred offset. The optimal number of cascaded stage is n = 3 and gain of each stage is between 2.5~3.

B. *Folding*

Folding architectures have been developed, in a piece-meal fashion, by circuit improvements over decades to suppress the complexity of flash ADC's [35] - [48]. They are derived here in a logical way that provides better insight.

A flash architecture consists of ZX generators (e.g. difference amplifier), ZX detectors (i.e. regeneration latch), and encoder to turn the thermometer codes into desired digital format. A straightforward way to reduce overall complexity by F times is to add an F-to-1 function preceding the ZX generators (ZG). A coarse MSB channel overhead is necessary to resolve the ambiguity introduced by the many-to-one mapping, while the original ADC is transformed into the fine LSB channel. Since the F-to-1 operation is in the signal path and the corresponding transfer characteristic must be folded F times, it is referred to as signal folding. The commonly used linear folding characteristics are sawtooth and triangular. It not trivial to realize such rectifier characteristic with negligible distortion especially at high input frequency. A compromise is made by moving the F-to-1 block across the ZG. Now, the ZG complexity switch back to the original, but the implementation of F-to-1 is much easier, since the F-to-1 deals with discrete ZXs and no linearity is required at all.

The F-to-1 operation on ZX's is nothing more than multiplexing of the ZX's. In the example shown in Figure 4, 12 parallel ZX generators are divided into three groups, i.e. F = 3. The group in which in the input falls is connected to the ZX detectors by an analog multiplexer, controlled by the MSB channels. This subranging architecture is *serial*, because the control signal from the MSB channel must be generated prior to multiplexing.

A common way to implement the multiplexer is by means of switching, as shown in Figure 5. Now suppose the switches and the MSB control signal are removed. Due to interference the rest of the cells some of the ZX's may be shifted. However, the net interference can be made zero if the signals from other cells are pegged at opposite levels. For this to happen, all we need to do is to have odd number of groups and to flip the ZX polarity of adjacent groups. Flipping the ZX polarity does not affect the quantization result, because the transition point of the thermometer code does not change within each group. Therefore, by eliminating the switches, the multiplexing becomes automatic; the merged transfer characteristic fed to each ZX detector becomes folding; and the serial subranging architecture becomes flash. It is named ZX folding to distinguish it from signal folding.

The derivation demonstrates that folding is in essence automatic multiplexing of ZX's. In fact folding characteristics inherently implies automatic multiplexing of ZX's. The number of folds corresponds to the number of multiplexed groups. Therefore, the implementation of folding is not necessarily limited to the method derived above, which is based on summation of odd number of ZX's of alternative polarity (Figure 6); any methods that result in the folding characteristics work. For example, one method is to directly implement automatic multiplexing based on rectifier self-switching characteristics [41]. The described summation can be realized to the same effect in two steps: subtotal first and then grand total, leading to the summation-based cascaded folding [34]. Since multiplication preserves the ZX points of both multiplier and multiplicand signals, it can be cascaded to obtain higher extent of folding [4]. By cascading, speed degradation from folding is alleviated because the associated slowing factors such as the merged loading at the summing node is distributed over several amplification stages. Also, the pre-amplification before each folding stage helps to peg the interfering signals in summation-based folding.

The folding characteristic multiplies the input frequency at the merging node, severely degrading the AC performance. T/H is necessary for Nyquist performance. In the pipelined multi-bit/stage architecture, T/H is built in for the sub-ADCs. This makes folding architecture a very suitable choice for sub-ADCs.

C. *Bit-synchronization*

The coarse channel decides in which group (or fold) the input sample falls. The fine channel detects on the fly where the input sample lies within the group. The input referred ZX points of the coarse channel must align with the group division of the fine channel; otherwise, the coarse channel may point to a wrong group. Since the misalignment is inevitable in reality, the so called bit-synchronization [42], [46], [47] must be executed to correct the possible one-group wrong decision by the coarse channel. Generally, when a ZX is redefined in another channel, bit-sync is necessary to resolve the conflict between the redefined ZX's.

A simple algorithm implementing bit-sync for the sub-ADC's is described. The fine channel is able to tell if the input sample lies in the upper or lower half of a group with positive or negative ZX polarity. The LSB of the coarse channel is also an indicator of the polarity, say 0 indicates negative, and 1 stands for positive. For a given input sample, the disagreement in polarity implies the misalignment causing erroneous quantization, and the fine channel information is used to correct the coarse channel output, since the fine channel defines the overall resolution. Given misalignment within 1/2 LSB of the coarse channel, there are two possibilities if the disagreement occurs. First, fine channel indicates that the input sample in the upper portion of the group, then the coarse channel output must be subtracted one LSB. Second, the input sample in the lower portion, the output must be added one LSB, corresponding to shifting the thermometer code down and up by one bit, respectively.

V. OFFSET AVERAGING

While the ZD block is made compact by the F-to-1 action, the ZG block is made small by averaging and interpolating. A lateral resistor R1 connected between adjacent outputs of a preamp array was proposed to average out the offsets at preamp inputs [49]. In fact, averaging evolves from interpolating when the interpolating resistor R1's are not buffered. The nominal ZX's are not altered if translational symmetry is maintained across preamp or ZG array within the input FS. Dummy preamps extending beyond the FS is necessary to preserve the symmetry. The idea of averaging is simple, however, using it at optimum is not that simple.

The preamp load resistor R0's and the averaging resistor R1's forms a spatial filtering network [50], and the preamp g_m stages generate current stimuli injected to the network (Figure 7). With the small-signal current flowing through R0's defined as its output, the filtering network is fully characterized with impulse response h(n) which is an exponential function of R0/R1 for this first-order (i.e. maximum lateral resistor span of one) network. The impulse response (IR) can be made rectangle-like with higher order network, where R2 connects the output nodes one preamp apart, R3 two preamp apart, and so on, with the lateral resistors labeled with their span of connection. To simplify discussion, h(n) is represented with its width W_{IR}, which is quantitatively defined based on certain criterion. The stimuli consists of "signal" and "noise". "Signal" $\Delta I_s(n) = g_m(n)\Delta V_{in}$ and "noise" $\Delta I_{os}(n) = g_m(n)V_{os}(n) + \Delta I_{tail}(n)$, where $g_m(n)$ is the differential transconductance, ΔV_{in} is an equivalent input voltage applied to cancel the effect of offset voltage $V_{os}(n)$ appearing at the input of a ZG generator, and ΔI_{tail} represents mismatch of amplifier tail currents plus any other long-range fluctuations due to factors other than $V_{os}(n)$ such as digital noises (Figure 3). Due to the clipped I_d-V_d characteristic, the signal currents assume non-zero value over a finite number of ZX-generators, equivalent to constant current sources windowed by $g_m(n)$. The window is approximately characterized with its width W_{ZX}. Noise current from V_{os} is windowed by $g_m(n)$ as well. Beyond the window, the additional noise term $\Delta I_{tail}(n)$ dominates and is usually comparable to the offset current within the signal window. Therefore, offset currents approximate white noise.

The input referred offset (rms) σ_{os} is minimized when the output SNR is maximized at the matched filtering condition $W_{IR} = W_{ZX}$ for a given signal window. To maintain translational symmetry, boundary condition $W_D = W_{IR}$ must be met, where W_D is the total number of dummy preamps. When the outputs of the differential preamps at the end of the array are cross connected with R1, the boundary condition is relaxed into $W_D = \min(W_{IR}, W_{ZX})$. The larger $\min(W_{IR}, W_{ZX})$, the more the averaging, and the smaller the σ_{os}; but on the other hand, more dummies are required. The dummies consumes not only hardware, but also voltage headroom, leading to smaller LSB. By noting that σ_{os} is proportional to the inverse square root of $\min(W_{IR}, W_{ZX})$, while LSB is a linear function of $\min(W_{IR}, W_{ZX})$, it can be shown that INL (rms), i.e. σ_{os}/LSB, is minimized when $W_D = 1/3\ W_{total}$, where W_{total} is the total number of preamps in the array (including the dummies). Under $\min(W_{IR}, W_{ZX}) = 1/3\ W_{total}$, the matched filtering condition is relaxed to $W_{IR} \leq W_{ZX}$. Therefore, the overall optimum condition minimizing INL is given by

$$W_D = W_{IR} = W_{tatal}/3 < W_{ZX} \quad , \tag{2}$$

If R0 is implemented with current source, i.e. R0 = infinity, the cross-connection at the ends is necessary to maintain translational symmetry of the impulse response [34]. Dummy preamps are still indispensable to maintain the translational symmetry of the stimuli. Now the impulse response degenerates into linear function and extends over the entire network: $W_{IR} = W_{total}$. To satisfy the boundary condition, W_{ZX} has to be smaller than W_{IR}, i.e. $W_{ZX} < W_{IR}$ (Figure 8), which must also be satisfied for averaging applied on folding amplifier array, where signal pegging requires $W_{ZX} < W_{total}/F = W_{IR}$. Therefore, with R0 = infinity, it is impossible to meet the matched filtering condition, and loss of "signal" and consequent degradation in speed is inevitable.

It is true that the R1's reduce the impedance at each preamp output node, but it should be noted that this does not necessarily means reduction in voltage gain, since the R1's also pass currents from other preamps. At optimum averaging, nothing happens to the gain and BW of the individual preamps, yet the INL of the array is improved to maximum.

Interpolating is a different form of averaging, where the averaging among distributive preamps are lumped into averaging within a big preamp. Fundamentally, averaging does not improve INL on top of interpolating. However, averaging is generally better than interpolating, because the former maintains individual preamp speed and gain, while the later suffers additional delay of the interpolating network.

For the 7b sub-ADC, a preamp is designed first, with the linear region (i.e. W_{ZX}) larger than 1/3 FS and R0 set to 6.8 K ohm to meet the specified gain of 3x. An array of 84 preamps are then formed with their threshold voltages spanning the maximum available voltage headroom. The neighboring preamps are both flipped to filter spatially correlated low frequency offsets. To overcome the KT/C noise, less than the optimum dummy is desirable; as a result, 20 out of 84 total preamps are used as dummies, with only 4% degradation in INL. R1 value is determined by reducing it from infinity to a point where the zero-crossings at the ends of FS start moving inward by more than 0.1 LSB. This results in R1/R0 ratio of 7 and $W_{IR} = W_D = 20$ with respect to the criterion of edge effect smaller than 0.1 LSB. Gain reduction is only 10%. Monte-Carlo simulations show that the implemented averaging lowers random offsets by 3.6x. Compared to simple flash without offset averaging, it translates into one-order-of-magnitude reduction in input transistor size, input capacitance and power dissipation.

VI. TRACK-AND-HOLD

Input T/H, functionally independent of the quantization, is responsible for the dynamic performance of ADC. The optimization of quantization architecture described above removes accuracy bottlenecks, but has little impact on the T/H, which still need to meet the overall linearity and speed requirement. Therefore, T/H poses the greatest challenge on circuit design.

A. closed-loop configuration

Op amp-based closed-loop feedback configuration implementing bottom-plate sampling is necessary to attain T/H linearity beyond 80 dB. The simple wrap-around configuration is used for its fast settling (Figure 9). All the switches are linearized with complementary MOSFET's, except for the NFET input switches, which are linearized using a bootstrap scheme to be described. Given 2 pF loading to T/H1 from the 6b sub-ADC and 6 pF (including 2 pF parasitics) to T/H2 from the reconstruction DAC, a compromise between power consumption and KT/C noise results in g_m of 2 6mS for both T/H1 and T/H2, and 2 pF and 3 pF sampling capacitances for T/H1 and T/H2 respectively.

The interstage switches are moved inside the feedback loop. First, this isolates the 6b sub-ADC loading to T/H1 in the reset mode and thus improves settling. Second, the sub-ADC is connected to the OTA only in hold phase and does not experience the common-mode jump between the two phases, improving OTA common-mode settling significantly. OTA output common-mode voltage is set to Vdd/2 = 1.7V to maximize the output swing in the hold phase, and is switched to the OTA input common-mode voltage, Vth + 2Veff, in the reset mode. Finally, the in-loop switches improve the hold-phase settling. The interstage switch and sampling capacitor form a pole close to the pole of the driving T/H. In the cascaded configuration, the overall BW degrades by about $\sqrt{2}$ -1. When moved to the feedback loop, this pole low-passes the feedback signal, enabling a fast start-up of the amplification in the absence of negative feedback at the beginning. Overshoot or ringing do not occur so long the feedback delay is much less than the

OTA delay. In the s-domain, the pole in the feedback path appears as zero when the loop is closed and cancels the closed-loop dominant pole when positioned close to it. On-resistance of feedback switch can be scaled up to the same effect, but it contributes to more kT/C noise.

B. Bootstrapping Circuit

The input T/H track-mode distortion at Nyquist input frequencies must be lower than -85 dBc. It is not enough to eliminate nonlinear charge injection with bottom-plate sampling in a fully balanced op amp-based SC T/H. At high input frequencies, nonlinearity in the MOSFET switch distorts capacitor charging currents. This distortion arises from the dependence of mobility on gate electric field, and from the nonlinear body effect. The nonlinearity in mobility has to do with surface effect, which is not well-behaved and therefore can not be accurately compensated for with complementary MOSFET. Simulations show that the minimum track-mode 3rd-order harmonic attainable with complementary MOSFET's is -65 dBc for single-ended full swing of 0.8 Vp-p at Nyquist input frequency of 25 MHz. The best size combination is 40 μm for NFET and 177 μm for PFET at minimum channel length of 0.6 μm. The switch size is limited by the nonlinearity in the junction capacitance.

The nonlinear switch on-resistance is reduced by bootstrapping its gate voltage substantially over Vdd (e.g, by one Vdd) [51], [52]. However, the process reliability rules require any gate voltage should not exceed Vdd by 10%. In this ADC, the NFET switch M1 is turned ON using the gate voltage on a replica FET M2 carrying a constant current, whose source is servoed to the analog input by feedback (Figure 9). The open loop gain-bandwidth product (GBW) or unity gain frequency of the op amp in this feedback loop is 280MHz so that the analog input bootstrapping gate voltage is delayed less than 8° at Nyquist relative to the switch input. A SC level-shift circuit bootstraps the gate voltage to a maximum of 3.6 V. All the PFET switches with their sources seeing this maximum voltage have their n-well tied to their sources. During the hold-mode, the op amp drives a dummy NFET switch through the other SC level-shift path. The dummy balances the op amp load. At the start of track mode, M1 is switched in right before the dummy is switched off so that the dummy can transfer charge to M1, facilitating the turn-on of M1.

The bootstrapping can also be implemented by simply connecting a pre charged capacitor across the gate and source of the NFET sampling switch during the track-mode. The constant Vgs removes the mobility nonlinearity but can not compensate for the body effect.

C. Operational Transconductance Amplifiers (OTA)

Some OTA topologies are tailored for implementation at low Vdd. Examples are folded cascode and cascaded two-stage [51], [52], [53]. It is of interest, however, to explore the lower limit that Vdd can go with conventional cascode topology [54]. For this ADC, the maximum signal swing is limited to 1.6V peak-to-peak differential by the switch gate headroom and the nonlinearity of the "linear capacitor" option available in the process [55]. It is not necessary to adopt topology for rail-to-rail swing. As a result, a multiple-Veff based design approach was adopted where all stacked transistors of the same type are assigned the same Veff and the voltage margin between the stacked transistors are saved (Figure 10). The required multiple-Veff biasing voltages are easily generated by scaling W/L based on the square law of the MOSFET current. The scaling is implemented by connecting variable number of unit transistors in series. The absence of voltage margin requires care in matching the layout of the biasing and biased transistors. Once matched, they tracks each other, immune to process and temperature variations. Therefore, this design is robust, especially, when used for this multibit/stage architecture, where the DC gain of OTA's are not that crucial. An overall DC gain of 80 dB is designed to ensure the required closed-loop linearity.

The maximum Veff available for the input NFET's is 0.4 V, not enough to set the T/H settling free of slew-rate limiting for the maximum input swing of 1.6Vp-p. The slew-rate causes dynamic nonlinearity or effectively reduces the available settling time, and 12b accuracy requires a settling time more than 10x the time-constant (ideally, 8.3x). The closed-loop BW of the T/H is designed to 250MHz, which guarantees 12b settling accuracy in half cycle of 75MHz clock.

The auxiliary amplifiers boosts the overall gain by about 30x. The GBW of the one driving the cascode FET is set to 300MHz based on the following considerations. It must be over 250MHz so that the zero-pole doublet corresponding to the auxiliary amplifier unity gain frequency is positioned out of the closed-loop band to suppress the slow settling component arising from the doublet. It should not be too high either, otherwise the auxiliary amplifier loop may become unstable. The level shifting source followers add a pole to the loop. This pole should not cause instability since it is about four times higher than the auxiliary GBW. The relative frequency can be told from the relative size of the transistor shown in the schematic. The four auxiliary amplifiers consume total power comparable to the main circuit.

Two small size analog input transistors with their drains cross-connected to the two differential legs are used to maintain the minimum bias necessary to keep the cascode transistors "alive" at the beginning of each mode when the large input swing would otherwise steer all the tail current to one leg. The cutoff of the cascode transistors are harmful not only because the recovery slows down the settling but also the auxiliary loop becomes unstable since the drastically reduced Cgs of the cascode transistor can not compensate the auxiliary amplifier.

The dual common-mode voltages are realized by two rotating level shifting switched capacitors pre charged to different voltages [31].

VII. ARCHITECTURAL BUILDING BLOCKS

A. *Residue amplifiers*

Residue amplification is an auxiliary function in A/D conversion, employed to overcome comparator offset. It equivalently amplifies input and input FS or LSB without running out of headroom. If implemented in CMOS SC circuits, it also automatically performs another auxiliary function, pipelining the subsequent stage.

Given the closed-loop configuration and gain, the time constant of *i*th stage is a function of the sampling capacitance Cs(i), the load capacitance Cs(i+1) from next stage, and the OTA transconductance Gm(i). The first stage sampling capacitance Cs(1) is determined by the reconstruction DAC, while the last stage load capacitance Cs(6) is determined by the 7b sub-ADC. The total amplifier power consumption is optimized with respect to {Cs(2), Cs(3), Cs(4), Cs(5)} and Gm(i) of each stage, under the constraint that overall settling accuracy better than 1.6%, for objective function sum(Veff(i) Gm(i)), where Veff(i) is determine in the headroom design. A program is written to find the optimal Cs(i) and Gm(i) for this nonlinear programming problem.

To take advantage of the small residue swing, the OTA of the first four stages are realized in double cascode topology, saving half of the power due to elimination of the auxiliary amplifiers. The last residue amplifier uses single cascode for FS swing, with increased channel length to compensate for the DC gain.

B. *Reconstruction D/A converter (DAC)*

Reconstruction DAC, together with residue-forming subtractor, constitutes an important auxiliary block that performs F-to-1 multiplexing necessary for subranging. The 6b DAC uniformly divides the input FS into 2^6 regions and multiplexes the residue in the region that the input sample falls to the following stages for fine quantization. In terms F-to-1 operation in signal path, this multiplexing is similar to signal folding, and therefore the residue characteristic is identical to signal folding characteristic. However, they are different in terms of implementation. The former is "serial", meaning the multiplexing is controlled by a previous quantization decision, while the later is flash or automatic, meaning the multiplexing is controlled directly by the signal itself - no need to wait for regeneration of ZX detectors.

Since the DAC multiplexes the sampled input signal directly, it must preserve the signal fidelity to the overall linearity of 12b, which is only attainable with capacitor matching in CMOS process. To reduce layout complexity, the DAC is realized in a two-step hybrid configuration, comprising a coarse 3b capacitor array and a fine 3b interpolating resistor ladder driven by the thermometer code, and the 1-of-8 code readily available from the 3b MSB and 3b LSB outputs of the 1st stage 6b folding sub-ADC, respectively (Figure 11). The first coarse tap above -FS/2 is formed by connecting the top plates (bottom plates in layout) of 7 unit capacitor Cp's on the positive side to -Vref = -FS/4 and 7 capacitor Cn's on the negative side to +Vref = +FS/4; the second tap corresponds to one of the 7 differential capacitor pairs flipping their connection to +/- Vref; and so on, till all the 7 pairs have flipped the connection. Two things to be noted. First, even though the circuit and the reference taps are differential, the errors in the taps are not differential. This is generally true for fully segmented DAC's controlled by thermometer code. Second, the differential circuit does not worsen the DAC matching by sqrt(2), instead, it improves the matching by sqrt(2) due to the averaging effect between the positive and negative side implied by residue gain expression, G = (Cn + Cp)/2Cf.

The 3b coarse DAC defines the overall DAC linearity, since the nonlinearity in the fine DAC weighs 3b less. In terms of spreading out the spurs arising from DAC nonlinearity, the 3b-3b two-step configuration is not as effective as a 6b fully segmented configuration. However, the reduced complexity eases layout matching; especially, the possibility of parasitic mismatch is greatly reduced. Assume systematic mismatch is completely eliminated by careful layout, the random mismatch requirement dictates the capacitor size. Based on published data on capacitor matching for similar process [56], the total capacitance of 4 pF (single side) is chosen to provide 200 ppm matching required for 3-sigma over-80dB SFDR. The matching specification is determined by Monte-Carlo simulations with random errors perturbing the DAC references of the programmed ADC which simulates the architecture of the real ADC. (The correlation between the tap errors was neglected for simplicity.) The *k*th tap error $\delta(k)$ (rms) is related to the random mismatch (rms) between a pair of capacitors, $\Delta C/C$(rms) by

$$\delta(k) \approx (\Delta C/C)\sqrt{(N_C-k)^2 k + k^2 (N_C-k)}/(2N_C) \quad , \tag{3}$$

where N_C is the total number of capacitor segments. The maximum error (rms) happens at the middle tap k = Nc/2, equal to quarter rms of the total capacitor mismatch. Due to the averaging effect between Cp and Cn, Ctotal should be doubled to 8pF.

Each capacitor consists of a polysilicon plate over thin oxide and a heavily doped n+ diffusion. The 3b coarse DAC layout (Figure 12) comprises two rows of differential unit capacitors, with the differential pairs in common centroid, addressed in scrambled order across the row to suppress the cumulative effect of gradient [57], which gives rise to second order INL. The differential common centroid layout obviates wire crossing inevitable in the conventional cross-connected common centroid layout, eliminating the possibility of mismatch in wiring parasitics. Shielding by grounded metal-3 on top of the array blocks long-range fringe capacitance [58]. Dummy capacitors extending over a distance of at least 50 um prevent uneven lithography at the boundaries from encroaching on the array.

C. Comparator amplifiers

Comparator preamps implement the first A/D conversion core function of generating zero-crossings. Amplification of zero-crossing, an auxiliary function, is built in to overcome the comparator latch offset.

The input differential transconductance stage consists of two differential pairs, with one comparing Vinp, the positive-side signal voltage, with Vrefp, the positive-side reference voltage, and the other comparing, Vrefn, the negative-side reference, with Vinn, with the negative-side signal. The generated differential currents are summed at the low impedance node provided by the cascode FET's. The summed current continues to pass the cascode FET's, and the load differential current sources [34] force it to flow through the resistor across the output nodes, generating a differential voltage ZX characteristic centered at the reference voltage. The differential resistor is implemented with a PFET biased in linear region using a master-slave circuitry. Half of the differential resistor is R0. The lateral averaging resistor R1 is implemented with un-silicided poly of sheet resistance 100 ohms.

The folding-by-3 amplifier (FA) are designed similarly. The difference is that three differential currents are generated and summed with the middle one flipped in polarity.

D. Comparator latch

Comparator latches implement the second A/D conversion core function of detecting the ZX's. They consist of a differential input Gm stage and a cascoded stage in the form of cross-coupled inverters. The amplified ZX signal is sampled on the input capacitors and generates a held differential current when the clock strobe signal turns off the input switch. In the meantime, the reset-to-Vdd switch releases the output nodes for regeneration. The differential current passes the the inverter NFET's in saturation, reaches the output nodes which are blocked from Vdd by the inverter PFET's in cutoff, and develops an output voltage by charging one output node and discharging the other, while the tail current pulls down the output common-mode voltage from Vdd by discharging both output nodes. The regeneration kicks in when one of the output nodes drops one threshold voltage (0.9V) below Vdd which turns on the PFET on the other side. The static input stage degenerates the NFET regeneration, though it avoids common-mode jump usually associated with dynamic latches at strobe [59].

The fully developed rail-to-rail digital signal is written to the subsequent RS flip-flop, and is held there by resetting the latch outputs to Vdd. Reset-to-Vdd enables use of RS flip-flop which exerts negligible kickback to the latch output nodes at strobe, though it worsens the latch dynamic performance in two ways. First, one PFET is never turned on during the entire regeneration, doubling the regeneration time-constant. Second, the common-mode excursion reduces the time available for regeneration. In this latch design, dynamic performance is traded for the detection accuracy which dictates the overall performance.

Regeneration time-constant is simulated to 0.15 ns which ensures bit-error-rate (BER) less than 10^{-11}. Monte-Carlo simulations show 7mV dynamic offset due to transistor mismatch in threshold voltage [19], [20]. The offset can be much higher than the simulated taking into account of unbalanced clock feedthrough due to mismatch in the switch FET size and the node capacitance receiving the fed through charge. The latch layout is therefore made as symmetrical as possible.

E. Encoder

This is the third and last core function of A/D conversion that converts the detected ZX's into a required digital format. The sum of the first-stage 6b sub-ADC output and the quantized residue output forms the overall 12b output. Since the residue is amplified by 2^5x before quantization, the second-stage 7b sub-ADC output should be divided by 2^5 before the summation. This is equivalent to shifting the 7b binary code by 5 bits towards the LSB direction and aligning its first bit (the MSB) with the 6th bit (the LSB) of the first-stage binary code for the summation. The commonly used name for this shift-and-add operation, "digital error correction," is confusing, since no error is involved. It is simply part of encoding.

Due to the half FS under-range in the second stage, the original all zero code is increased to a code corresponding to 1/2 LSB of the first-stage, causing a same amount of offset in the overall code. This offset can be corrected either by digital subtraction or equivalently by shifting all the comparator threshold voltages and DAC reference voltages by positive half LSB of the 6b.

Concatenating 3b the coarse and 3b fine binary codes forms the outputs folding sub-ADC's. However, the coarse code may contain one LSB error corresponding to mis-detecting one ZX group by the coarse channel. The correction must be done on the thermometer code, which is the required format to switch the reconstruction DAC capacitors. The thermometer code is transformed by XOR gates into one-of-n code, which is then encoded into binary code by a ROM encoder. The implementation of the bit-sync algorithm described in IV-C is shown in Figure 13.

F. Peripheral Circuits

All clocks are generated on-chip from an external low phase noise balanced sinewave. On-chip clock circuitry synthesizes two non-overlapping phases and distribute them all over the chip. The front-end clock receiver consists of three differential amplifiers which sharpen the received 1 Vp-p sinewave into a single-ended rail-to-rail square wave with 0.3 ps rise/fall time. A subsequent stage comprising a three- and a two-inverter chain in parallel generates complementary clocks. The CMOS inverters are sized so that the skew between the complementary does not exceed 40 ps for all process corners [60]. The clock jitter generated from these circuits must be smaller than 1 ps to meet the 12b SNR requirement at Nyquist [31], [61]. The power supply and ground bounce of clock buffers contributes negligible sampling jitter because they are synchronized with the clock. The zero-crossing is sharpened just enough to make the thermal noise contribution to jitter negligible. (Local clock buffers provide the required rise/fall edge.) Therefore, the front-end receiver stage dominates the clock jitter and are designed carefully.

The input Veff (i.e. Vgs - Vth) of the first amplifier is maximized to 0.44 V for minimum current clipping. The R and C are chosen so that the 3-dB corner frequency equals to the clock frequency and zero-crossing slope gain is maximized without being two much deviated from the AC gain at the clock frequency, resulting in slope gain of 7. Since the slope is maximized with minimum bandwidth and DC gain, the noise (rms) to slope ratio, i.e. the jitter (rms) is minimized. The cascaded amplifiers decreases successively in DC gain but increases in unity gain bandwidth to accommodate the increasingly sharpened zero-crossings. The simulated noise at the second stage output is 2.16 mV (rms), which, divided by the zero-crossing slope, gives rise to a jitter of 0.44 ps (rms).

Unlike the clock buffers, output drivers cause harmful signal-correlated noise. Differential current mode output drivers injects much less noise to the substrate and the power supply but drains too much current. CMOS drivers only consume dynamic power but are too noisy. To compare their effect on SNR and SFDR, both are made available on-chip and the whole chip shares a single power Vdd and ground Gnd with the substrate tied to Gnd. Analog, digital and output driver Vdd's and Gnd's are separable by FIB. Total 2nF decoupling poly-diffusion capacitors are laid out in unused areas all over the chip.

Reference voltages are applied externally. The master bias current of each circuit block such as the bootstrap circuits, the T/H's, the RA's and sub-ADC's are adjustable on board for improved testability.

VIII. PERFORMANCE

The chip (Figure 14) is fabricated through MOSIS in a 0.6μm, triple-metal, single-poly standard digital CMOS process with analog option of linear-capacitor and unsilicided poly resistor. The active chip area is 4x4 mm^2 and the ADC dissipates 850 mW excluding the power of output driver. The differential output drivers are used by default for the performance test.

Code density test yields 1.3 LSB INL and 0.8 LSB DNL (Figure 15). The 8 segments fairly identifiable of the measured INL plot reflect the mismatch among the 8 unit capacitors of the reconstruction DAC. The INL and DNL peaks are contributed by the 7b sub-ADC.

This ADC achieves over 80 dB SFDR for signal frequencies up to 75 MHz at 50 MS/s (Figure 16). The SFDR rolloff at 80 MHz is due to the finite bandwidth of the feedback loop bootstrapping the gate voltage of the input sampling switch. The SNR rolloff at higher input frequencies are attributed to the clock jitter.

Sampling jitter is measured by locked-histogram test in which the clock and the input signal are synchronized such that the sampling always occurs at the vicinity of the input zero-crossing. To amplify the jitter effect, input frequency is set to 4x of the clock frequency. The output histogram is fitted with binned Gaussian distribution, yielding 0.61 ps jitter (rms) (Figure 17). Perfect fitting can not be achieved unless the bin sizes are adjusted with the measured DNL.

The Vdd tolerance test confirms that the multiple-Veff based design is viable, since the SFDR and SNR for low input frequency (fin = 1 MHz) maintains almost constant in a wide Vdd range from 3.05 V to 3.6 V (Figure 18). The SFDR degradation at lower Vdd for high input frequency (fin = 20 MHz) is due to the increased nonlinearity of the input sampling switches.

The effect of CMOS buffer on SNR and SFDR are measured at 50 MS/s. The bounce exhibits itself by SNR rolloff at input frequencies higher than Nyquist. SFDR is not affected within Nyquist frequency. The rolloff corresponds 1.8 ps sampling jitter. It seems that the power and ground bounce noise are only picked up by the clock inverters while rejected by the all-differential analog circuitry.

The programmed ADC is also used to process the measured data and ensures consistency among them. Figure 19 compares the measured spectrum with the one reconstructed from the INL data measured under the same condition. The reconstructed spectrums are obtained from the FFT of the programmed 12b ADC output with the quantization thresholds perturbed by the measured INL data. The similarity between the spectrums confirms that INL dominates SFDR and that the superb SFDR is not a result of the possible dithering effect from the inherent thermal noise. In terms of SNDR, the reconstructed spectrum is 3.5 dB better than the measured (67.5 dB vs. 64.0 dB), suggesting a random noise energy of about 1.24x the INL distortion. The random noise is also directly measured by grounded input test, from which a noise (rms) of 0.66 LSB (i.e. 257μV) is extracted. Simulations on the programmed ADC show that the reconstructed spectrum degrades in SNR by 3.5 dB when this noise is added at the input, and degrades further by 4 dB at 200 MHz input frequency when the 0.61 jitter (rms) is added at the clock - consistent with the measured SNR rolloff shown in Figure 16.

The results are summarized in Table 2

IX. CONCLUSIONS

The key to high SFDR is to randomize the ADC error characteristic by statically segmenting it more and dynamically perturbing it. For each bit increment in the first-stage resolution of a multi-stage architecture, the interstage gain error and reconstruction DAC nonlinearity are segmented one time more, leading to 2~3 dB improvement in SFDR for the same SNDR. This 6b-7b two-stage pipelined ADC achieves over 80 dB SFDR for signal frequencies up to 75 MHz at 50 MS/s without the need for trimming, calibration and dithering. The BW bottleneck due to increased interstage gain is removed by pipelining five wideband 2x SC stages. Folding, averaging and interpolating architecture suppresses the complexity of the 6b and 7b sub-ADC's. Bootstrapping using replica source follower Vgs removes NFET sampling switch nonlinearity in both mobility and the body effect, without any node running substantially above Vdd. Multiple-Veff based design save precious headroom and proves a viable solution for low-Vdd design. Whole-chip simulation methodology helps in success on first silicon.

X. REFERENCES

[1] T. Gratzek, B. Brannon, J. Camp, and F. Murden, "A new paradigm for base station receivers: high IF sampling + digital filtering," 1997 IEEE Radio Frequency Integrated Circuits Symposium, pp. 143 - 146, June 1997.

[2] T. Gratzek, B. Brannon, J. Camp, and F. Murden, "ADCs for digital receivers: the whole world tunes in," 1996 IEEE MTT-S Digest, pp. 1335 - 1338.

[3] F. Murden and R. Gosser, "12b 50MSample/s Two-Stage A/D Converter," ISSCC Dig. Tech. Papers, San Francisco, CA, Feb. 1995, pp. 278-279.

[4] P. Vorenkamp and R. Roovers, "A 12-b, 60-MSamples/s Cascaded Folding and Interpolating ADC," IEEE J. Solid-State Circuits, vol. 32, pp. 1876-1886, Dec. 1997.

[5] D. Birdsall and A. Kuckreja, "A 12-bit, 65 MSPS BiCMOS ADC for cellular base-station applications," Proceedings of the 1998 Bipolar/BiCMOS Circuits and Technology Meeting, pp. 43 - 46, Sept. 1998.

[6] W. R. Bennett, "Spectra of quantized signals," Bell Syst. Tech. J., vol. 27, pp. 446-472, July 1948.

[7] A. Gersho, "Principle of Quantization," IEEE Trans. Circuits Syst., vol. CAS-25, pp. 427 - 436, July 1978.

[8] A. G. Clavier, P. F. Panter, and D. D. Grieg, "Distortion in a pulse count modulation system," AIEE Trans., vol. 66, pp. 989 - 1005, 1947.

[9] R. M. Gray, "Quantization Noise Spectra," IEEE Trans. Inform. Theory, vol. 36, pp. 1220 - 1244, Nov. 1990.

[10] W. T. Colleran, "A 10-bit, 100 MS/s A/D converter using Folding, Interpolating, and Analog Encoding," Ph.D. dissertation, UCLA, Dec. 1993.

[11] H. Pan, "A 3.3-V, 12b, 50-MS/s A/D converter in 0.6-μm CMOS with over 80-dB SFDR," Ph.D. dissertation, UCLA, Dec. 1999.

[12] L. Schuchman, "Dither signals and their effects on quantization noise," IEEE Trans. Commun. Technol., vol. COM-12, pp. 162 - 165, Dec. 1964.

[13] H. S. Fetterman, D. G. Martin, and D. A. Rich, "CMOS pipelined ADC employ dither to improve linearity," Proceedings of the IEEE 1999 Custom Integrated Circuits Conference, pp. 109 - 112, May, 1999.

[14] L. R. Carley, "A noise shaping coder topology for 15+ bit converters," IEEE J. Solid-State Circuits, vol. 24, pp. 267 - 273, April 1989.

[15] B. H. Leung and S. Sutarja, "Multibit D S A/D converter incorporating a novel class of dynamic element matching techniques," IEEE Trans. Circuits Syst. II, vol. 39, pp. 35 - 51, Jan. 1992.

[16] R. T. Baird and T. S. Fiez, "Linearity enhancement of multibit ΔΣ A/D and D/A converters using data weighted averaging," IEEE Trans. Circuits Syst. II, vol. 42, pp. 753 - 762, Dec 1995.

[17] R. Radke, A Eshraghi and T. S. Fiez, "A spurious-free delta-sigma DAC using rotated data weighted averaging," Proceedings of the IEEE 1999 Custom Integrated Circuits Conference, pp. 125 - 128, May, 1999.

[18] R. Adams, K. Q. Nguyen, and K. Sweetland, "A 113-dB SNR oversampling DAC with segmented noise-shaped scrambling," IEEE J. Solid-State Circuits, vol. 33, pp. 1871 - 1878, Dec. 1998.

[19] M. J. M. Pelgrom, A. C. J. Duinmaijer and A. P. G. Welbers, "Matching Properties of MOS Transistors," IEEE J. Solid-State Circuits, vol. 24, pp. 1433-1440, Oct. 1989.

[20] M. J. M. Pelgrom and M. Vertregt, "CMOS technology for mixed signal ICs," Solid-State Electronics, vol.41, pp. 967 - 974, July 1997

[21] S. H. Lewis and P. R. Gray, "A pipelined 5-Msample/s 9-bit analog-to-digital converter," IEEE J. Solid-State Circuits, vol. 22, pp. 954 - 961, Dec. 1987.

[22] S. H. Lewis *et al*, "A10-b 20-Msample/s analog-to-digital converter," IEEE J. Solid-State Circuits, vol. 27, pp. 351- 358, March 1992.

[23] H.-S. Lee, D. A. Hodges, and P. R. Gray, "A self-calibrating 12 bit CMOS A/D converter," IEEE J. Solid-State Circuits, vol. 19, pp. 813 - 819, Dec. 1984.

[24] Y. M. Lin, B. Kim, and P. R. Gray, "A 13-b 2.5-MHz self-calibrated pipelined A/D converter in 3-μm CMOS," IEEE J. Solid-State Circuits, vol. 26, pp. 628 - 636, Dec. 1991.

[25] S.-H. Lee and B.-S Song, "Digital-domain calibration of multistep analog-to-digital converters," IEEE J. Solid-State Circuits, vol. 27, pp. 1679 - 1688, Dec. 1992.

[26] E. G. Soenen and R. L. Geiger, "An architecture and an algorithmn for fully digital correction of monolithic pipelined ADC's," IEEE Trans. Circuits Syst. II, vol. 42, pp. 143 - 153, Dec. 1995.

[27] I. E. Opris, L. D. Lewicki, and B. C. Wong, "A single-ended 12-bit 10 Msample/s self-calibrating pipeline A/D converter," IEEE J. Solid-State Circuits, vol. 33, pp. 1898 - 1903, Dec. 1998.

[28] B.-S. Song, M. F. Tompsett and K. R. Lakshmikumar, "A 12-bit 1-Msample/s capacitor error-averaging pipelined A/D converter," IEEE J. Solid-State Circuits, vol. 23, pp. 1324 - 1333, Dec. 1987.

[29] L. A. Singer and T. Brooks, "A 14-bit 10-MHz calibration-free CMOS pipelined A/D converter," 1996 Symposium on VLSI Circuits Digest of Technical Papers, pp. 94 - 95, June, 1996.

[30] T. L. Brook, D. H. Robertson, D. F. Kelly, A. D. Muro, and S. W. Hartson, "A cascaded sigma-delta pipeline A/D converter with 1.25 MHz signal bandwidth and 89 dB SNR," IEEE J. Solid-State Circuits, vol. 32, pp. 1896 - 1997, Dec. 1997.
[31] K. Y. Kim, N. Kusayanagi and A. A. Abidi, "A 10-b, 100-MS/s CMOS A/D converter," IEEE J. Solid-State Circuits, vol. 32, pp. 302 - 311.
[32] M. Flynn and B. Sheahan, "A 400MSample/s 6b CMOS Folding and Interpolating ADC," ISSCC Dig. Tech. Papers, San Francisco, CA, Feb. 1998, pp. 150-151.
[33] A. G. W. Venes and R. J. van de Plassche, "An 80-MHz, 80-mW, 8-b CMOS Folding A/D Converter with Distributed Track-and-Hold Preprocessing," IEEE J. Solid-State Circuits, vol. 31, pp. 1846-1853, Dec. 1996.
[34] K. Bult and A. Buchwald, "An Embedded 240-mW 10-b 50MS/S CMOS ADC in 1-mm^2," IEEE J. Solid-State Circuits, vol. 32, pp. 1887-1895, Dec. 1997.
[35] B. D. Smith, "An Unusual Analog-Digital Conversion Method," IRE Trans. Instrum. Meas., pp. 155-160, June, 1956.
[36] J. D. Edson and H. H. Henning, "Broadband Codecs for an Experimental 224 Mb/s PCM Terminal," Bell Syst. Tech. J., vol. 44, pp. 1887-1940, Nov. 1965.
[37] A. Arbel and R. Kurz, "Fast ADC," IEEE Trans. Nucl. Sci., vol. NS-22, pp. 446-451, Feb. 1975.
[38] U. Fieldler and D. Seitzer, "A High-Speed 8 Bit A/D Converter Based on a Gray-Code Multiple Folding Circuit," IEEE J. Solid-State Circuits, vol. SC-14, pp. 547-551, June, 1979.
[39] R. J. van de Plassche and R. E. J. van de Grift, "A High-Speed 7 Bit A/D Converter," IEEE J. Solid-State Circuits, vol. SC-14, pp. 938-943, Dec. 1979.
[40] R. E. J. van de Grift and R. J. van de Plassche, "A Monolithic 8-Bit Video A/D Converter," IEEE J. Solid-State Circuits, vol. SC-19, pp. 374-378, June, 1984.
[41] R. E. J. van de Grift, I. W. J. M. Rutten, and M. van der Veen, "An 8-bit Video ADC Incorporating Folding and Interpolation Techniques," IEEE J. Solid-State Circuits, vol. SC-22, pp. 944-953, Dec. 1987.
[42] R. J. van de Plassche and P. Baltus, "An 8-bit 100-MHz Full-Nyquist Analog-to-Digital Converter," IEEE J. Solid-State Circuits, vol. 23, pp. 1334-1344, Dec. 1988.
[43] J. van Valburg and R. J. van de Plassche, "An 8-b 650-MHz Folding ADC," IEEE J. Solid-State Circuits, vol. 27, pp. 1662-1666, Dec. 1992.
[44] W. T. Colleran and A. A. Abidi, "A 10-b, 75-MHz Two-Stage Pipelined Bipolar A/D Converter," IEEE J. Solid-State Circuits, vol. 28, pp. 1187-1199, Dec. 1993.
[45] R. Roovers and M. Steyaert, "Design of CMOS A/D Converters with Folding and/or Interpolating Techniques," Advanced A-D and D-A Conversion Techniques and Their Applications, 6-8 July 1994, Conference Publication No.393, pp. 76-81.
[46] B. Nauta and A. G. W. Venes, "A 70-MS/s 110-mW 8-b CMOS Folding and Interpolating A/D Converter," IEEE J. Solid-State Circuits, vol. 32, pp. 1302-1308, Dec. 1995.
[47] M. P. Flynn and D. J. Allstot, "CMOS Folding A/D Converters with Current-Mode Interpolation," IEEE J. Solid-State Circuits, vol. 31, pp. 1248-1257, Sept. 1996.
[48] A. G. W. Venes and R. J van de Plassche, "An 80MHz 80mW 8b CMOS Folding A/D Converter with Distributed T/H Preprocessing," in ISSCC Dig. Tech. Papers, San Francisco, CA, Feb. 1996, pp. 252-253.
[49] K. Kattmann and J. Barrow, "A Technique for Reducing Differenctial Nonlinearity Errors in Flash A/D Converters," ISSCC Dig. Tech. Papers, San Francisco, CA, pp. 170-171 , Feb. 1991.
[50] J. L. White and A. A. Abidi, "Active Resistor Networks as 2D Sampled Data Filters," IEEE Trans. Circuits Syst. I: Fundamental Thoery and Applications, vol. 39, pp. 724-733, Sept. 1992.
[51] T. B. Cho and P. R. Gray, "A 10 b, 20 Msample/s, 35 mW Pipeline A/D converter," IEEE J. Solid-State Circuits, vol. 30, March 1995.
[52] A. M. Abo and P. R. Gray, "A 1.5-V, 10-bit, 14.3-MS/s CMOS pipeline analog-to-digital converter," IEEE J. Solid-State Circuits, vol. 34, pp. 599 - 606, May 1999.
[53] B. K. Ahuja, "An improved frequency compensation technique for CMOS operational amplifiers," IEEE J. Solid-State Circuits, vol. 18, pp. 629 - 33, Dec. 1983.
[54] K. Bult and G. J. G. M. Geelen, "A fast-settling CMOS op amp for SC circuits with 90-dB DC gain," IEEE J. Solid-State Circuits, vol. 25, pp. 1379 - 1384, Dec. 1990.
[55] T.-I. Liou and C.-S. Teng, "n^+-poly-to-n^+-silicon capacitor structures for single-poly analog CMOS and BiCMOS processes," IEEE Transactions on Electron Devices, vol. 36, pp. 1620 - 1628, Sept. 1989.
[56] H. P. Tuinhout, H. Elzinga, J. T. Brugman, and F. Postma, "The floating gate measurement technique for characterization of capacitor matching," IEEE Transactions on semiconductor manufacturing, vol. 9, pp. 2 - 8, Feb. 1996.
[57] C. S. G. Conroy, W. A. Lane, and M. A. Moran, "Statistical design techniques for D/A converters," IEEE J. Solid-State Circuits, vol. 24, pp. 1118 - 1128, Aug. 1989.

[58] M. J. McNutt, S. LeMarquis, and J. L. Dunkley, "Systematic Capacitance Matching Errors and Corrective Layout Procedures," IEEE J. Solid-State Circuits, vol. 29, pp. 611 - 616, May 1994.
[59] T. Kobayashi, K. Nogami, T. Shirotori, and Y. Fujimoto, "A current-controlled latch sense amplifier and a static power-saving input buffer for low-power architecture," IEEE J. Solid-State Circuits, vol. 28, pp 523 - 527, April 1993.
[60] M. Shoji, "Elimination of process-dependent clock skew in CMOS VLSI," IEEE J. Solid-State Circuits, vol. 21, pp. 875 - 879, Oct. 1986.
[61] B. E. Boser and B. A. Wooley, "The design of sigma-delta modulation analog-to-digital converters," IEEE J. Solid-State Circuits, vol. 23, pp. 1298 - 1308, Dec. 1988.

Table 1: SFDR vs. First-Stage Resolution.

1st-stage resolution	Gain Error	SNDR (dB)	SFDR (dB)
1	2^{-11}	72.1	79.6
2	2^{-10}	71.3	79.6
3	2^{-9}	71.0	81.5
4	2^{-8}	70.9	83.1
5	2^{-7}	70.9	86.3
6	2^{-6}	70.9	88.3

Table 2: Performance Summary

Resolution	12 bits
INL / DNL @Fin = 10 MHz	1.3 LSB / 0.8 LSB
Analog Input	1.6 Vp-p differential
Input Capacitance (T/H)	2 pF
Latency	4
Clock Jitter	0.61 ps
Bit Error Rate @Fs = 100 MHz	< 10^{-8} (no error measured so far)
Conversion Rate	50 MS/s
SFDR @Fin < 75 MHz	> 80 dB
SNR @Fin = 1 MHz / 200 MHz	64 dB / 60 dB
Power Dissipation	850 mW
Min. Power Supply Voltage	3.05 V
Technology	3.3-V, 0.6-µm CMOS (1p3m + lin. cap.)
Active/Total Die Area	4x4 mm^2 / 5.5x4.8 mm^2

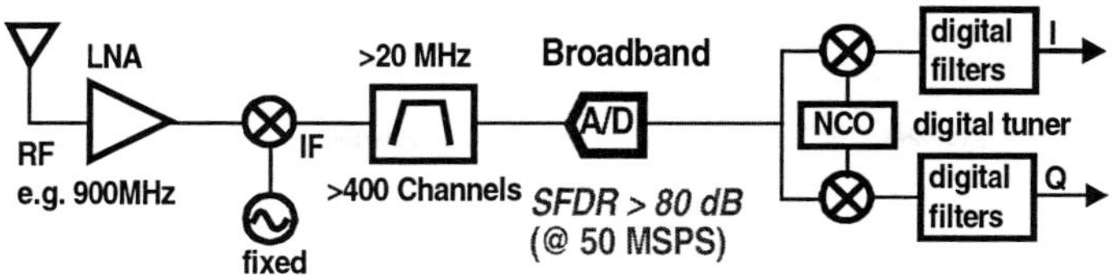

Figure 1: Wideband digital receiver.

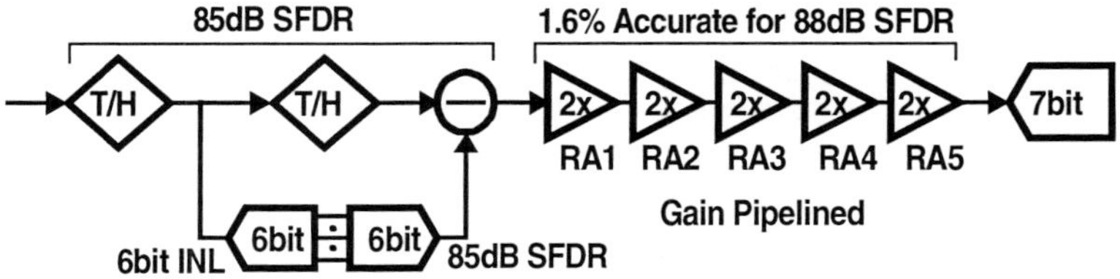

Figure 2: The 12b two-stage ADC.

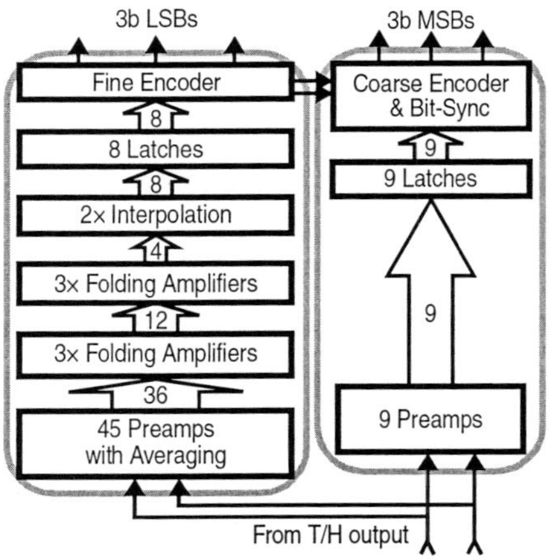

Figure 3: The 6b sub-ADC.

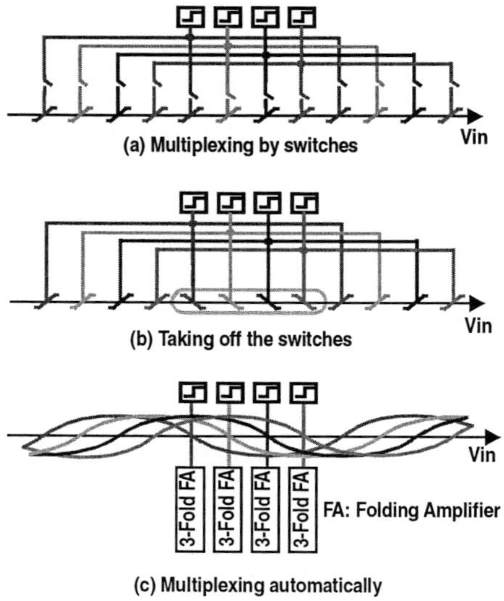

Figure 5: Automatic multiplexing of ZX's.

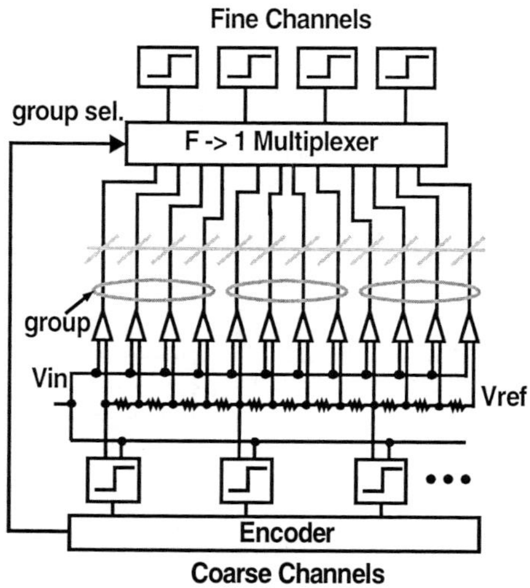

Figure 4: Serial Multiplexing of ZX's.

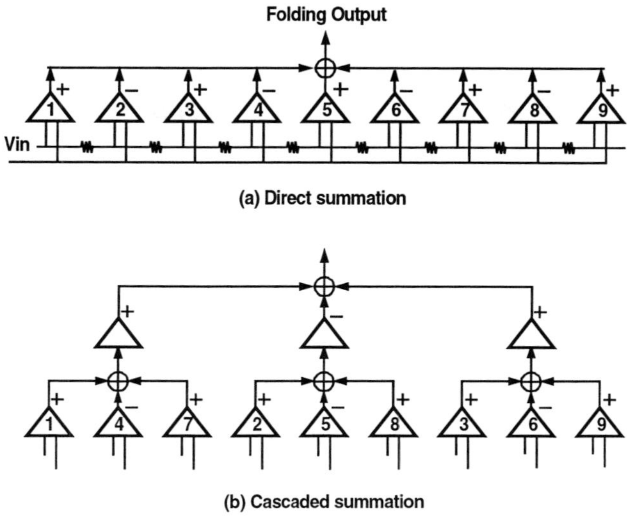

Figure 6: Folding by summation.

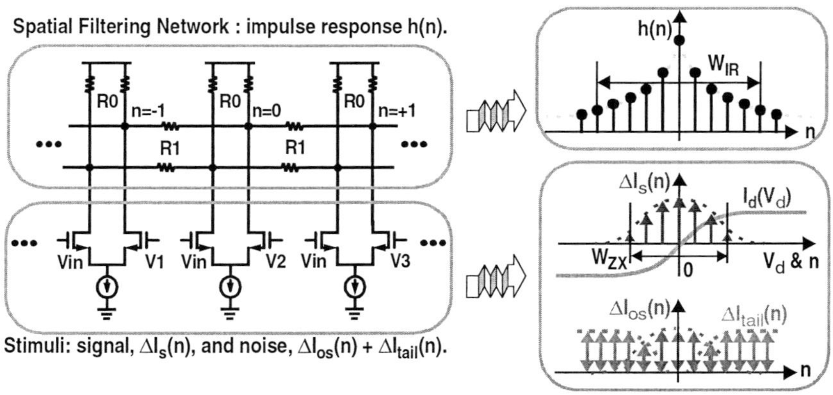

Figure 7: Offset averaging modeled as spatial filtering.

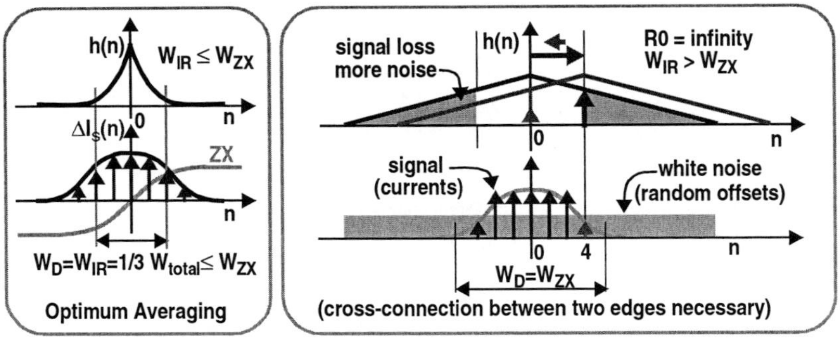

Figure 8: Optimal averaging vs. R0 = infinity.

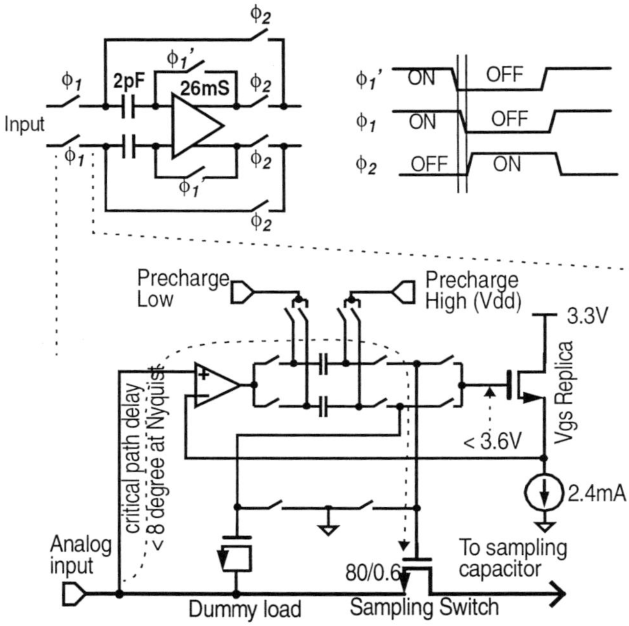

Figure 9: T/H configuration and the bootstrap circuit linearizing the sampling switch.

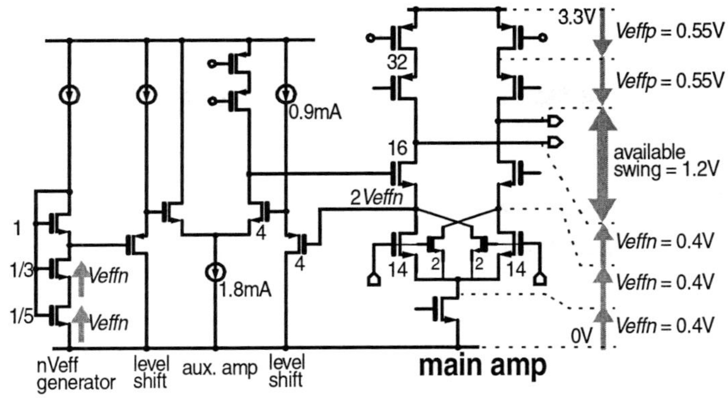

Figure 10: Multiple-Veff based design of T/H OTA.

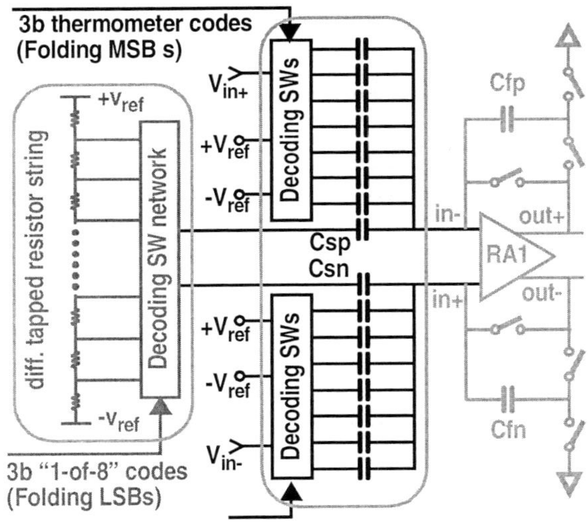

Figure 11: Two-step hybrid reconstruction DAC.

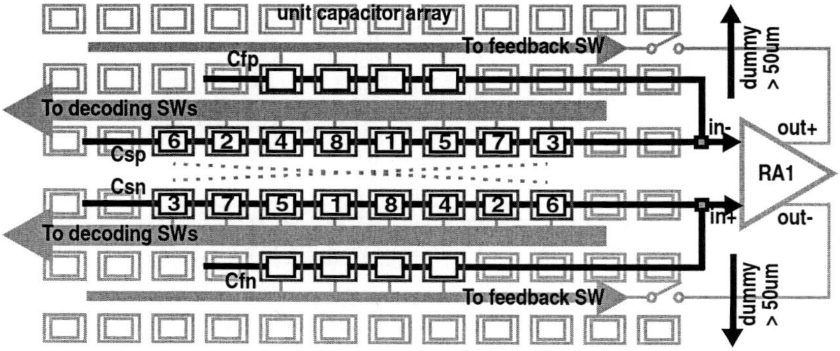

Figure 12: DAC capacitor layout.

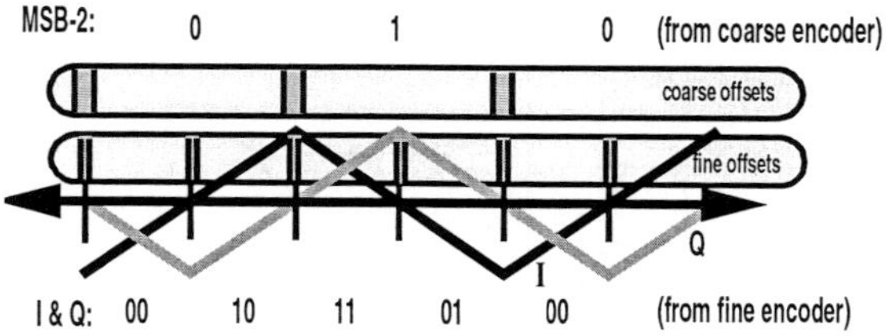

Figure 13: Bit-sync.

MSB-2	I	Q	code shift
0	0	0	same
0	0	1	down
0	1	0	same
0	1	1	up
1	0	0	up
1	0	1	same
1	1	0	down
1	1	1	same

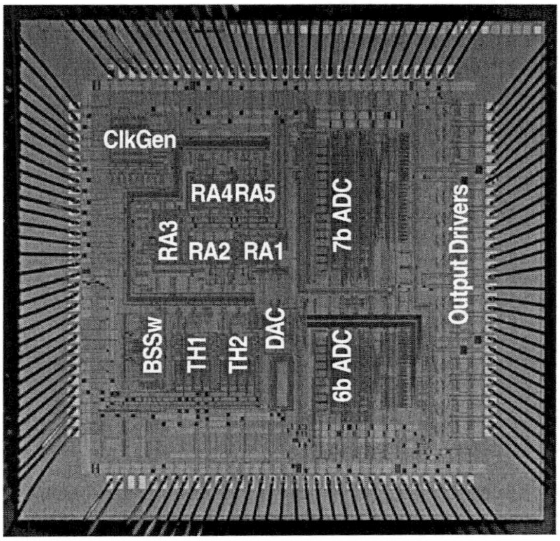

Figure 14: The 12b ADC die photograph.

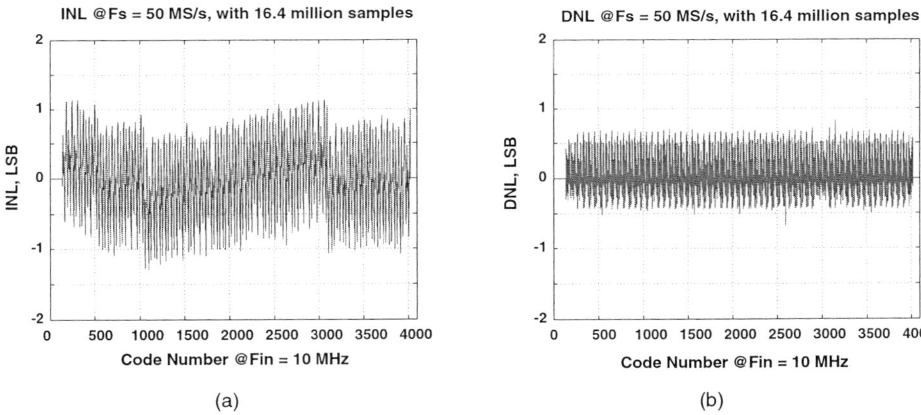

Figure 15: Code density test results.

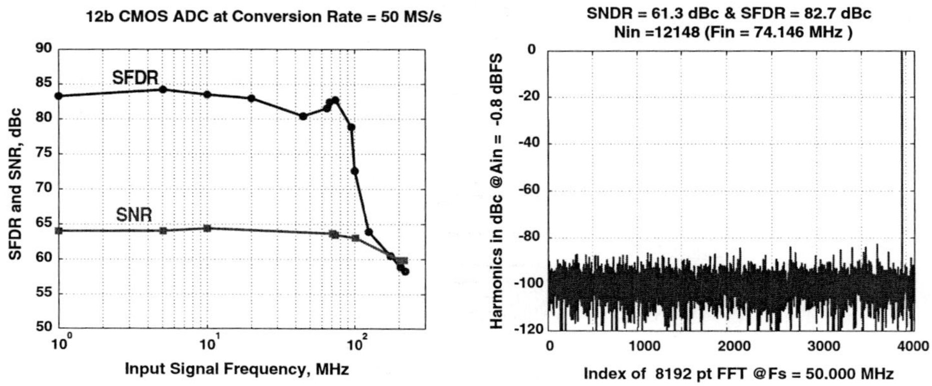

Figure 16: The measured dynamic performance.

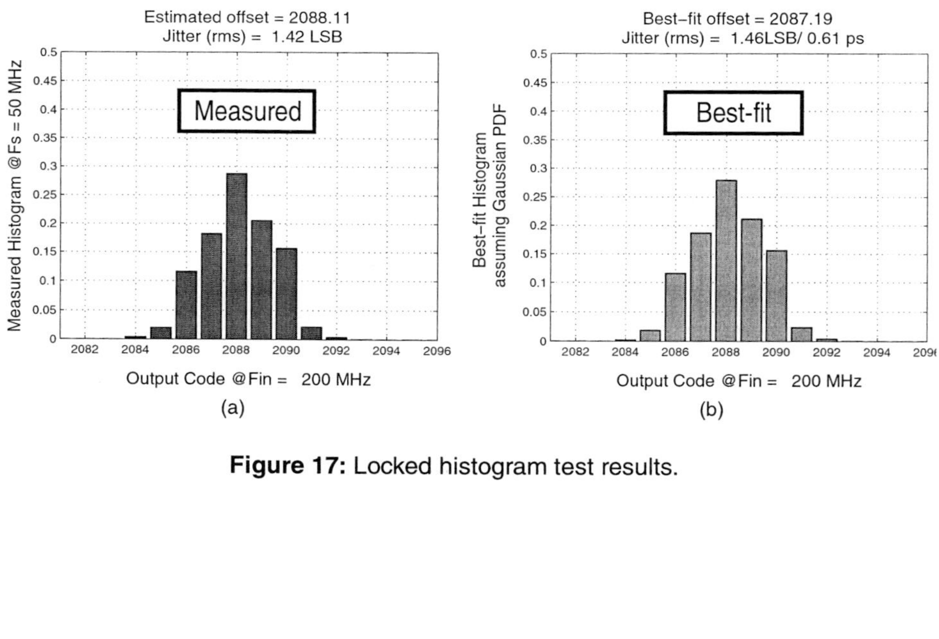

Figure 17: Locked histogram test results.

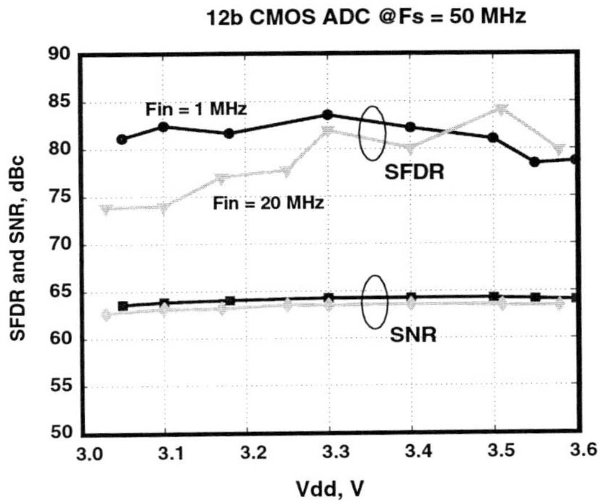

Figure 18: Vdd Tolerance Test.

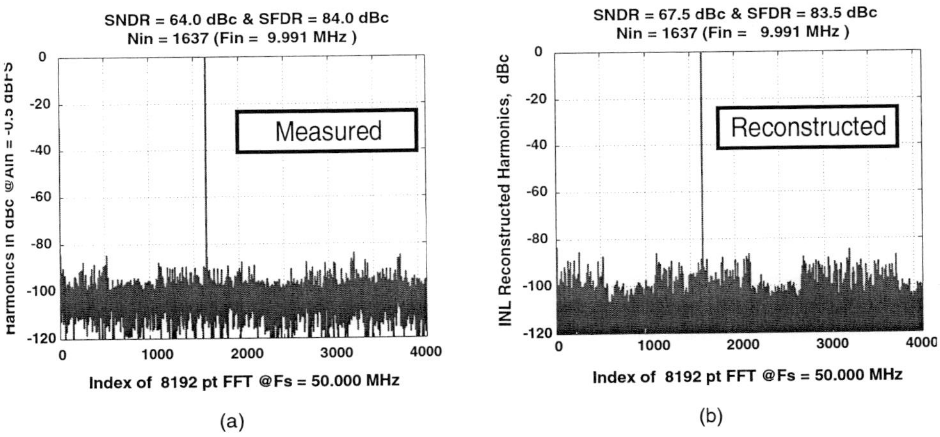

Figure 19: Directly measured spectrum vs. the INL reconstructed.

A 10-bit, 20-30 MSPS CMOS Subranging ADC with 9.5 Effective Bits at Nyquist

Brian Brandt and Joseph Lutsky
National Semiconductor
Salem, New Hampshire, USA

Abstract

A CMOS subranging ADC incorporates several features to enhance performance and reduce power dissipation. The combination of an extended settling period for the fine references, absolute-value signal processing, and interpolation in the comparator banks alleviates the principal speed-limiting operation. A front-end sample-and-hold amplifier (SHA) provides sustained dynamic performance at high input frequencies and performs single-ended to differential conversion with a signal gain of two and with low distortion. The SHA holds its differential output for a full clock cycle while it simultaneously samples the next single-ended input, thereby allowing it to drive two comparator banks on consecutive clock phases. The remaining analog circuits are implemented in a fully differential manner. The use of pipelining allows every input sample to be processed by the same channel, thereby avoiding the use of ping-pong techniques, while providing a conversion latency of only two clock cycles. The dynamic performance with a single-ended input approaches that of an ideal 10-bit ADC, typically providing 9.7 effective bits for low input frequencies and 9.5 bits at Nyquist. This performance level is comparable to the best reported for 10-bit CMOS ADCs with differential inputs and significantly better than those with single-ended inputs. The typical maximum DNL is ±0.4 LSB and the maximum INL is ±0.55 LSB without trimming or calibration. At 20 MSPS, the ADC power is 55mW and the SHA power is 20mW from a 5V supply. The active area is 1.6 sq. mm in a 0.5-μm double-poly, double metal CMOS technology.

R. J. van de Plassche et al. (eds.), Analog Circuit Design, 75–94.

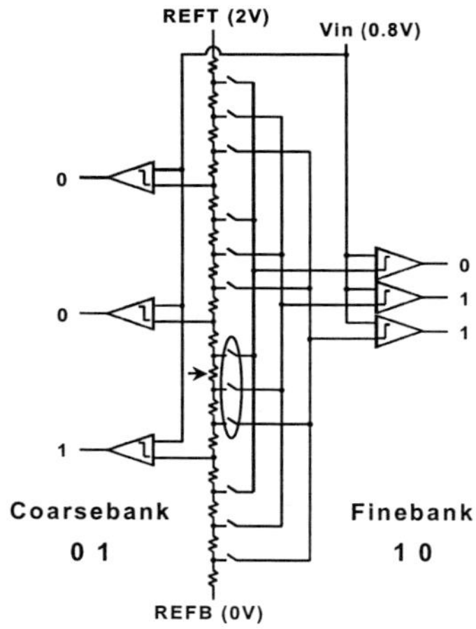

Fig. 1. Simple 4-bit CMOS subranging ADC.

I. Introduction

Many imaging, video, and digital communication applications require low-cost, low-power analog-to-digital converters (ADC) with 8-10 bits of resolution and sampling rates between 10 and 50 MSPS. The two-stage CMOS subranging ADC (CSA) introduced by Dingwall and Zazzu [1, 2] and developed by several others [3-7] has found widespread commercial use in these applications. The simple example of a 4-bit CSA shown in Fig. 1 consists of a 2-bit coarse comparator bank (coarsebank), a 2-bit fine comparator bank (finebank), a reference resistor ladder, and a 3-output analog multiplexor (AMUX). The coarsebank determines in which of the four coarse reference segments the analog input lies. This information is encoded into the two most-significant bits (MSBs) of the 4-bit ADC output and it is used to signal the closing of the proper set of three switches in the AMUX. The AMUX outputs, which serve as the reference voltages for the fine comparators, are thereby connected to the appropriate internal tap points within the selected coarse segment. The fine comparators then determine in which of the four bins of the chosen coarse segment the input lies. This information is encoded into the two least-significant bits (LSBs) of the ADC output.

The number of comparators in a CSA is considerably lower than in a flash ADC and is proportional to $2^{N/2}$, where N is the resolution.

For example, only 62 comparators are nominally required for a 10-bit ADC. Thus, a significant reduction in power, die area, and input capacitance are provided by this architecture without the precise linear subtraction or high-speed, high-gain amplifiers that are required by other two-stage and multi-stage architectures. While reference resistor matching is relatively important, requirements for capacitor matching and linearity are not stringent. This allows CSAs to achieve at least 10-bit integral and differential linearity without trimming or calibration. Other features of CSAs include low conversion latency (comparable to flash and folding converters) and a minimal requirement for digital support logic.

Along with the features of CSAs come several limitations that have been addressed in the implementation reported here. First, the number of switches in the AMUX is approximately 2^N. Thus, for a 10-bit CSA, the parasitic capacitance of over 1000 switches must be swung over the entire input range by the resistor ladder as the AMUX outputs follow the ADC input from one sample period to the next. This operation usually limits the speed of CSAs because the fine references typically must settle to the resolution of the ADC within half a clock cycle. The timing described in Section IV provides a full clock cycle for the fine references to settle. In addition, the absolute value signal processing discussed in Section III reduces the maximum swing of the fine references by half. Lastly, the use of interpolation in the comparator banks, which is described in Section VI, reduces the required number of fine references by 75%.

With some exceptions [8,9], previously reported CSAs have typically exhibited significantly reduced dynamic performance for input frequencies near the Nyquist limit. The front-end sample-and-hold amplifier (SHA) presented in Section V provides sustained dynamic performance for high input frequencies and also buffers the ADC input from the large switched input capacitance of the comparator banks. The SHA also converts the single-ended ADC input to the differential form used internally. In contrast to earlier single-ended CSAs, the differential architecture presented in Section II is less sensitive to digital noise and it suppresses even-order harmonic distortion and the effects of charge injection. Many previous CSAs interleave the operation of multiple coarse and fine comparator banks to increase their throughput to one conversion per clock cycle. As in [7], the architecture presented in Section II uses pipelining to allow every input sample to be processed by the same signal channel. Measured results that characterize the performance of the CSA are presented in Section VII.

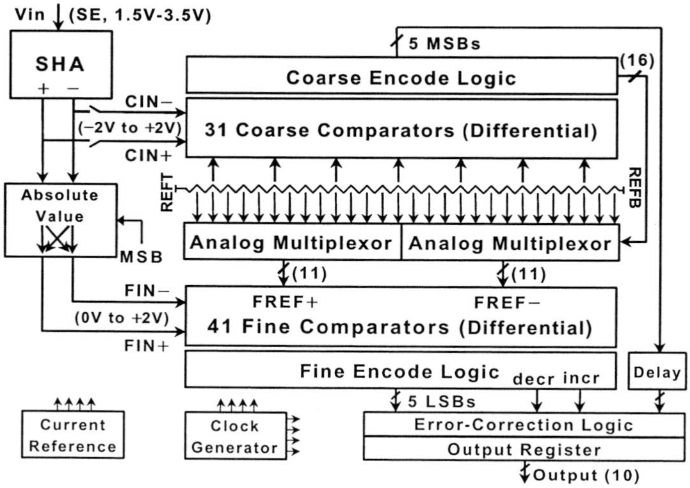

Fig. 2. Architecture of the CSA.

II. Architecture

Fig. 2 shows the architecture of a CSA that addresses the limitations described in Section I. The 10-bit CSA includes a 5-bit coarsebank and a 5-bit finebank. Each of the 31 comparators in the coarsebank compares the differential output of the SHA to a unique differential coarse reference voltage derived from the resistor ladder that spans horizontally across the center of the ADC. The finebank has five additional comparators on each end of the 31 nominal comparators. These ten additional comparators generate signals used by the error-correction logic to increment or decrement the five MSBs in the event of an error in the coarse conversion [3]. In contrast to the simple example in Fig. 1, there are two analog multiplexors in the architecture of Fig. 2. The left AMUX selects a set of taps from the left half of the resistor ladder to serve as the positive components of the differential references for the finebank, while the right AMUX selects a set from the right half to serve as the negative components.

The single-ended input to the ADC, Vin, is sampled by an integrated sample-and-hold amplifier. A unique feature of this SHA is that it converts the single-ended ADC input to differential outputs that swing between the reference voltages REFB and REFT, nominally set at 1.5V and 3.5V respectively. A nominal 2-Vpp ADC input between 1.5V and 3.5V is converted to a 4-Vpp differential signal between –2V and +2V, thereby increasing the internal LSB size from 2mV to 4mV. A second unique feature of this SHA is its ability to drive both the coarsebank and the finebank on two succeeding half-period clock intervals while simultaneously sampling the next ADC input, as described in Section V.

III. Absolute Value Signal Processing

The differential output of the SHA is sampled directly by the coarsebank but it is rectified by an absolute value (ABSVAL) circuit before being passed to the inputs of the finebank. The ABSVAL circuit is a set of switches that directly passes positive differential inputs and swaps negative differential inputs. The sign of the ABSVAL input is indicated by the MSB of the ADC, which is determined by the coarsebank. As an example, single-ended inputs in the top half of the ADC input range (nominally 2.5V to 3.5V) are mapped to positive differential signals (0V to 2V) by the SHA and result in an MSB of 1. In this case, the ABSVAL circuit passes the SHA outputs directly to the finebank inputs. Single-ended inputs in the bottom half of the ADC input range (1.5V to 2.5V) are mapped to negative differential signals (-2V to 0V) by the SHA, resulting in an MSB of 0. An MSB of 0 signals the ABSVAL to swap the SHA outputs before passing them to the finebank.

Since the input to the finebank is always a positive differential signal, only positive differential fine references need to be supplied from the resistor ladder by the AMUXs. For example, a single-ended ADC input of 3.5V will be presented as a +2-V differential input to both the coarsebank and the finebank. The two AMUXs will select sets of taps from the two ends of the resistor ladder to serve as the differential fine references. As the ADC input is progressively reduced from 3.5V to 2.5V, the differential input to both comparator banks will be correspondingly reduced from +2V to 0V and the AMUXs will select sets of taps progressively closer to the center of the resistor ladder. As the input is further reduced from 2.5V to 1.5V, the differential input to the coarsebank will decrease from 0V to -2V, but the input to the finebank will be *increased* from 0V to +2V because of the absolute value processing. Accordingly, the AMUXs will again select sets of taps progressively closer to the two ends of the resistor ladder.

Instead of the two AMUXs each spanning the entire reference ladder, as would be required in a straightforward differential implementation, each need only span half because of the absolute value processing. This results in 50% fewer switches, or the same total number as in the single AMUX of a single-ended implementation. However, compared to a single-ended implementation, the parasitic capacitance on each set of AMUX outputs is reduced by half. Furthermore, each set need only swing half as far, in worst case, from one sample period to the next. Since the settling of the AMUX outputs often limits the speed of CSAs, these reductions are especially important.

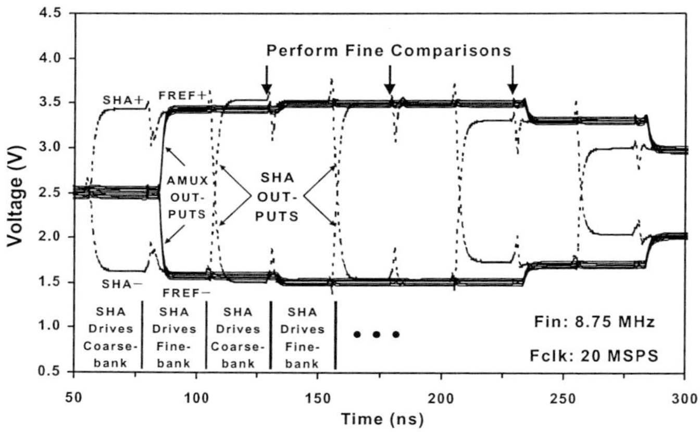

Fig. 3. Simulated waveforms that demonstrate the reduced swing of the fine references (AMUX outputs).

The reduced swing of the fine references is demonstrated in Fig. 3, which shows simulations of the ADC's operation for a sinusoidal input. The outputs of the SHA swing back and forth between positive and negative differential values corresponding to the 8.75-MHz input frequency, which is near the Nyquist limit for a sampling rate of 20 MSPS. Fig. 3 also shows the two sets of AMUX outputs that serve as the positive and negative components of the differential references for the fine comparators. With some delay, the fine references follow the magnitude of the SHA differential output. However, the references do not follow the sign of the SHA output and therefore they need only swing over half the reference range instead of the full range as in a conventional CSA.

A further benefit of the absolute value processing is that the worst-case transient charge that is switched onto the SHA outputs when the finebank samples it (via the ABSVAL switches) is reduced by half. This effectively reduces the maximum output step of the SHA by 50% during this critical settling period.

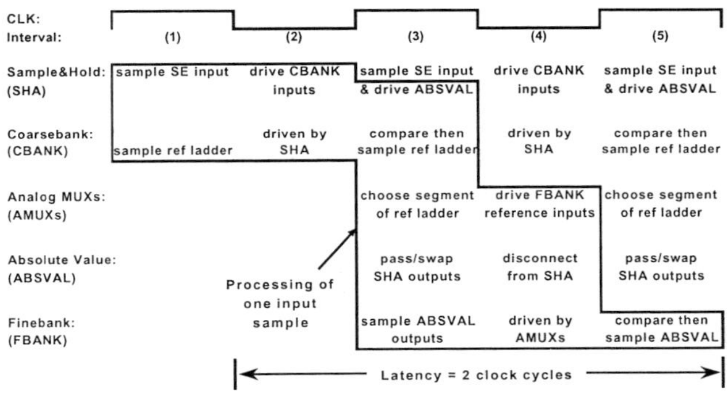

Fig. 4. Timing diagram.

IV. Timing

The timing diagram presented in Fig. 4 details the processing of one ADC input sample by the five main analog blocks of the CSA. During clock interval 1, the SHA samples the ADC input while the coarsebank differentially samples the reference ladder. The ADC input sampling instant is defined by the falling edge of the clock, which defines the end of interval 1. The SHA then holds its output and drives the coarsebank during interval 2. The comparator amplifiers in the coarsebank amplify the difference between their individual differential coarse references sampled in interval 1 and the differential output of the SHA. Once the amplification is complete, the coarse comparators are strobed at the beginning of interval 3 and then the coarsebank immediately returns to sampling the reference ladder in preparation for the next coarse conversion. Once the coarse comparators are strobed at the beginning of interval 3, the analog multiplexors select the proper segments from the top and bottom halves of the resistor ladder. The absolute value circuit passes or swaps the outputs of the SHA to provide the rectified input for the finebank, which is in sampling mode throughout interval 3. During interval 4, the AMUXs drive the differential reference inputs of the finebank. The differences between those references and the ABSVAL output sampled during interval 3 are amplified in the fine comparators. The fine comparators are strobed at the beginning of interval 5 and then the finebank immediately returns to sampling the ABSVAL outputs in preparation for the next fine conversion.

The digital outputs of the ADC change at the start of interval 6, just two clock cycles after the sampling instant at the end of interval 1. While the conversion latency is low, the timing presented in Fig. 4 actually provides more time for the fine references to settle than the half clock cycle provided in conventional CSAs. The fine references have a full clock cycle to settle from the time they start to change at the beginning of interval 3 to when the fine comparisons are performed at the beginning of interval 5. This extended settling period, along with the 50% reduction in the swing and capacitance of the fine reference lines described in Section III, alleviates the principal speed-limiting operation in CMOS subranging ADCs.

The processing of the second ADC input sample begins in interval 3, one clock cycle after beginning the processing of the first input sample. Thus, through the use of pipelining, a throughput of one output per clock cycle is achieved while allowing each input sample to be processed by the same channel. This processing avoids the deleterious effects of interleaving ("ping-pong") techniques that are often used in CSAs.

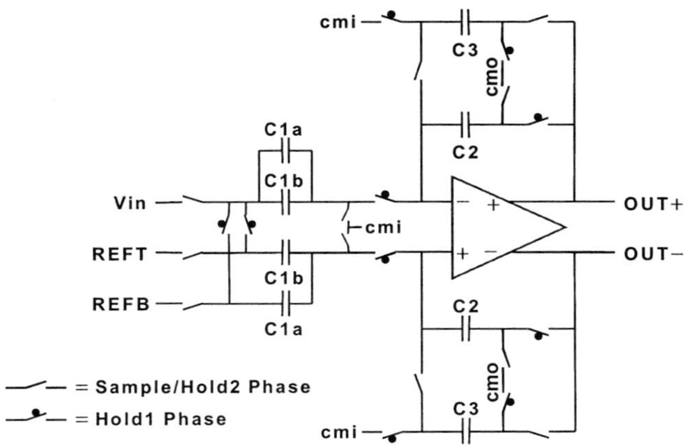

Fig. 5. SHA and single-ended to differential converter.

V. Sample-and-Hold Amplifier

The front-end SHA included in Fig. 2 enables the pipelined operation described in Section IV. The switched-capacitor implementation of the SHA depicted in Fig. 5 converts a single-ended input to differential form with better than -70 dB THD in a manner similar to [10]. However, the topology in Fig. 5 is capable of holding its output for a full clock cycle so that it can drive the coarsebank during interval 2 and then, during interval 3, drive the finebank while it simultaneously samples the next ADC input. While similar in some respects to previously published non-resetting SHA topologies, the SHA in Fig. 5 does not use interleaving techniques as in [11] so that every input sample is processed by the same circuit elements. Furthermore, the SHA does not depend on capacitor matching to cancel its previous output as in [12, 13].

The SHA operates on a two-phase, non-overlapping clock scheme. During the sample phase, which coincides with the second half of the hold operation, the switches without black dots in Fig. 5 are closed. The right sides of the four sampling capacitors (C1) are connected to cmi, a low-impedance bias that sets the common-mode input voltage of the fully-differential output transconductance amplifier (OTA). The left sides of the upper pair of sampling capacitors track the single-ended input, Vin. The left sides of the lower pair of sampling capacitors effectively sample the midpoint of the input range (i.e. the common mode of the references) by capacitively interpolating between REFT and REFB. The C3 capacitors provide feedback around the OTA and force its inputs to virtual ground so that both C2 capacitors are reset, with their left sides at the cmi potential and their right sides at the cmo potential. Cmo is a bias voltage that defines the common-mode output voltage of the OTA and it is set midway between REFT and REFB, nominally 2.5V. The sampling instant is defined by opening the switches that connect the right sides of the four sampling capacitors to cmi. A short time later, the rest of the switches are opened and the other set of switches (those denoted with black dots) are closed to configure the circuit into the first hold phase. Note that all the switches that operate near the cmi potential are NMOS-only switches and open slightly ahead of the other switches, which are CMOS switches, on their respective phases.

During the first hold phase, the left sides of all four sampling capacitors (C1) are shorted together, thereby forcing their differential charge onto the C2 capacitors, which are configured in feedback around the amplifier. Since all the capacitors in Fig. 5 have the same value (1 pF), the charge transfer from the C1 capacitors to the C2 capacitors causes a voltage gain of two. A 2-Vpp single-ended input is converted to a 4-Vpp differential output, which drives the coarsebank input during this phase. The C3 capacitors are reset with their left sides connected to cmi and their right sides connected to cmo.

The second hold phase coincides with the next sample phase. The act of resetting the C2 capacitors described earlier forces their charge onto the C3 capacitors. Since the C2 and C3 capacitors have the same value, the differential output of the SHA remains unchanged, enabling it to drive the finebank input. The sampling network is isolated from the amplifier at this time and can therefore sample the next single-ended input, Vin. This pipelining in the SHA allows three operations to be performed in two clock intervals without resorting to ping-pong techniques. The full-cycle hold capability of the SHA is evident in the simulated waveforms presented in Fig. 3.

Any size mismatch between the C2 and C3 capacitors results in a small change in the differential output of the SHA during the second half of the hold operation. Essentially, the sample-to-hold gain of the SHA is C1/C2 during the first half of the hold operation and C1/C3 during the second. However, only the gain during the second half is critical because that is when the finebank samples the SHA. If C1/C2 is not exactly equal to C1/C3, an error can occur in the coarse conversion, but this error is corrected by the error-correction comparators and logic included in Fig. 2. In addition, the offset voltage of the OTA causes a change in the SHA output during the second half of the hold operation. The magnitude of this change is equal to the offset voltage, which is only a few millivolts because of the large devices used in the OTA. Coarse conversion errors that result from the offset voltage are also corrected by the error-correction comparators and logic.

The topology of the SHA in Fig. 5 allows the common-mode input and output voltages of the OTA to be set independently. Furthermore, the common-mode input voltage of the amplifier is well controlled and independent of the ADC input voltage, Vin, except for second-order effects. Consequently, a telescopic cascode OTA was chosen because it provides high-speed low-distortion settling for less than half the power of an equivalent folded-cascode amplifier. Switched-capacitor feedback circuitry was employed in a standard manner to maintain the common-mode output voltage of the amplifier.

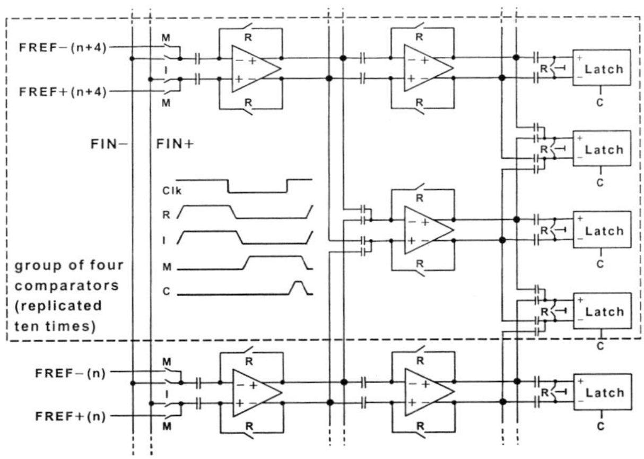

Fig. 6. Section of the fine comparator bank.

VI. Interpolated Comparator Banks

The fine comparator bank, a section of which is depicted in Fig. 6, compares its differential input to pairs of differential references chosen from the resistor ladder by the AMUXs. The top and bottom rows in Fig. 6 are conventional multistage, offset-canceling comparators consisting of two high-speed gain stages followed by a regenerative latch. However, the inner latches and amplifier share the outputs of the amplifiers in the top and bottom rows through capacitive interpolation. As described in [7], interpolation provides several benefits, including a 75% reduction in the number of fine references (AMUX outputs) and in the number of taps in the reference ladder (256 instead of 1024). Note that only one pair of fine references is required for every four comparators in Fig. 6. Other benefits include a reduction in the number of amplifiers, lower power and die area, and significantly lower comparator bank input capacitance. Lastly, interpolation attenuates front-end sources of DNL, such as charge injection mismatch in input sampling switches and resistor mismatch in the reference ladder, by 75%.

In contrast to the single-ended implementation based on inverter gain stages presented in [7], the interpolation scheme in Fig. 6 is fully differential to provide greater immunity to digital noise and better charge injection cancellation. In addition, only two stages of gain are used in front of the latches to reduce the delay through the comparator bank. Consequently, pipelining is not necessary in the amplifier stages, which all operate on the same clock signals as indicated in Fig. 6.

Interpolation is also used in the coarsebank. However, because error correction relaxes the accuracy required in the coarsebank, only one stage of gain is used to reduce the input-referred offset of the coarse latches. As a result, the coarsebank uses interpolation by two instead of four. It also uses 75 fF input sampling capacitors instead of the 150 fF sampling capacitors used in the finebank.

Each differential interpolation amplifier has a gain of about seven and consists of an NMOS differential pair, an NMOS tail current source, PMOS diode-connected loads, and NMOS reset switches. A regenerative latch similar to [14] was used in both comparator banks. Common-centroid layout techniques and non-minimum device sizes were used to reduce the offset voltage of the regenerative latch to less than 6 mVrms.

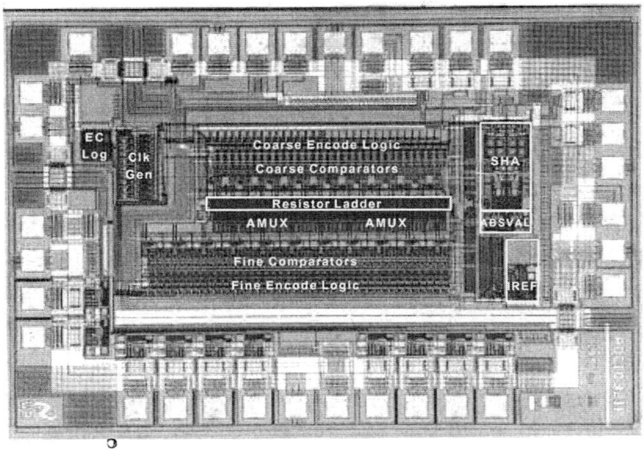

Fig. 7. Die photograph.

VII. Measured Results

Fig. 7 shows the CSA as implemented in a 0.5μm double-poly, double-metal CMOS technology. Note that because capacitor matching and linearity are not critical in CSAs, double-poly capacitors were not required to implement the core of the ADC but were available at low cost and resulted in a small decrease in power and die area. As in many switched-capacitor circuits, the distortion performance of the SHA does depend on capacitor linearity. Capacitors formed between metal interconnect layers could be used to implement the SHA in a single-poly process at a small increase in overall ADC power and die area.

The reference resistor ladder in the center of the die is a continuous piece of polysilicon with 256 kelvin taps and without bends or corners. Since both comparator banks access the resistor ladder in a differential manner, the reference range is effectively doubled. Furthermore, linear process gradients in the polysilicon are cancelled. Both factors improve the static and dynamic performance of the ADC. Also, because of the absolute value signal processing and extended settling period for the fine references, the settling time of the reference ladder is not critical and a relatively large ladder resistance (1.3 K-Ohm) was used to reduce power dissipation.

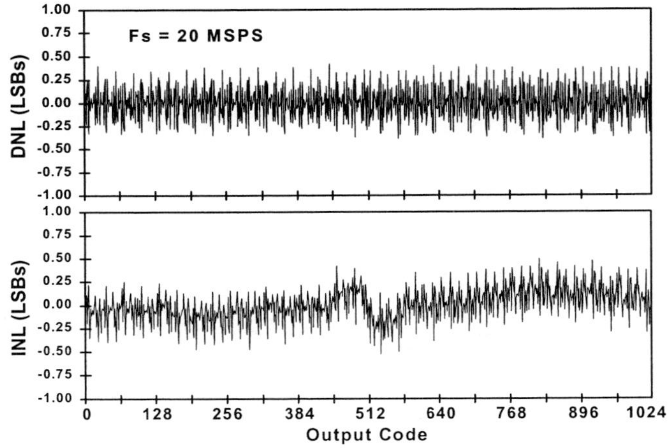

Fig. 8. Typical measured dc performance.

Fig. 8 presents the measured DC performance of a typical part at a sampling rate of 20 MSPS. The maximum DNL is less than 0.4 LSB and the maximum INL is 0.5 LSB. The disturbance near midrange in the INL data has a magnitude less than half an LSB and is not seen in all parts

All dynamic performance results were obtained using a low-phase-noise sinusoidal signal source to drive the ADC input through low-distortion low-pass filters. The dc component of the sinusoidal input was centered between REFT and REFB, which were held by external buffer amplifiers at 3.5V and 1.5V, respectively. However, the performance did not change noticeably either upon altering the dc component of the input or upon varying the center of the references from 1.9V to 3.4V. Digital output data was acquired into a high-speed buffer and then transferred to a computer for analysis. The frequency of the sinusoidal input source was locked to a low-jitter clock source so that a prime number of sinewave periods was acquired and data windowing was unnecessary. Frequency domain analysis was performed to extract the distortion and noise components separately.

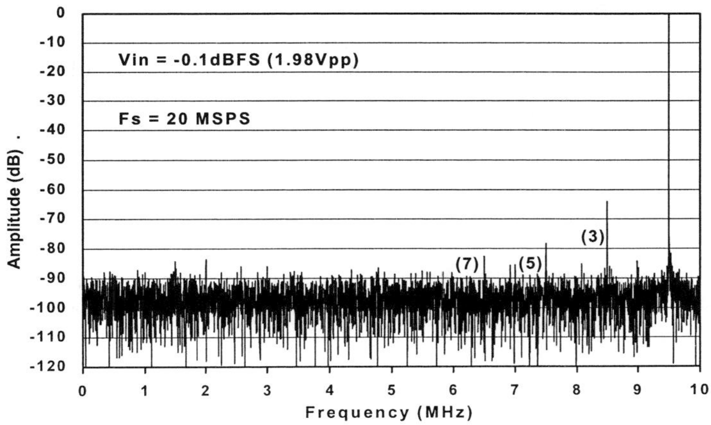

Fig. 9. Typical measured spectrum for a 9.5 MHz input.

Fig. 9 presents a typical measured output spectrum for an input frequency near the Nyquist limit. The 9.5 MHz fundamental is evident along with the third, fifth, and seventh harmonics, which have been aliased down into the signal band. The third harmonic is 64 dB below the –0.1 dBFS (1.98 Vpp) input. The fifth and seventh harmonics are about 80 dB below the input.

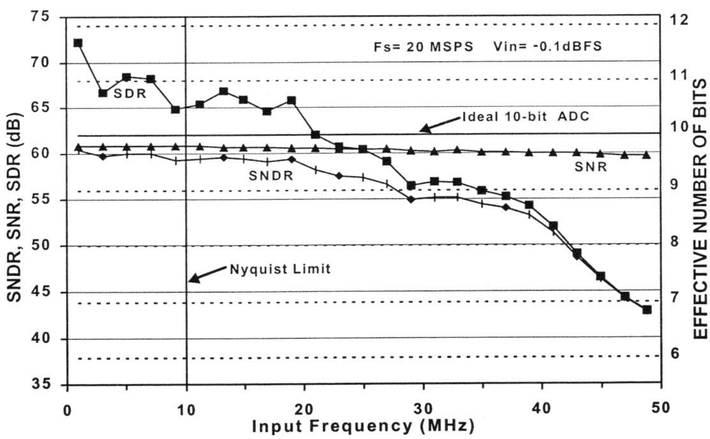

Fig. 10. Dynamic performance as a function of input frequency.

Good dynamic performance is maintained for input frequencies beyond the Nyquist limit, as demonstrated in Fig. 10. This figure shows the typical measured signal-to-distortion ratio (SDR), signal-to-noise ratio (SNR), and signal-to-(noise+distortion) ratio (SNDR) as a function of the input frequency. For low input frequencies, the distortion performance approaches the 12-bit level. For input frequencies below two times the Nyquist limit, both the SNR and SNDR approach that of an ideal 10-bit converter. Furthermore, the harmonic distortion does not contribute significantly to the SNDR despite the use of a single-ended input in this ADC. Earlier publications [10] have reported difficulty in implementing video-rate switched-capacitor single-ended to differential conversion with low distortion.

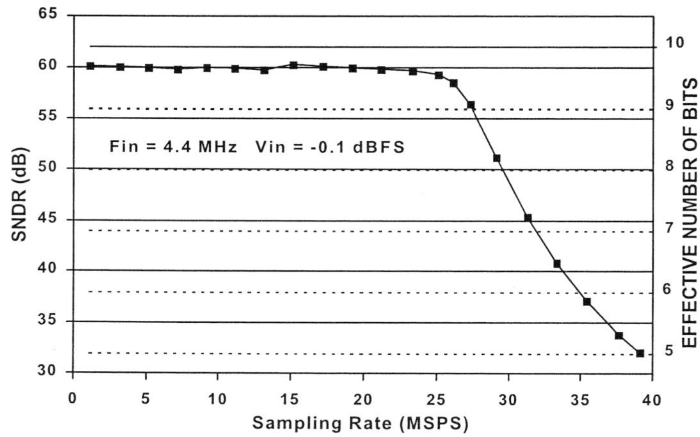

Fig. 11. Dynamic performance vs. sampling frequency.

Above two times the Nyquist limit the SNDR is limited by distortion, which is thought to arise from the input sampling network of the SHA. However, at least 9-bit performance is maintained for frequencies nearly three times the Nyquist limit and 8-bit performance is achieved at over four times Nyquist. Note that the SNR is maintained at near-ideal levels for inputs at least as high as 50 MHz. Fig. 11 presents the performance of the ADC as a function of the sampling rate. For a video-frequency (4.4 MHz) input, the sampling rate of the converter can be extended to 27 MSPS while maintaining better than 9-bit performance.

Max Conversion Rate for Video Inputs:	27 MSPS
Max Conversion Rate for Nyquist Inputs:	20 MSPS
Conversion Latency:	2 clock cycles
DNL:	±0.4 lsb
INL:	±0.55 lsb
Effective Bits at Fin=1, 4.4, 10 MHz:	9.7, 9.6, 9.5 bits
SNDR at Fin=1, 4.4, 10 MHz:	60, 59.5, 58.7 dB
SNR at Fin=1, 4.4, 10 MHz:	60.5, 60.5, 60.5 dB
THD at Fin=1, 4.4, 10 MHz:	−71, −67, −64 dB
SFDR at Fin=1, 4.4, 10 MHz:	73, 69, 64 dB
Supply Voltage:	5 V
Nominal Input Range (single-ended):	2 Vpp
Reference (REFB, REFT) Range:	0.9 - 4.4 V
Input Capacitance (incl. ESD & pad):	4 pF
Core ADC Power (no SHA):	55 mW
Total Power (with SHA):	75 mW
Active Area:	1.6 mm^2
Technology:	0.5μ CMOS (2-poly, 2-metal)

Table I. Typical performance at 20 MSPS.

Table I lists the typical measured performance parameters of the CSA at a sampling rate of 20 MSPS. Of particular note are the low DC and dynamic nonlinearity and the sustained dynamic performance at high input frequencies, all of which are provided in a small die area, at low power consumption, and without trimming or calibration. While initially this product was intentionally targeted at 20-MSPS video and imaging applications, the architectural features described earlier result in the important advantage that the speed of this CSA is not limited by the settling of the fine references. Therefore, the speed can be increased by allocating more power to the SHA and comparators, both of which were only given sufficient power for 20-MSPS operation. Table II lists the typical measured performance parameters at 30 MSPS when the power is increased from 75 mW to 120 mW. Note that nearly identical performance is maintained at 30 MSPS at only a slightly higher ratio of power to sampling rate.

DNL:	±0.4 lsb
INL:	±0.45 lsb
Effective Bits at Fin=1, 4.4, 15 MHz:	9.7, 9.4, 9.3 bits
SNDR at Fin=1, 4.4, 15 MHz:	60, 58.5, 58 dB
SNR at Fin=1, 4.4, 15 MHz:	60, 59, 59.5 dB
THD at Fin=1, 4.4, 15 MHz:	–74, –68, –63 dB
SFDR at Fin=1, 4.4, 15 MHz:	77, 70, 66 dB
Nominal Input Range (single-ended):	1.75 Vpp
Core ADC Power (no SHA):	80 mW
Total Power (with SHA):	120 mW

Table II. Typical performance at 30 MSPS.

While the first implementation of the CSA is for a 5V power supply, the architecture can be migrated to lower supply voltages through straightforward redesign of the SHA and interpolation amplifiers. In this implementation, the 5-V supply provides the flexibility of a wide range of reference voltages (0.9V to 4.4V). In addition, while the first implementation is a discrete ADC product, the relatively low power and die area, along with the differential topology, will facilitate use of this CSA as a cell in highly-integrated system chips.

VIII. Conclusion

A CMOS subranging ADC has been reported that addresses several limitations found in earlier implementations. This is the first CSA to employ an integrated front-end SHA. In addition to extending good dynamic performance to input frequencies well beyond the Nyquist limit, the SHA buffers the ADC input from the large (switched) input capacitance of the comparator banks. It also performs single-ended to differential conversion with a gain of two and with low distortion. The output of the SHA is held for a full clock cycle without employing ping-pong techniques, allowing it to drive both the coarsebank and the finebank on consecutive clock phases.

The combination of an extended settling period for the fine references, absolute-value signal processing, and interpolation in the comparator banks results in the important benefit that the speed of this CSA is not limited by the settling of the fine references. Interpolation also provides several additional advantages and it has been implemented in a fully differential manner to provide increased noise immunity and to double the internal signal swing. By accessing the reference ladder differentially, the linearity of the ADC is improved by doubling the reference range and by canceling linear process gradients. This feature may help extend the usefulness of the CMOS subranging architecture to applications requiring 12-bit resolution. Lastly, pipelining allows every input sample to be processed by the same signal channel while preserving a conversion latency of only two clock cycles.

As a result of the features listed above, the performance of the CSA with a single-ended input approaches that of an ideal 10-bit ADC without trimming or calibration, providing 9.7 effective bits for low input frequencies and 9.5 bits at Nyquist. While dynamic performance at this level has been reported for 10-bit CMOS ADCs with differential inputs [15], the performance of reported 10-bit CMOS ADCs with single-ended inputs is at least 0.5 effective bits lower.

Acknowledgement

The authors gratefully acknowledge the contributions of the members of the Data Conversion Systems group, especially J. Margolycz, R. Cullen, S. Kwong, M. Courcy, D. Birdsall, C. Hinojosa, T. Nguyen, R. Taft, P. Holloway, R. Subrahmanyan, D. Boisvert, P. Tucci, and D. Archer.

References

[1] A. Dingwall and V. Zazzu, “An 8-MHz CMOS subranging ADC”, *ISSCC Dig. Tech. Papers*, pp. 72-73, Feb. 1985.

[2] A. Dingwall and V. Zazzu, “An 8-MHz CMOS subranging 8-bit A/D converter”, *IEEE J. Solid-State Circuits*, pp. 1138-1143, Dec. 1985.

[3] T. Matsuura, T. Tsukada, S. Ohba, E. Imaizumi, H. Sato, S. Ueda, “An 8b 20MHz CMOS half-flash A/D converter”, *ISSCC Dig. Tech. Papers*, pp. 220-221, Feb. 1988.

[4] N. Fukushima, T. Yamada, N. Kumazawa, Y. Hasegawa, M. Soneda, “A CMOS 40MHz 8b 105mW two-step ADC”, *ISSCC Dig. Tech. Papers*, pp. 14-15, Feb. 1989.

[5] S. Hosotani, T. Miki, A. Maeda, N. Yazawa, “An 8-bit 20-MS/s CMOS A/D converter with 50-mW power consumption”, *IEEE J. Solid-State Circuits*, pp. 167-172, Feb. 1990.

[6] M. Ito, T. Miki, S. Hosotani, T. Kumamoto, Y. Yamashita, M. Kijima, T. Okuda, K. Okada, "A 10 bit 20 Ms/s 3V supply CMOS A/D converter", *IEEE J. Solid-State Circuits*, pp. 1531-1536, Dec. 1994.
[7] K. Kusumoto, A. Matsuzawa, K. Murata, "A 10-b 20-MHz 30-mW pipelined interpolating CMOS ADC", *IEEE J. Solid-State Circuits*, pp. 1200-1206, Dec. 1993.
[8] *ADC1175 Datasheet*, National Semiconductor, 1998.
[9] *ADC1175-50 Datasheet*, National Semiconductor, 1999.
[10] W. Song, H. Choi, S. Kwak, B. Song, "A 10-b 20-Msample/s low-power CMOS ADC", *IEEE J. Solid-State Circuits,* pp. 514-521, May 1995.
[11] B. Rothenberg, S. Lewis, P. Hurst, "A 20-Msample/s switched-capacitor finite-impulse response filter using a transposed structure", *IEEE J. Solid-State Circuits,* pp. 1350-1356, Dec. 1995.
[12] G. Nicollini, P. Confalonieri, D. Senderowicz, "A fully differential sample-and-hold circuit for high-speed applications", *IEEE J. Solid-State Circuits,* pp. 1461-1465, Oct. 1989.
[13] G. Uehara, P. Gray, "A 100 MHz A/D interface for PRML magnetic disk read channels", *IEEE J. Solid-State Circuits,* pp. 1606-1613, Dec. 1994.
[14] G. Yin, F. Op't Eynde, W. Sansen, "A high-speed CMOS comparator with 8-b resolution", *IEEE J. Solid-State Circuits,* pp. 208-211, Feb. 1992.
[15] S. Lewis, H. Fetterman, G. Gross, R. Ramachandran, T. Viswanathan, "A 10-b 20-Msample/s analog-to-digital converter", *IEEE J. Solid-State Circuits,* pp. 351-358, Mar. 1992.

A 2.5MHz Output-Rate Delta-Sigma ADC with 90dB SNR and 102dB SFDR

Ichiro Fujimori, Lorenzo Longo[1], Armond Hairapetian[1], Kazushi Seiyama[2], Steve Kosic, Jun Cao[1], and Shu-lap Chan

AKM Semiconductor, San Diego, CA, [1]Newport Communications, Irvine, CA, [2]Asahi-Kasei Microsystems, Kanagawa, Japan

Abstract

A 16-b, 2.5MHz output-rate ADC for wireline communications and high-speed instrumentation has been developed. A 2-1-1 cascaded delta-sigma modulator employing 4-b quantizers in all stages makes quantization noise sources negligible at 8X oversampling ratio. Data weighted averaging with bi-directional rotation eliminates tones generated by multibit DAC nonlinearity to increase SFDR. Switched-capacitor design techniques using low-threshold transistors reduce front-end sampling distortion. The 24.8mm^2 chip in 0.5-um CMOS also integrates the decimation filter and voltage reference. The ADC achieves 90dB SNR in the 1.25MHz bandwidth and 102dB SFDR with 270mW power dissipation.

1. Introduction

The growing trend in wire-line communications and high-speed instrumentation is to shift more signal processing from the analog to the digital domain. This implies that the analog-to-digital converter (ADC) is moved towards the front-end of the system with less analog pre-processing, which makes the performance requirement more stringent. For example, in systems like echo-cancelled xDSL modems, the ADC must simultaneously cope with the desired (weak) signals and the interfering (strong) signals. Wide dynamic range is needed to avoid clipping by the interfering signal, while having a noise floor low enough to recover the desired signal. High linearity is required to prevent masking the desired signal by intermodulation distortion products of the interfering signal. The target performance of an ADC aimed at such application is 90dB signal-to-noise ratio (SNR) and

R. J. van de Plassche et al. (eds.), Analog Circuit Design, 95–117.

100dB spurious-free dynamic-range (SFDR) with a bandwidth exceeding 1MHz. This paper presents a 16-b, 2.5-MHz output-rate ADC, which achieves these performance goals with low power dissipation.

This ADC development involves two key design issues. One is a delta-sigma modulator (DSM) topology feasible at a low oversampling-ratio (OSR) of eight. This is important for integration with the decimation filter. The resulting sampling clock of 20MHz makes digital switching noise easier to manage, and a cost effective single chip solution possible. Also a lower sampling clock will relax speed requirements in the analog circuits, and hence reduce power dissipation. In this development, an architectural approach that combines merits of cascaded DSM structures and multibit quantization makes all quantization noise sources negligible at 8X OSR. The other key issue is the design for high SFDR. At the system level, a dynamic element matching (DEM) algorithm for the multibit digital-to-analog converter (DAC) which reduces tones has been developed. In the analog design, switched-capacitor (SC) techniques for low sampling distortion have been explored. Details describing both key issues will be covered in this paper.

2. Architecture Level Design

2-1. Conventional Delta-Sigma Architectures for Low OSR

Cascaded DSM structures realize high-order noise-shaping by cascading delta-sigma stages of second order or lower to avoid instability, and are suited for high-speed ADC's with low OSR. A common choice is fourth-order noise-shaping implemented as a 2-1-1 cascaded DSM as shown in Fig. 1 [1]-[4]. In a cascaded structure, the quantization noise of the preceding stage is extracted and fed to the following stage. As a result, the digital output of the following stage includes information of quantization noise for both the preceding stage and itself. By replicating the noise transfer function of the preceding stage in the noise cancellation logic (NCL), the quantization noise can be cancelled using the digital representation. Ideally, quantization noise of only the final stage suppressed by the total noise-shaping order appears at the DSM output, which is usually referred to as theoretical quantization noise (TQN).

Conventionally, 1-b quantization has been used in the cascaded stages because of its inherent linearity. In this ADC, a signal-to-quantization-noise ratio (SQNR) substantially lower than the 90dB SNR target is required. A 2-1-1 cascade DSM with 1-b quantizers can achieve an SQNR set by inband TQN of lower than 90dB at 24X OSR [2], but the resulting sampling clock of 60MHz is too high for single-chip integration. Furthermore, because of the rapid signal growth of two serial integrators, the output of a second-order stage must be scaled prior to the next stage to avoid overload at signals near full-scale. A scaling factor smaller than 1/4 is often required, which results in a 12dB degradation of SQNR set by TQN [1][5].

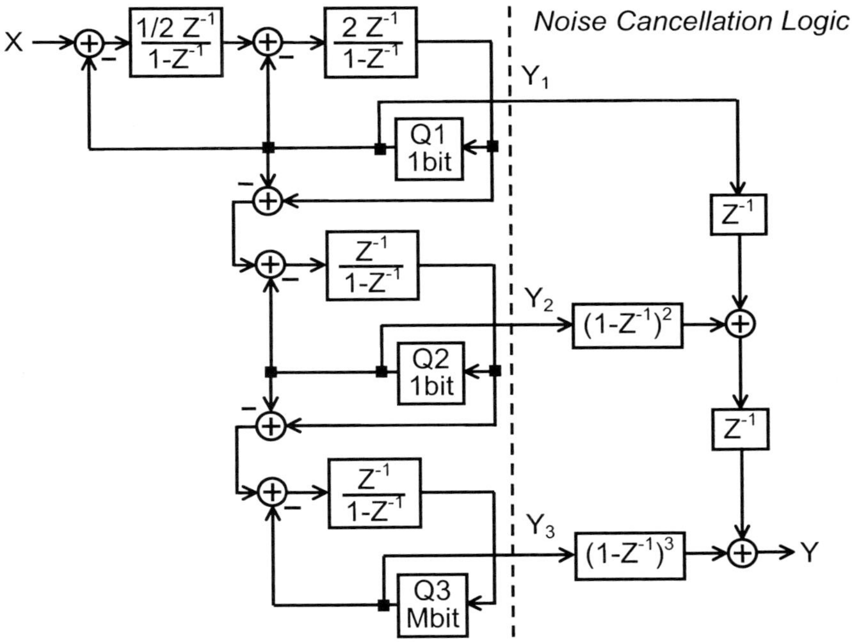

Figure 1: Conventional 2-1-1 cascaded delta-sigma modulator

One way to achieve further reduction of TQN is to use a multibit quantizer in the final stage [4][6]. This is a useful method, because the nonlinearity of the multibit DAC is suppressed by third-order noise-shaping, and can be made negligible. However, quantization noise leakage (QNL) from the other stages becomes the limiting factor. This leakage is caused by errors between the ideal noise cancellation logic and the imperfect analog circuits. With a 2-1-1 cascade using 1-b quantizers in the first two stages, the QNL limits the OSR to 16X, even with a 4-b quantizer in the final stage [4].

The cascaded pipeline approach [7] can reduce both TQN and QNL so that operation at 8X OSR is feasible. However, the relatively high 12-b resolution required in the pipeline converter stage results in extra design complexity and increase of die area.

2-2. Cascaded Multibit Delta-Sigma Approach

The architecture proposed in Fig. 2 overcomes the limitation of conventional 2-1-1 cascaded DSM's by using multibit quantizers in the first two stages [8]. This simple extension has several advantages that make a low OSR of 8X feasible and reduce power. Increasing the

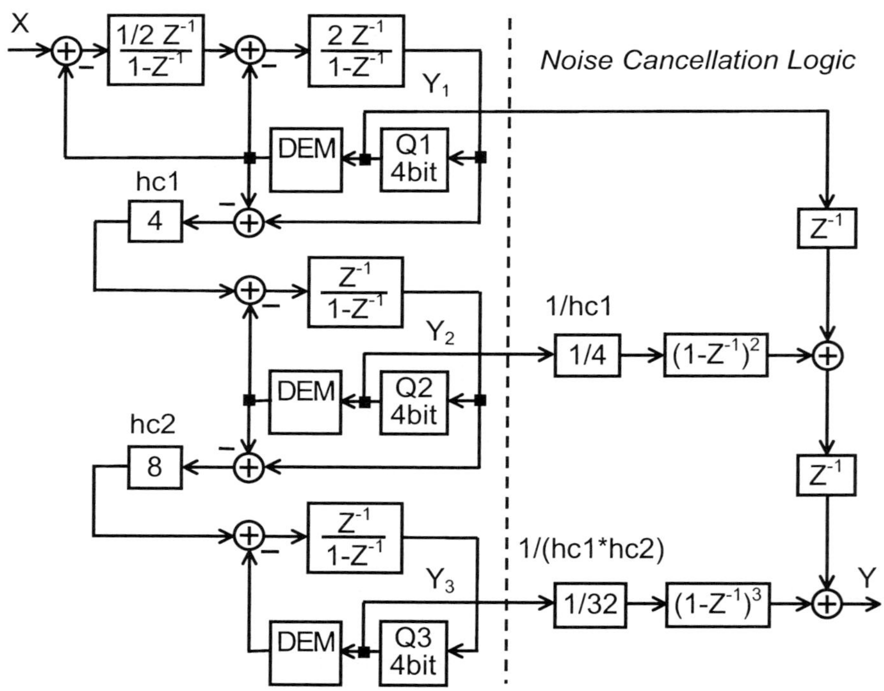

Figure 2: A 2-1-1 cascaded multibit delta-sigma modulator

quantizer resolution in the first and second stage does not directly reduce TQN, because of the noise cancellation. However, the smaller quantizer error extracted from the preceding stage allows insertion of an interstage gain to utilize the full dynamic range of the next stage. In case of 4-b quantizers, interstage gain as large as 15 will not cause overload. A three-stage implementation allows use of two interstage gains h_{C1}=4 and h_{C2}=8. Because of the gain insertion prior to the input, the output of the next stage must be scaled by inverse of the gain for noise cancellation. At the final stage, the output is scaled by the total interstage gain. As the result, TQN is reduced by factors h_{C1} and h_{C2}.

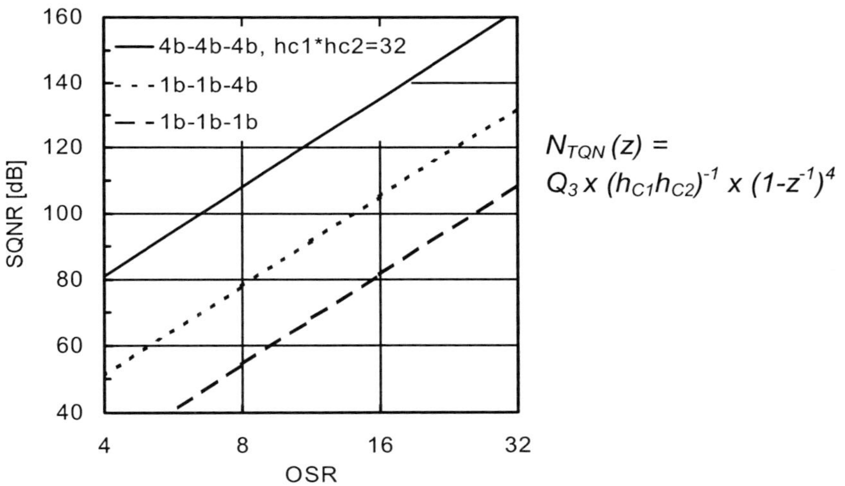

Figure 3: Theoretical quantization noise vs. OSR

In this architecture, TQN is determined by three factors: Noise shaping order, final stage quantizer resolution, and interstage gains h_{C1} and h_{C2}. Figure 3 shows SQNR set by TQN vs. OSR with contributions of the three factors illustrated. At 8x OSR, fourth-order noise shaping gives 54dB SQNR. Four-bit quantization in the last stage gives another 24dB. Total interstage gain of 32 provides the final 30dB, totaling in 108dB. Unlike classical DSM's, noise shaping provides only half of the total SQNR in the new architecture.

Another advantage of multibit quantization in the earlier stages is the reduced QNL. Figure 4 shows simulated SQNR set by QNL vs.

OSR. The dominant source for QNL is the first stage leakage caused by integrator pole errors, because of the inadequate suppression by first-order noise-shaping [9]. Coefficient mismatch errors and integrator gain errors are suppressed by at least second-order noise-shaping, and have minor effect. The integrator pole error is determined by the finite opamp DC gain A_{DC}, and the integrator feedback factors f_{FB1} and f_{FB2} for the first and second integrators respectively. In this simulation, the A_{DC} of all opamps is 80dB. In case of using 1-b quantizers in the first two stages, the SQNR at 8X OSR is only 87dB. The proposed architecture achieves 110dB SQNR. An OSR of 32X would have been required to achieve the same SQNR with 1-b quantizers.

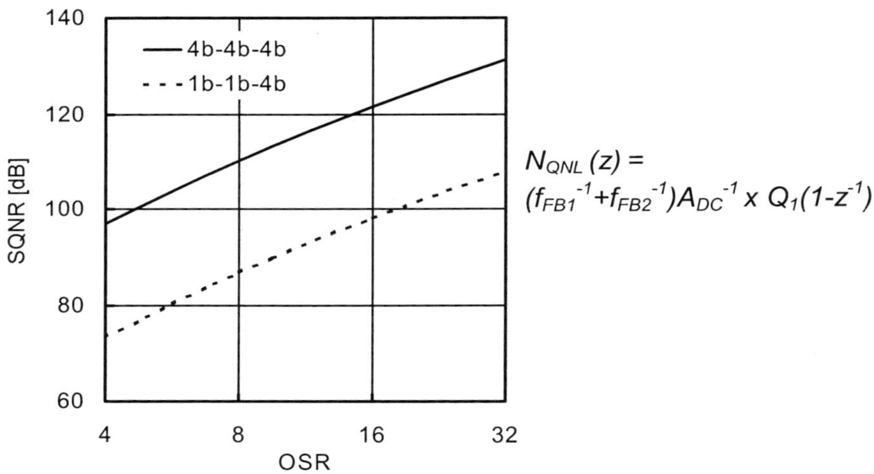

Figure 4: Quantization noise leakage vs. OSR

One drawback of the interstage gain is the reduced feedback factor in the integrator of the following stage. As a result, the QNL of the latter stages may become significant even though the noise shaping order is higher for those stages. Note that at 8X OSR, an additional noise shaping order provides only 10dB better suppression. To keep this effect minor, h_{C2} is made larger than h_{C1} within the total interstage gain. The latter stage is less susceptible to pole error because of the increased noise shaping of QNL.

2-3. Linearization of Multibit DAC

The non-linearity of the first-stage multibit DAC directly affects the performance without suppression by noise shaping. However, improvement of capacitor matching in modern process technology and development of DEM algorithms has made this a viable choice. The analog CMOS process used for this ADC employs a reversed process sequence for poly-silicon layers to optimize capacitor matching [10]. In this process, first, both bottom and top plate poly-silicon layers are deposited. Next, only the top plate is etched and then guarded. Finally, the bottom plate is etched. Since the deposition and etching of the top plate is done while the surroundings are completely flat, mismatch lower than 0.1% can be achieved with reasonable capacitor sizes.

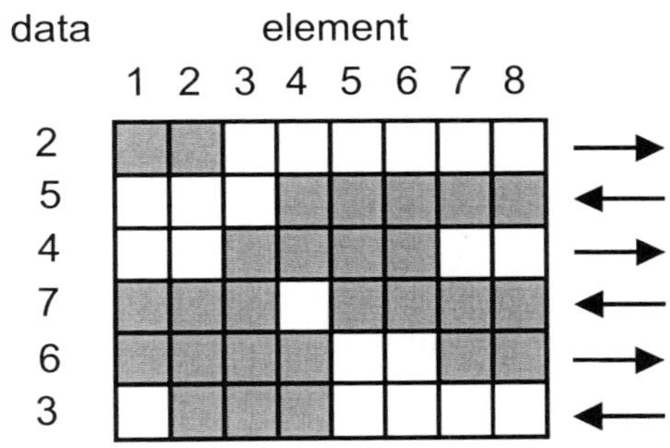

Figure 5: Concept of Bi-DWA

A common DEM algorithm, data weighted averaging (DWA), achieves first order mismatch shaping, but can introduce signal dependent tones [11]. This degrades SFDR and makes the inband mismatch induced noise power dependent on signal level [12]. One method to avoid these tones is to rotate the DWA index pointer by a predetermined sequence [13], but the DEM processor hardware becomes prohibitively large for a quantizer resolution of four bits or larger. For this design, DWA with bi-directional rotation was used to avoid these tones. A 3-b DAC example of bi-directional DWA (Bi-DWA) is shown in Fig. 5. The difference from conventional DWA is that the direction of rotation alternates every cycle. Here, the odd cycles are right rotation and the even cycles are left rotation. Both directions have separate index pointers, and rotate independently.

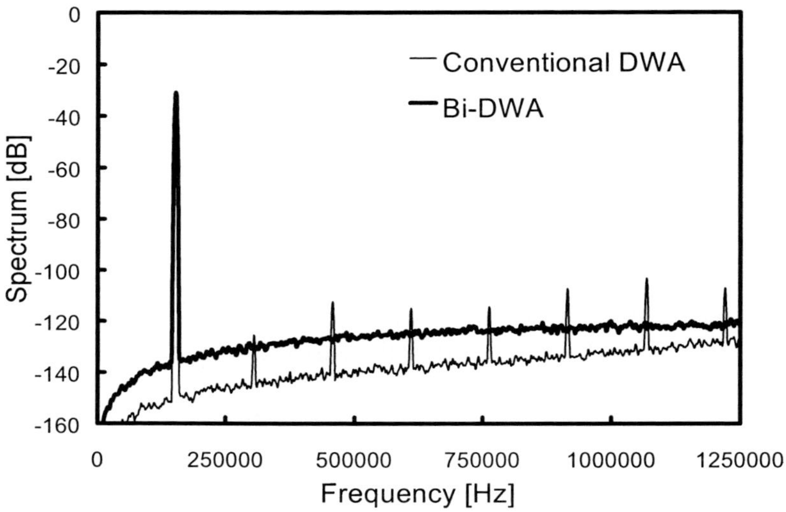

Figure 6: Simulated FFT for Bi-DWA with 0.1% mismatch

Figure 6 compares simulated FFT spectrums for both DWA and Bi-DWA with 0.1% mismatch. The simulation includes quantization noise and mismatch induced noise, but no analog device noise. In conventional DWA, because of its periodic nature, the mismatch error will be translated to signal dependent tones. This degrades SFDR and must be avoided. In Bi-DWA, the jump between the separate index pointers is a nearly random sequence. This eliminates tones while preserving the noise shaping property by rotation, which is clearly demonstrated in the simulation.

The cost of Bi-DWA is the rise of the noise floor because of the lost efficiency of rotation. This degrades the signal-to-noise plus distortion ratio (SNDR) set by mismatch-induced noise and distortion. Figure 7 shows simulated SNDR vs. input level in the 1.25MHz bandwidth with a 150kHz signal. The capacitor mismatch is 0.05%, which corresponds to actual capacitor sizes in the design. Conventional DWA shows variation to signal level. The out-of-band tones generated by DWA may fold back to inband due to modulation by delta-sigma limit cycles, which makes the inband noise power dependent on signal level. A dip at –24dBFS indicates the worst case for tone aliasing. On the other hand, for Bi-DWA, the tones are low enough so that the aliased signals are not significant. This results in a straighter but lower line because of the higher noise floor. The degradation of SNDR compared to the worst case of conventional DWA is only 1dB. This implies that if the mismatch noise is minor within the noise budget, it is beneficial to use Bi-DWA and improve SFDR.

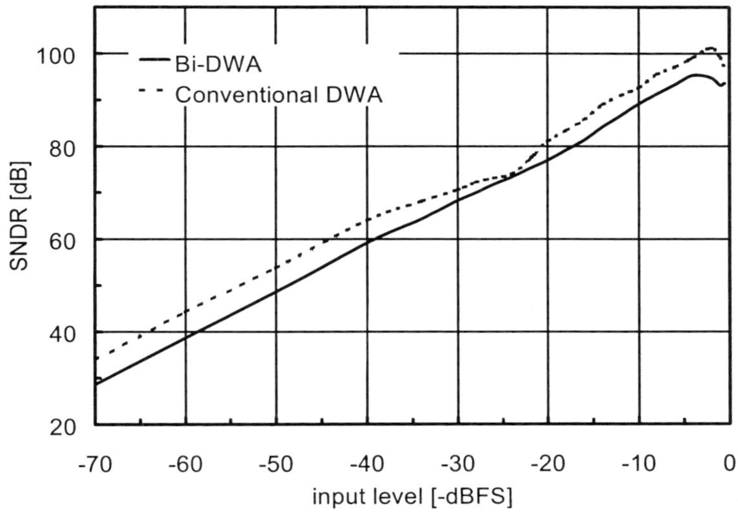

Figure 7: Simulated SNDR vs. input level for Bi-DWA

As shown in Fig. 8, the Bi-DWA decoder is implemented by modifying the general mismatch shaping structure of [14]. This structure realizes arbitrary mismatch shaping by using a vector quantizer in a feedback structure similar to a DSM. First-order mismatch shaping can be implemented by insertion of a single delay in the feedback path. In case of the rotation algorithm, barrel shifters with edge selectors can replace the general vector quantizer to save area. A bi-directional vector selector can be implemented by two sets of barrel shifters, which are time interleaved. This allows a common vector in the feedback loop to pass index pointers for both right and left rotation to further save area.

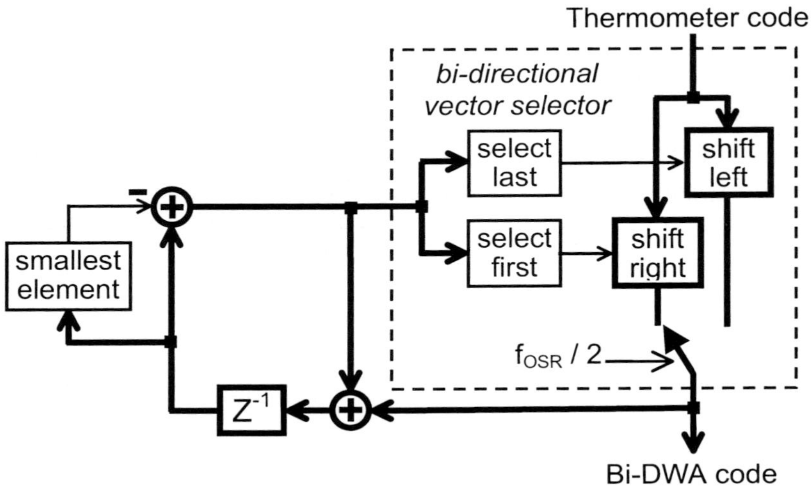

Figure 8: Bi-DWA decoder implementation

2-4. Noise Allocation

Table 1: ADC noise allocation

Noise Sources	SNR	Comments
Theoretical Quant.	108dB	
Quant. noise leakage	110dB	
Element mismatch	99dB	0.05% (sigma)
Digital filter	98dB	16-b rounding
Analog device noise	93dB	kT/C noise
Total	91dB	

The ADC noise allocation is shown in Table 1 to describe the power efficiency of the new architecture. The cascaded multibit approach makes both TQN and QNL negligible in the overall noise budget. The use of Bi-DWA and capacitor sizes decided by kT/C noise constraints makes the modulator first stage mismatch noise a minor contribution. Applying Bi-DWA to all three stages makes mismatch of the latter stages completely negligible. The noise generated in the digital filter is limited by the 16-b output rounding, and is also a minor contribution. This leaves most of the noise budget for analog device noise. In addition, the insertion of interstage gain reduces the contribution of latter stages, which otherwise would have been significant with limited noise shaping at 8X OSR. As a result, the noise budget of this ADC is dominated by analog noise of the first stage, which in this case is mainly kT/C noise. In this situation, the smallest capacitors possible decided by the physical noise limit to meet the target SNR can be used in the first stage, and aggressive power scaling can be done in the latter stages. This strategy has been proven most power efficient to obtain wide dynamic range in past high OSR designs such as audio converters [15].

3. Analog Circuit Implementation

3-1. Switched-Capacitor Integrator Design

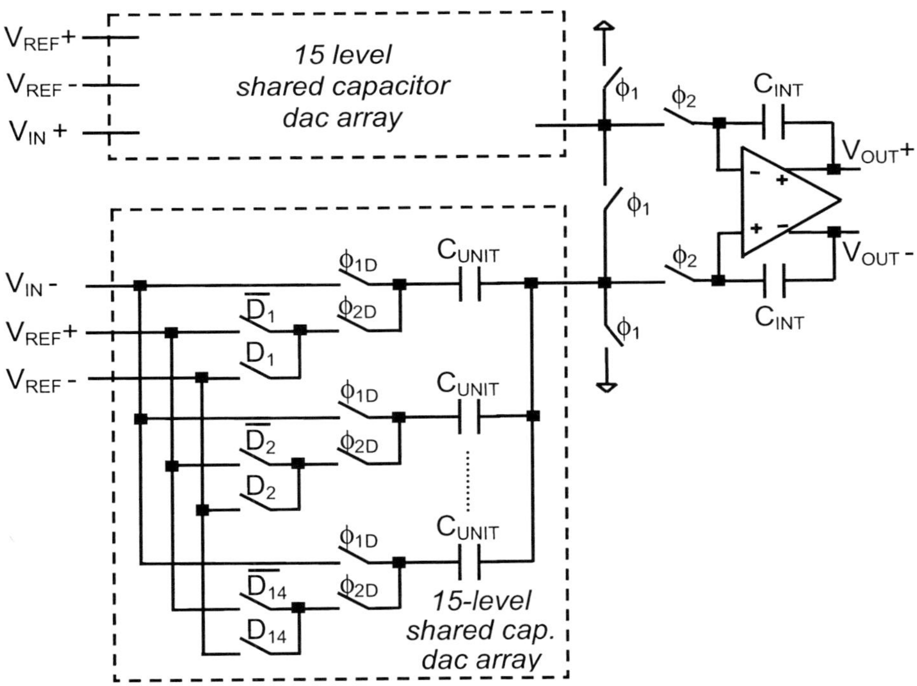

Figure 9: Shared-capacitor integrator + DAC

Fully-differential switched-capacitor circuits were used for the DSM implementation. The power supply voltage is 5-volts, and the sampling clock is 20MHz. Figure 9 shows the circuit diagram of the front-end SC integrator. In this circuit, input sampling and the 4-b DAC function share a common set of capacitors. During the sampling phase $\phi 1$, a cap array of 15 unit capacitors is tied in parallel, and the bottom plates are connected to the analog input. During the integrate phase $\phi 2$, the bottom plate of each unit is connected to the reference voltage that corresponds to its digital input code, in order to implement the DAC function.

Eliminating one set of switched-capacitors results in a 3dB reduction of kT/C noise, smaller AC load, and smaller integrator pole error. Another advantage of the shared capacitor implementation for use at the front-end integrator is the relaxed ADC input drive requirement. All capacitors are pre-charged to the previous quantizer result, which closely tracks the input in an oversampled multibit DSM. The disadvantage of this scheme is the signal-dependent charge load on the reference, which can generate second-order harmonic-distortion. Lowering the reference impedance with external bypass capacitors reduces the effect of this non-linearity. The reference driver amplifier is stable with a 10uF load capacitance. The ringing caused by the resonator formed by bonding inductance and the sampling capacitors also generate distortion. An internal RC damping network in parallel to the capacitors reduces this effect.

3-2. Switched-Capacitor Sampling Distortion Reduction

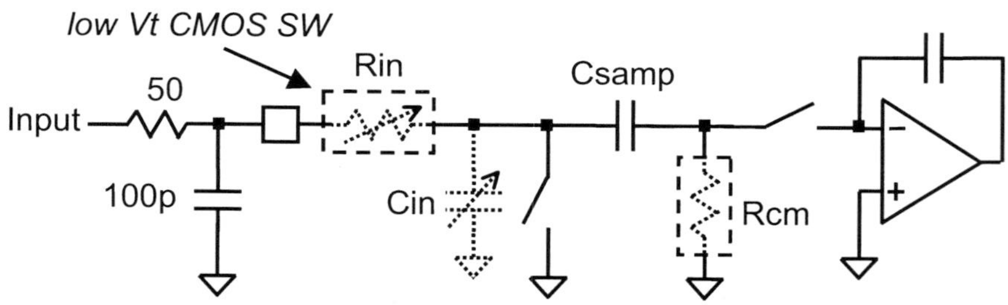

Figure 10: Front-end SC sampling network

The main concern in analog design is the SC sampling distortion at the front-end. In this design, use of low-threshold (±0.3V) transistors and optimum switch sizing achieves the SFDR target without employing special techniques such as clock bootstrapping [7]. At the front-end integrator, the continuous input signal is sampled by the SC network modeled in Fig. 10. The dominant source of sampling distortion is the signal dependent settling inaccuracy due to variation of RC time constants. Signal dependent sampling timing variation may also cause distortion, but is eliminated by four-phase non-overlap clocking with delayed bottom plate switching [16].

The RC variations are mainly determined by input switch on-resistance R_{IN} and switch junction capacitance C_{IN}. The resistance of the sampling switch R_{CM} remains constant because it is tied to the common mode voltage. For the input switches, low-threshold CMOS transmission gates reduce R_{IN} by 40% compared to normal-threshold (±0.8V) switches. Also, C_{IN} is reduced by roughly the same amount because of the smaller size needed to realize the same resistance.

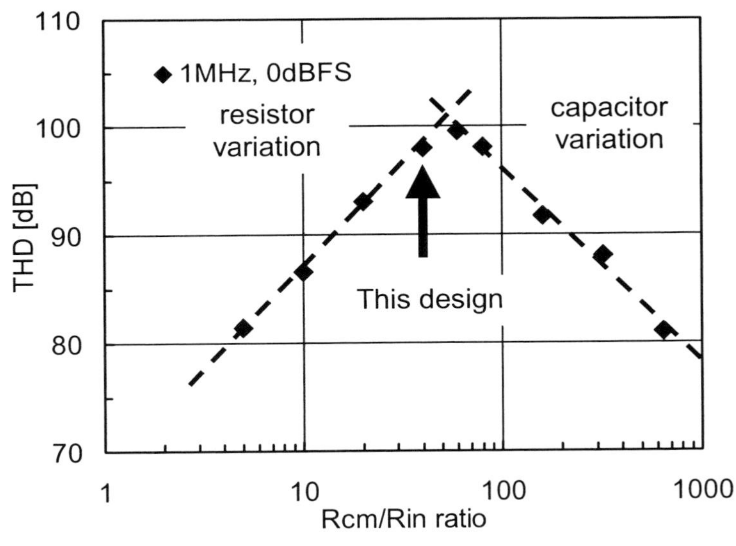

Figure 11: Simulated THD vs. Rcm/Rin ratio

The size ratio between the input and sampling switches was found very critical to reduce sampling distortion. Figure 11 shows the simulated distortion of the SC sampling network of Fig. 10. The actual circuit used for simulation is fully differential. In the ADC application circuit, an external first-order low pass filter prior to the input is required for anti-alias filtering, and for decoupling the signal source from the signal dependent charging event. This is also included in the simulation. The switch sizes were changed to vary the R_{CM} to R_{IN} ratio from 5 to 600. The total resistance, $R_{CM} + R_{IN}$ is kept to a constant value required by static settling. Total harmonic distortion (THD) up to the fifth harmonic was summed for a 1MHz, 2Vpp signal.

The THD first improves with increasing R_{CM} to R_{IN} ratio, which means with larger input switch size. In this region, THD is decided by the variation of the input switch resistance R_{IN}. The non-linear junction capacitance C_{IN} is still small compared to the sampling capacitor. The THD peaks to –100dB at a ratio of 80, and then starts to degrade. The right side of the peak is the region where THD is decided by C_{IN} variation. Now, R_{IN} is much smaller than the input source resistance, and C_{IN} becomes comparable to the sampling capacitor. For this design, an input switch with resistance 40 times smaller than the sampling switch was used. This size is half of the peak and has 1dB worse THD. The trade-off was made to reduce the amount of signal dependent charge injection generated by the input switch. In case of delayed bottom plate switching, this charge will be absorbed by the external input filter, which may cause distortion. For switches of the integrate phase, minimum sizes allowed by static settling requirements for QNL avoidance is used to prevent any excessive kT/C noise due to opamp noise sampled onto the capacitor [2].

3-3. Opamp Design

The circuit diagram of the first integrator opamp is shown in Fig. 12. For this opamp, DC gain higher than the pole error reduction requirement was implemented to avoid distortion caused by gain dependency on voltage swing. A folded-cascode topology with gain boost achieves high dc gain with the necessary bandwidth. A regulated cascode [17] using a single low-threshold transistor increases gain with minimum current overhead, and with swing reduction of only 0.6V compared to a cascode load biased in high-swing condition [18]. The simulated dc gain is 96dB. Bandwidth requirements for low distortion are greatly relaxed by multibit quantization. This is because the settling nonlinearity due to slew-rate limiting is reduced by smaller step transitions. The closed loop dominant pole is 90MHz, which is only 4.5 times the sampling frequency. The QNL caused by the resulting settling error is negligible because of multibit quantization in the first stage.

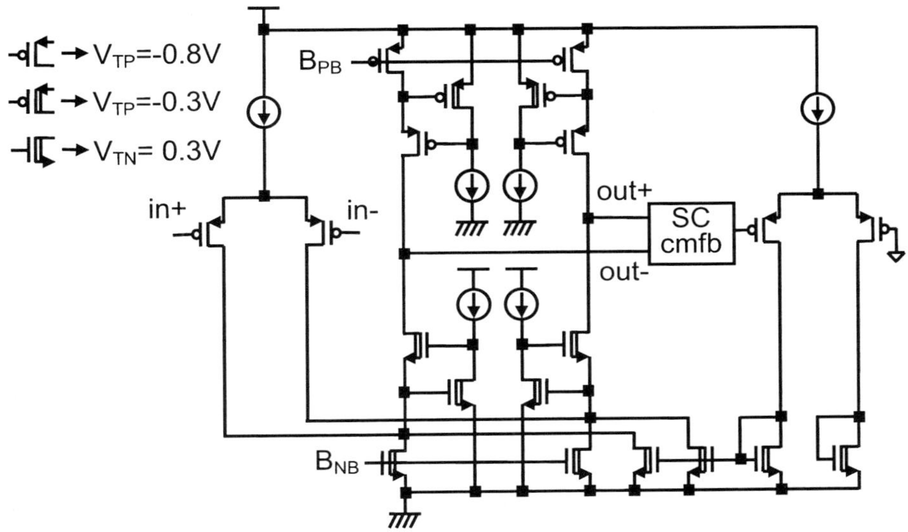

Figure 12: Opamp circuit diagram

3-4. Four-bit Quantizer

The 4-b quantizer is implemented by a straight flash architecture. Although the clock rate is only 20MHz, fast conversion is required to allow the digital switching noise of the immediately following DEM process to settle before the next critical SC sampling edge. A comparator circuit consisting of a preamplifier and a latch with boot-strapped capacitive feedback was employed for fast conversion [19]. The comparator achieves 3nsec regeneration time with ±1mV difference at the input.

One advantage of the cascaded structure is the immunity to flash converter nonlinearity, such as comparator offset and hysteresis. In a single loop, the effect of non-linearity is only suppressed by the amount of noise shaping provided by the loop, and may exhibit undesirable tones, which degrade SFDR [20]. In case of the 2-1-1 cascade, the first stage quantizer non-linearity only appears at the modulator output as second-order noise-shaped leakage, which is completely negligible. As a result, neither auto-zero capacitors nor large input device sizes are required to reduce offset.

4. Chip Implementation

The block diagram of the ADC is shown in Fig. 13. The digital filter consists of three half-band FIR stages to perform 8X decimation and low pass filtering. A direct-form poly-phase implementation is adopted to take advantage of coefficient symmetry and zero-value even-tap coefficients [21]. The stop-band attenuation is 85dB, and the pass-band ripple is ±0.004dB. The first FIR stage can be bypassed to configure a 4X decimation mode, resulting in an ADC operating at a 5MHz output-rate.

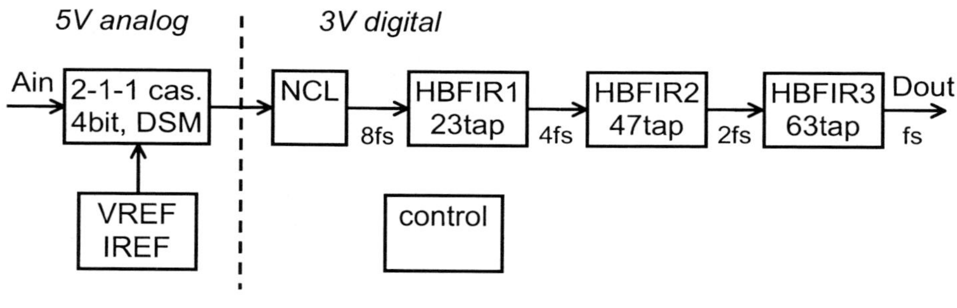

Figure 13: ADC block diagram

The chip micrograph is shown in Fig. 14. The chip is fabricated using triple-metal, double-poly, 0.5-um CMOS. The die dimensions are 4.6 x 5.4mm^2. The analog circuit including the voltage reference occupies 30% of the core area. The area of the DSM is 2.9mm^2. In the digital, the DEM overhead is 12% of the core.

One key issue in the chip implementation is to avoid digital switching noise coupling to the critical analog circuitry. Digital operation at 3-volts directly reduces the amount of switching noise. The low 20MHz sampling clock and digital filter design with short delay paths allow quiet time to place the critical analog sampling edge. The data output buffers are designed with minimum strength and slew rate control, and have designated power supply pins. For power supply decoupling purposes, a total of 3nF gate capacitance is placed on-chip. A lightly-doped substrate with separate substrate ties for analog and digital achieves good isolation [22].

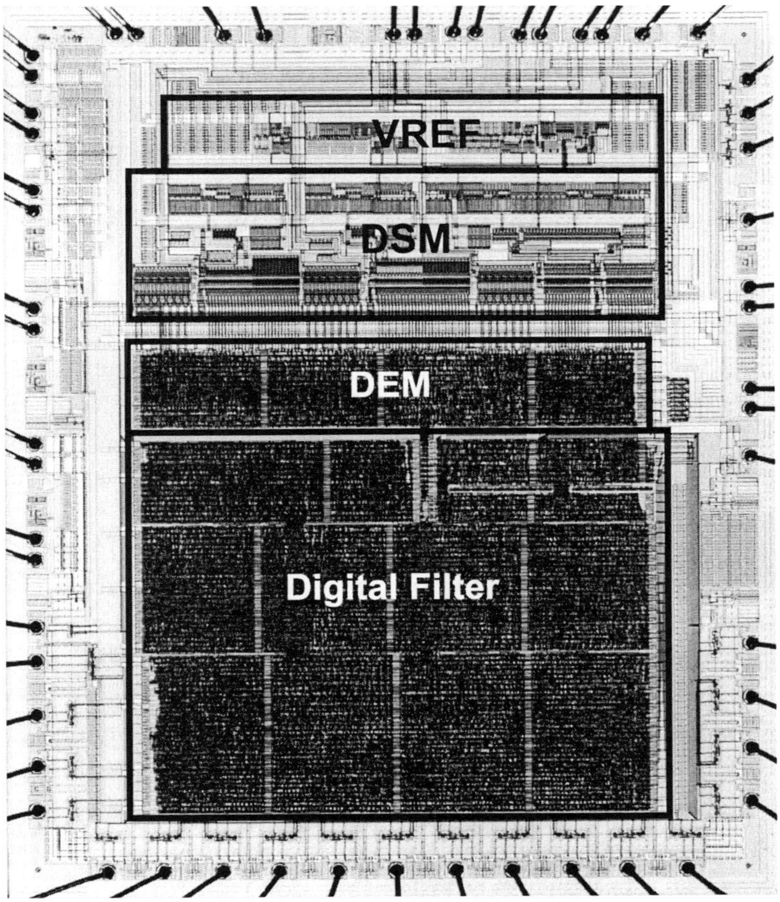

Figure 14: Chip micrograph

5. Measured Results

All measurements were taken with 5V analog and 3V digital power supplies. A 1.28GHz signal was divided by an RF prescaler IC to generate a low jitter 20MHz clock. A passive bandpass filter module was placed between the input signal source and the ADC. Figure 15 shows a 32768 point FFT spectrum with a -0.1dBFS, 100kHz sine wave input. The peak spurious is the third harmonic, and the SFDR is 102dB. The THD is -98dB. These results indicate the effectiveness of the low distortion switched-capacitor design techniques. Figure 16 shows SNDR and SFDR vs. input level for both 100kHz and 500kHz signals. The measurement bandwidth is 1.25MHz. The peak SFDR for a 100kHz signal is 110dB at -10dBFS. The SFDR at 500kHz is approximately 15dB lower compared to the 100kHz case, because of the higher second harmonic. The SFDR flattens to 100dB above -10dBFS. For SNDR, the performance near full-scale is limited by the noise floor instead of distortion components. The SNR is 90dB for both frequencies. Figure 17 shows a FFT spectrum with two high-frequency tones at 1MHz and 975kHz, both at –6.5dBFS. The resulting third-order intermodulation distortion products (IM3) are 92dB below the amplitude of the two tones. In the plot, the skirt around the 1MHz signal is the residue of the signal source noise after bandpass filtering. The tones at lower frequencies are artifacts of the stand-alone DAC used to generate the 975kHz signal.

The performance summary of the ADC is shown in Table 2. A SNR of 90dB has been achieved with DSM power dissipation of 105mW. The standard deviation of SNR is 0.3dB for 100 samples on 2 wafers. This data indicates the power efficiency and robustness of the new architecture. The total power of 270mW is less than half compared to a previously reported ADC with comparable performance [7]. The SNR at the 5MHz output-rate with a 2.5MHz bandwidth is 80dB, which in this case is dominated by quantization noise.

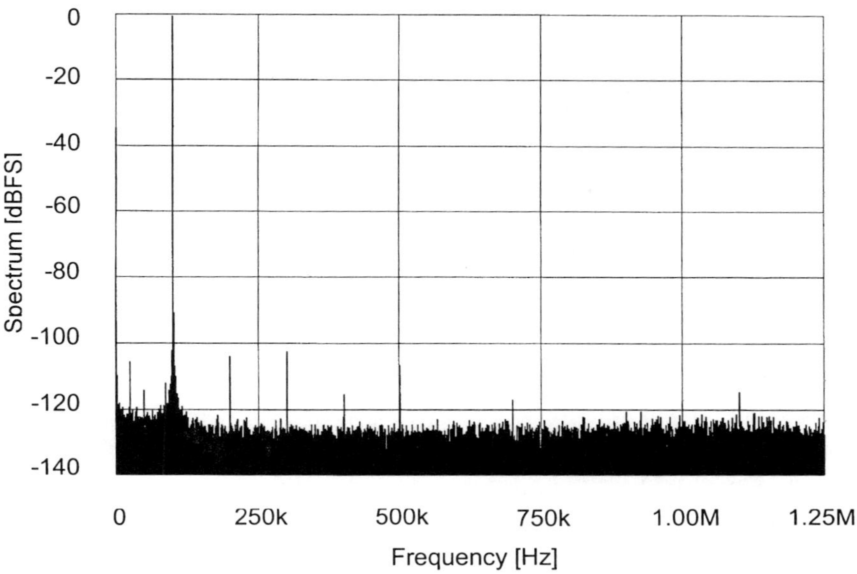

Figure 15: Measured FFT spectrum: -0.1dBFS, 100kHz input

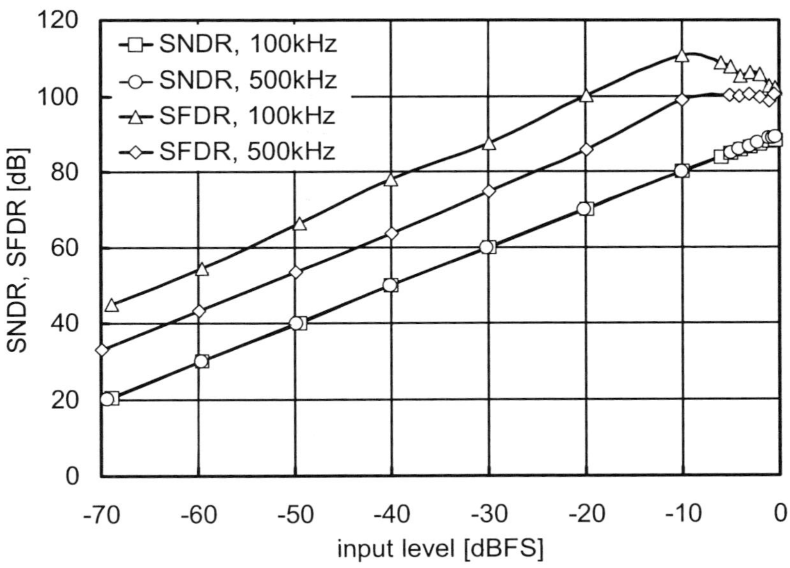

Figure 16: SNDR and SFDR vs. input level at 100kHz and 500kHz

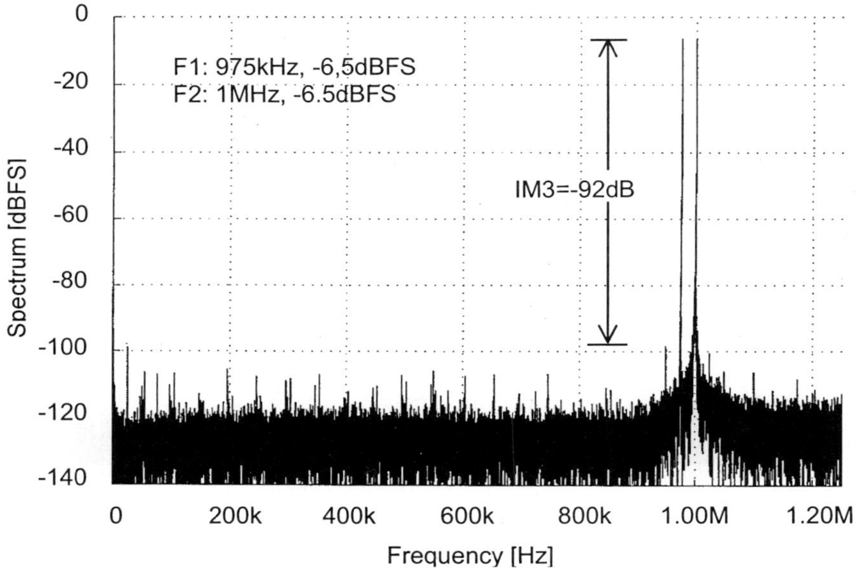

Figure 17: Measured FFT spectrum: dual-tone input

Table 2: ADC performance

Resolution	16 bits
Output rate/Sample rate	2.5MHz / 20MHz
Input swing	+/- 2.0Vpp differential
SNR (1.25MHz BW)	90dB (sigma=0.3dB)
THD (-0.1dBFS, 100kHz)	98dB (sigma=1.2dB)
SFDR(-0.1dBFS, 100kHz)	102dB
SFDR (-10dBFS, 100kHz)	110dB
IM3 (975kHz/1MHz, -6.5dBFS)	-92dB
Power supplies	5V analog, 3V digital
Power dissipation	270mW (DSM:105mW)
Technology	0.5u DPTM CMOS
Chip size	4.6 x 5.4 mm^2

6. Conclusions

A 16-b ADC in 0.5-um CMOS achieves 90dB SNR in the 1.25MHz bandwidth with 270mW power dissipation. The cascaded multibit delta-sigma architecture used in this ADC employs multibit quatization in the early stages of a cascaded DSM, which allows insertion of interstage gain to reduce TQN, and reduction of QNL from the first stage. A 4-b, 2-1-1 cascaded DSM using this approach achieves SQNR set by TQN and QNL of 108dB and 110dB, respectively at 8X OSR. The recent advancement of high-order DEM algorithms [20][23] make the proposed architecture a strong candidate for future high-speed ADC's with even lower OSR. In this ADC, relatively good capacitor matching allowed a simple modification (bi-directional rotation) to the existing DWA algorithm to reduce tones with minor sacrifice of SNR. Considering the improvement of capacitor matching in modern process technology, the proposed Bi-DWA algorithm is also a practical choice to increase SFDR. Switched capacitor sampling distortion has been reduced through use of low-threshold transistors and optimal sizing of the switches without any special circuit techniques. The measured SFDR of 102dB and IM3 of –92dB demonstrates the effectiveness of the design techniques. The ADC also successfully integrates the decimation filter. No SNR degradation due to digital switching noise has been observed, proving the robustness of the noise coupling avoidance techniques in conjunction with the low master clock of 20MHz.

7. References

[1] G. Yin and W. Sansen, "A high-frequency and high-resolution fourth-order delta-sigma A/D converter in BiCMOS technology," *IEEE J. Solid-State Circuits*, vol. 29, pp. 857-865, Aug. 1994.

[2] A. Marques, V. Peluso, M. Steyaert, and W. Sansen, "A 15-b resolution 2-MHz nyquist rate delta-sigma ADC in a 1-um CMOS technology," *IEEE J. Solid-State Circuits*, vol. 33, pp.1065-1075, Jul. 1998.

[3] Y. Geerts, A. Marques, M Steyaert, and W. Sansen, "A 3.3-V, 15-

bit, delta-sigma ADC with a signal bandwidth of 1.1MHz for ADSL applications," *IEEE J. Solid-State Circuits*, vol. 34, pp. 927-936, Jul. 1999.

[4] F. Medeiro, B. Perz-Verdu, and A. Rodriquez-Vazquez, "A 13-bit, 2.2MS/s, 55mW multibit cascade sigma-delta modulator in CMOS 0.7-um single-poly technology," *IEEE J. Solid-State Circuits*, vol. 34, pp.748-760, Jun. 1999.

[5] D. Ribner, R. Baertsch, S. Garverick, D. McGrath, J. Krisciunas, and T. Fujii, " A third-order multistage sigma-delta modulator with reduced sensitivity to nonidealities," *IEEE J. Solid-State Circuits*, vol. 26, pp. 1764-1774, Dec. 1991.

[6] B. Brandt and B. Wooley, "A 50-MHz multibit sigma-delta modulator for 12-b, 2-MHz A/D conversion," *IEEE J. Solid-State Circuits*, vol. 26, pp. 1746-1756, Dec. 1991.

[7] T. Brooks, D. Robertson, D. Kelly, A. Del Muro, and S. Harston, "A cascaded sigma-delta pipeline A/D converter with 1.25MHz signal bandwidth and 89dB SNR," *IEEE J. Solid-State Circuits*, vol. 32, pp. 1896-1906, Dec. 1997.

[8] I. Fujimori, L. Longo, A. Hairapetian, K. Seiyama, S.Kosic, J. Cao, and S. Chan, "A 90dB SNR, 2.5MHz output-rate ADC using cascaded multibit delta-sigma modulation at 8x oversampling ratio," *ISSCC Digest of Technical Papers*, pp.338-339, Feb. 2000

[9] M. Rebeschini, N. van Bavel, P. Rakers, R. Greene, J. Caldwell, and J. Haug, " A 16-b 160-kHz CMOS A/D converter using sigma-delta modulation," *IEEE J. Solid-State Circuits*, vol. 25, pp. 431-440, Apr. 1990.

[10] S. Kayanuma and K. Iki, "Method for fabrication of semiconductor device having polycrystalline silicon and metal silicides," U.S. Patent 5 397 729, Mar. 1995.

[11] R. Baird and T. Fiez, "Linearity enhancement of multibit delta-sigma A/D and D/A converters using data weighted averaging," *IEEE Trans. on Circuits and Systems II*, vol. 42, pp. 753-762, Dec. 1995.

[12] I. Fujimori, A. Nogi, and T. Sugimoto, "A multi-bit delta-sigma audio DAC with 120dB dynamic range," *ISSCC Digest of Technical Papers*, pp.152-153, Feb. 1999.

[13] R. Rudke, A. Eshraghi, and T. Fiez, "A spurious-free delta-sigma DAC using rotated data weighted averaging," in *proc. IEEE 1999*

CICC, pp. 7.5.1-4, May 1999.

[14] R. Scherier and B. Zhang, "Noise-shaped multibit D/A converter employing unit elements, *Electronic Letters*, vol. 31, pp. 1712-1713, Sep. 1995.

[15] I. Fujimori, K. Koyama, D. Trager, F. Tam, and L. Longo, "A 5-V single-chip delta-sigma audio A/D converter with 111dB dynamic range," *IEEE J. Solid-State Circuits*, vol. 32, pp.329-336, Mar. 1997.

[16] D. Haigh and B. Singh, "A switching scheme for switched capacitor filters which reduces the effect of parasitic capacitances with switch control terminals," in *Proc. IEEE 1983 ISCAS*, pp. 586-589, May. 1983.

[17] B. J. Hosticka, "Improvement of the gain of MOS amplifiers," *IEEE J. Solid-State Circuits*, vol. SC-14, pp. 1111-1114, Dec. 1979.

[18] R. Gregorian and G. Temes, *Analog MOS Integrated Circuits for Signal Processing*, New York: Wiley Interscience, 1986, ch. 4.

[19] K. Kim, N. Kusayanagi, and A. Abidi, "A 10-bit, 100Ms/s CMOS A/D converter, *IEEE J. Solid-State Circuits*, vol. 32, pp. 302-311, Mar. 1997.

[20] E. Fogleman, I. Galton, W. Huff, and H. Jensen, "A 3.3V single-poly CMOS audio ADC delta-sigma modulator with 98dB peak SINAD," in *Proc. IEEE 1999 CICC*, pp. 7.4.1-4, May 1999.

[21] S. Norsworthy, R. Schreier, and G. Temes, *Delta-Sigma Data Converters: Theory, Design, and Simulation*, New York: IEEE Press, 1997, ch. 13.

[22] B. Stanisic, N. Verghese, R. Rutenbar, R. Carley, and D. Allstot, "Addressing substrate coupling in mixed-mode IC's: Simulation and power distribution synthesis," *IEEE J. Solid-State Circuits*, vol. 29, pp. 226-238, Mar. 1994.

[23] A. Yasuda, H. Tanimoto, and T. Iida, "A third-order delta-sigma modulator using second-order noise-shaping dynamic element matching," *IEEE J. Solid-State Circuits*, vol. 33, pp. 1879-1897, Dec. 1998.

A 13-bit bandpass sigma delta modulator for 10.7MHz digital IF with a 40MHz sampling rate

Jurgen van Engelen

Broadcom Corporation
16215 Alton Parkway
Irvine, CA 92619
e-mail: engelen@broadcom.com

Abstract

This paper presents a sixth-order continuous-time bandpass sigma delta modulator (SDM) for analog-to-digital conversion of 10.7MHz intermediate frequency (IF) signals in AM/FM broadcast radio receivers. Sampled at 40MHz, the single-loop one-bit SDM achieves 67dB SNDR in 200kHz and 80dB in 9kHz. The third order intermodulation (IM3) is at -82dBc for a -10dBFS input level. The 0.5μm CMOS chip occupies 0.9x0.4mm^2 and consumes 60mW at 3.3V (digital) and 5.0V (analog). The sample frequency is variable and can be set from 30MHz to 80MHz.

1 Introduction

Due to scaling of the CMOS technology, more and more stages of (broadcast) radio receivers can be integrated onto a single die; adding intermediate frequency (IF) functionality to the baseband receiver IC [1]. Implementing (IF) signal processing stages in the digital domain takes full advantage of the increased speed, providing more flexibility, better noise immunity and potential improvements in performance and power by the technology scaling. The required analog-to-digital (A/D) conversion of the IF signal also alleviates problems due to DC-offset, flicker (1/f) noise and I/Q channel mismatch as occur in conventional analog heterodyne receivers [2].

R. J. van de Plassche et al. (eds.), Analog Circuit Design, 119–138.

Bandpass sigma-delta modulation [3–6] is well suited for A/D conversion of narrow band signals modulated on a carrier, as occur in AM/FM receivers. Through the use of oversampling and negative feedback, the quantization errors of a coarse quantizer are suppressed in a narrow signal band in the output of the modulator. Figure 1 shows an example of a single-loop multi-bit bandpass sigma delta A/D converter and the corresponding output frequency spectrum.

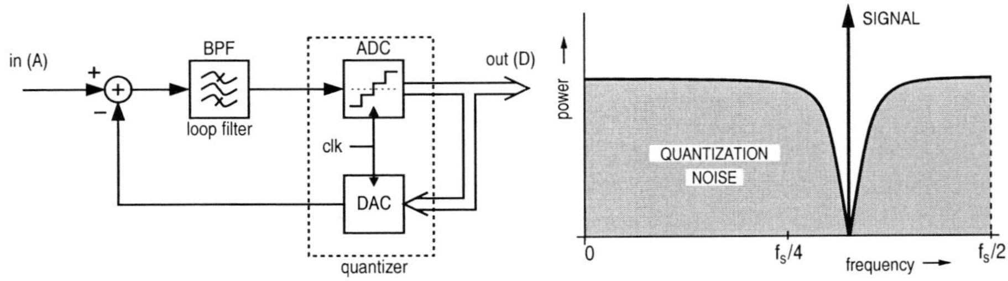

Figure 1: Example of a bandpass ΣΔ modulator (left) and a typical output spectrum (right).

A typical digital IF architecture for a radio receiver [7] using a sigma delta modulator (SDM) is shown in Fig. 2. The receiver consists of a low noise amplifier (LNA) followed by a wideband bandpass filter. Using a tunable local oscillator (LO) and a mixer, the FM signals (88-108MHz) are downconverted and the AM signals (520-1650kHz) are upconverted to the 10.7MHz IF frequency. This configuration allows a single (ceramic) SAW filter for channel selection for both AM and FM signals. The filter is followed by an automatic gain control (AGC) amplifier which feeds the signal to the SDM ADC. A digital signal processing (DSP) unit performs the demodulation, final channel selection and various baseband functions. A digital-to-analog converter (DAC) is used to output the baseband signal.

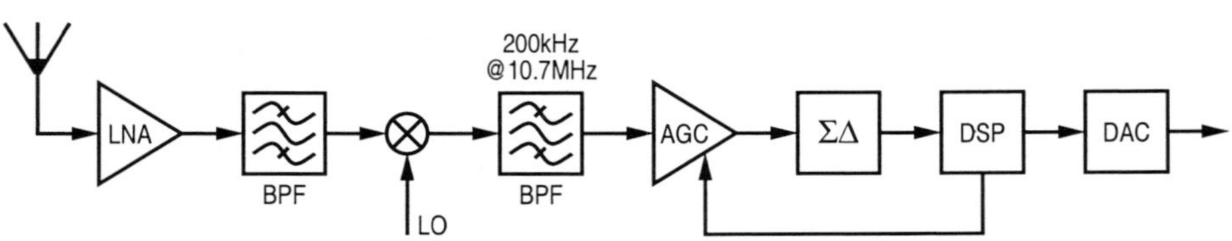

Figure 2: A typical digital IF radio receiver using a ΣΔ modulator.

2 Requirements

The requirements for the SDM are determined by its input signal characteristics. For FM/AM receivers, a dynamic range (DR) of 65dB and 90dB respectively is desirable. Preferably, the SDM should achieve this DR at low input signal levels to alleviate the requirements for the AGC. The ceramic filter has a fixed bandwidth of approximately 200kHz, corresponding to a single channel for FM signals. For AM signals, the typical carrier spacing is 9kHz and the filter feeds several channels to the AGC and SDM. The A/D conversion of the *multi-channel* AM signals places stringent requirements on the linearity of the SDM in order to avoid intermodulation distortion of adjacent channels. A third order intermodulation product (IM3) below -90dB is desirable. For linearity, the most critical parts of the modulator are the DAC, the summation point and the first stage of the loop filter. This can be easily seen by calculating the effects of nonlinear behavior in the various parts of the modulator *back* to the input (see Fig. 1).

3 Architecture

In order to achieve a dynamic range (DR) of 90dB in 9kHz several sigma-delta modulator architectures can be considered. A modulator employing a multi-bit quantizer can be used to obtain such an DR, but suffers from non-linearity caused by mismatch in the quantizer DAC levels. By using advanced dynamic element matching (DEM) techniques, this linearity problem can be overcome at the cost of additional hardware [8]. For a one-bit quantizer, which has two DAC levels, the mismatch only results in gain mismatch and a dc offset which are linear deviations. So, for reason of linearity, a single bit quantizer will be used.

The high DR required can be achieved by using a multi-stage (MASH) [9] architecture, but mismatches in the loop and recombination filter transfer functions cause imperfect cancellation of the quantization errors of the first low order loop. A part of the low-order noise-shaped quantization errors leaks to the output of the SDM, thus reducing the dynamic range. Therefore, a single loop, one-bit modulator architecture is chosen here. For such a modulator, the theoretically achievable DR depends on the order of the loop filter and the oversampling ratio[1]. Here, the SDM uses a sixth-order bandpass loop filter, giving an effective third order noise shaping of the quantization errors. The loop filter uses continuous-time (CT) circuitry, which has several advantages over a switched capacitor (SC) implementation. First, the clock feed through and settling errors of the sample & hold (S/H) are suppressed by the feedback loop as the S/H is placed between the loop filter and ADC inside the loop. Second, a CT circuit can operate at higher frequencies, as the sample frequency f_s is not limited by charge transfer accuracy requirements. Third, the tuning frequency of a CT filter does not depend on the sample frequency. As a result, the sample frequency can be set by the system requirements for the DSP unit, but a tuning mechanism is needed for the loop filter. A disadvantage is the increased sensitivity to clock jitter of the feedback DAC.

In the case of such a high-order one-bit SDM, the design of the CT loop filter transfer function $G(p)$ is mainly determined by the signal-dependent stability of the SDM [10]. As the loop of the SDM is sampled by the quantizer, the stability of the loop can be analyzed in discrete-time (DT) domain by applying the invariant impulse response transformation (see Fig. 3). The (sampled) response of the CT loop filter $G(p)$ to the pulse shape $R(p)$ of the quantizer DAC can be replaced by an equivalent discrete-time (DT) loop filter $G_{eq}(z)$, and the loop filter design can be done in the DT domain [11, 12]. Note that an appropriate CT loop filter prototype should be chosen such that it provides sufficient degrees of free-

[1]The oversampling ratio (OSR) equals the ratio between half the sampling frequency and the signal bandwidth: $OSR = f_s/2 \cdot BW$.

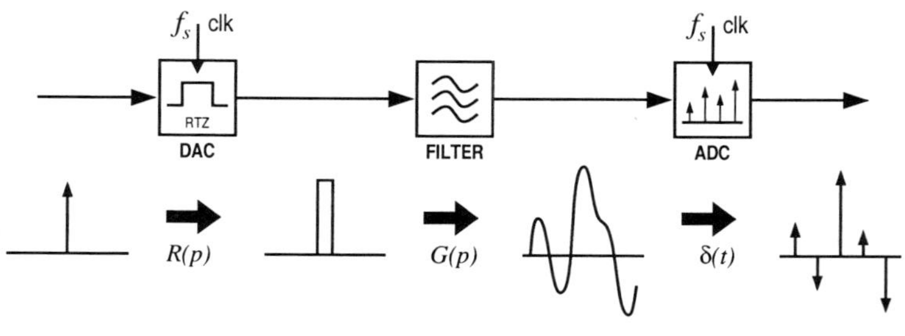

Figure 3: Sampled impulse response of the pulse shaping DAC followed by a CT loop filter.

dom for the desired placement of the poles and zeroes of the equivalent DT loop filter transfer function [13].

4 Stability

The design of the DT loop filter is a trade-off between performance and stability. The poles of the DT loop filter transfer function are placed at the intended tuning frequency to ensure a high in-band gain and a proper noise shaping action. The zeroes of the transfer function determine the SDM's stable operating range. Locating the zeroes near the poles results in a loss of performance, but placing the zeroes too far away from the poles causes unstable behavior, even at low input amplitudes. As an example, Fig. 4 shows the estimated in-band quantization noise, the maximum stable input amplitude and the actual SNR as a function of the loop filter parameter a of a fourth order low pass SDM. The DT loop filter transfer function is given by:

$$G(z) = \frac{(1 - a \cdot z^{-1})^4}{(1 - z^{-1})^4} - 1 \tag{1}$$

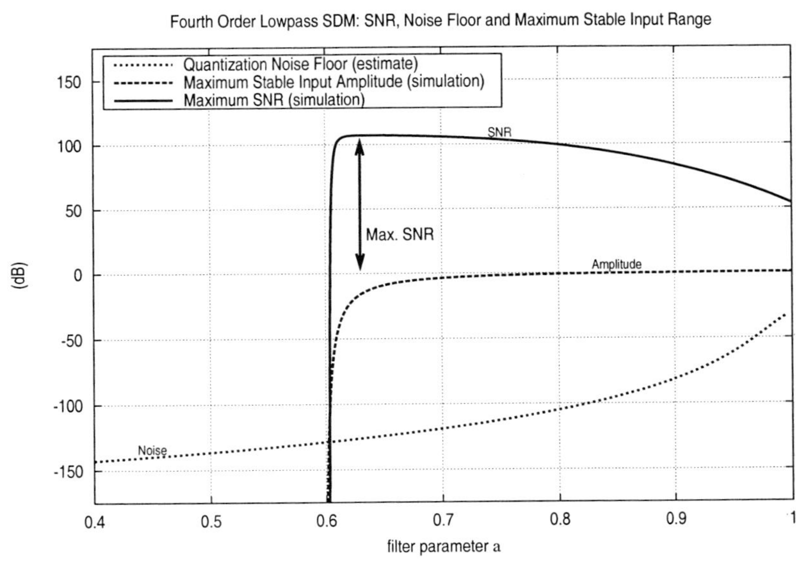

Figure 4: Performance of a fourth order low pass SDM with a loop filter defined by (1) as a function of the zero locations (OSR=100).

The resulting noise transfer function (NTF) equals:

$$NTF(z) = \frac{1}{G(z)+1} = \frac{(1-z^{-1})^4}{(1-a\cdot z^{-1})^4} \tag{2}$$

The parameter a determines the loop filter zero locations. When $a = 0$, the zeroes are located far from the poles, and the estimated in-band quantization noise [10] is at a minimum, but the modulator is unstable, even at very low input signals. At $a \approx 0.6$, the maximum input signal amplitude for which the modulator is stable increases rapidly, thereby also increasing the SNR. For $a > 0.65$, the maximum input signal amplitude hardly increases any further, but the in-band quantization noise rises due to a less effective noise shaping. The value for a at which a maximum SNR is reached lies near the stability boundary for a (in this case $a \approx 0.6$). In order to find this boundary without the use of extensive simulations, the stability of the SDM is analyzed.

The stability analysis of a SDM is complicated by the non-linearity of the quantizer (ADC and DAC). By using the describing function method, the quantizer can modeled by a quasi-linear, signal-dependent transfer function, and stability analysis techniques for linear systems (such as root locus analysis) can be applied. Often, the non-linear element is modeled by a single, signal dependent gain only. Here, the quantizer model consists of a gain λ *and* a phase shift $\alpha\Delta\phi$ (see Fig. 5). The gain results from the variable input and fixed output amplitude of the one-bit quantizer. The phase shift is in fact a phase uncertainty and represents the inaccuracy in time with which the zero-crossings of the quantizer input signal are detected. Sampling of the input signal causes a zero crossing to be detected by the sample moment following this crossing. As the zero crossing could have occurred anywhere within the previous sample period, an uncertainty in the phase transfer of the signal is introduced. Figure 6 shows an exam-

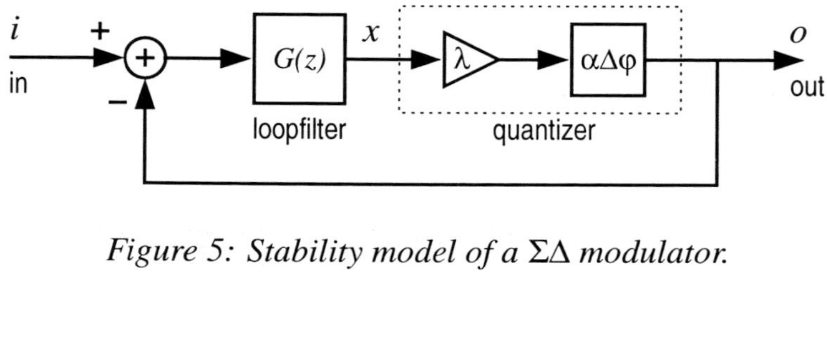

Figure 5: Stability model of a ΣΔ modulator.

ple in which an input signal with frequency $f_s/4$ (solid-line) is shifted in phase without changing the output samples of the quantizer (denoted by impulses).

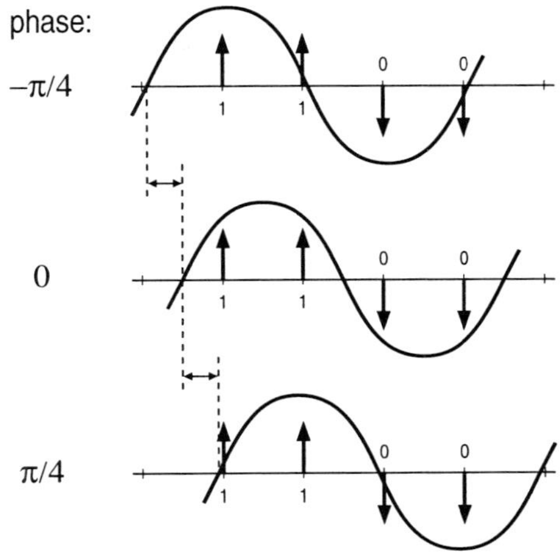

Figure 6: Phase Uncertainty of a single bit quantizer with $f_{in} = f_s/4$.

Adding this phase uncertainty to the quantizer model improves the root locus stability analysis and allows calculation of the small input-signal amplitude stability-boundary for the location of the loop filter zeroes [10]. In Fig. 7, the calculated and experimentally determined stability boundary for the loop filter parameter a is shown for several SDMs as a function of the tuning frequency. The loop filters of these SDMs belong to the class of tunable bandpass loop filters described by:

$$G(z) = \frac{(1-ae^{j\theta_0}z^{-1})^N(1-ae^{-j\theta_0}z^{-1})^N}{(1-e^{j\theta_0}z^{-1})^N(1-e^{-j\theta_0}z^{-1})^N} - 1 \quad (3)$$

in which θ_0 is the tuning frequency, $2N$ is the order of the loop filter and a controls the location of the loop filter zeroes. The experimental values have been determined by lengthy simulations. For most of the tuning range, the stability analysis using the extended model for the quantizer provides an accurate estimate of the stability boundary. For the sixth- and higher order SDMs, the calculated values are even within 5% of the experimentally determined values for more than 60% of the tuning range.

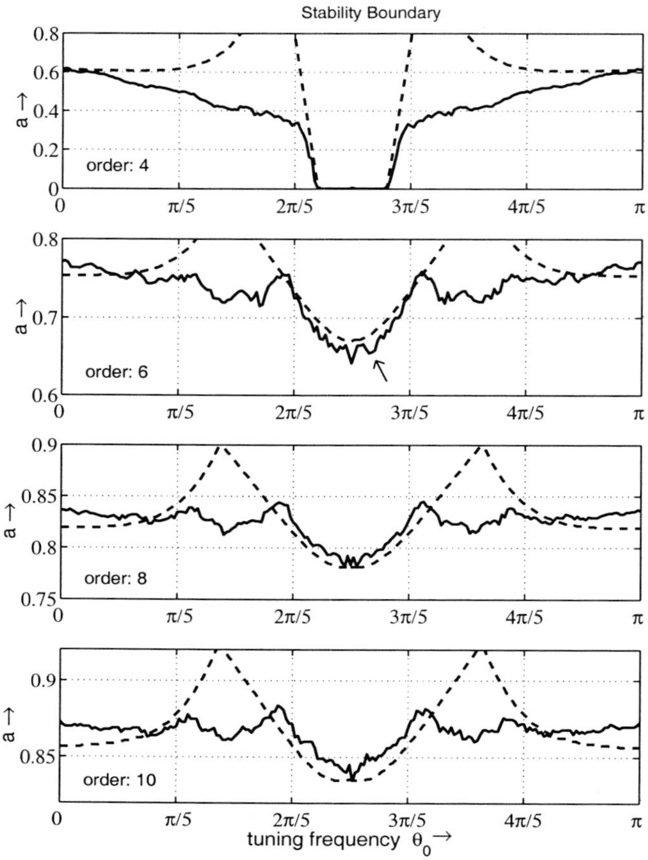

Figure 7: Theoretically (dashed) and experimentally (solid) determined minimum values for a of a 4th, 6th, 8th and 10th order bandpass SDM, for which it is stable for small signals.

5 Loop Filter Design

The equivalent DT loop filter transfer function of the sixth order SDM is also chosen from the class described by (3). With a tuning frequency of 10.7MHz, the sample frequency is set to 40MHz, giving an oversampling ratio of 100 for a bandwidth of 200kHz (FM) and a relative tuning frequency of $\theta_0 = 1.681$. According to Fig. 7, the loop filter parameter a should satisfy $a > 0.68$ when $\theta_0 = 1.681$ and the order is six ($N = 3$). In order to allow a stable operation at a larger input amplitude and provide a safety margin for implementation tolerances, a is set to $a = 0.75$, at the cost of a small reduction in SNR (see Fig. 8). For these values, the maximum stable input amplitude equals $A_{max} = -3.6$dBFS and the peak SNR equals 93dB in a bandwidth of 200kHz.

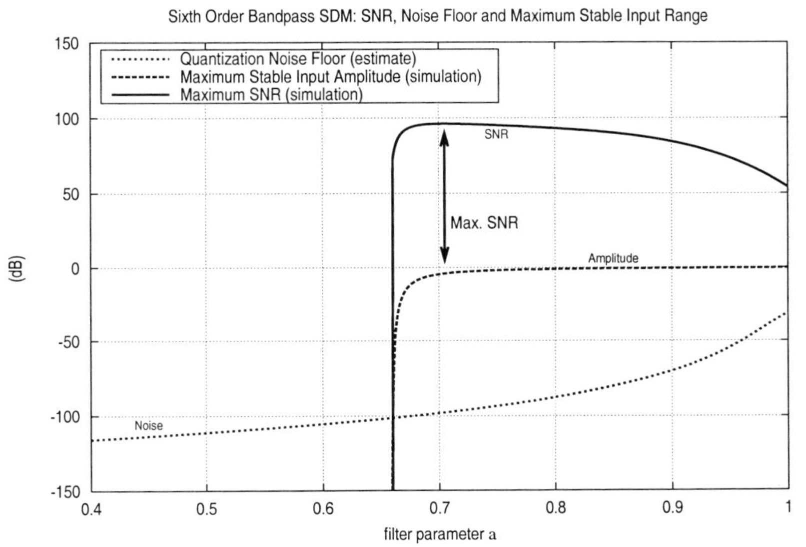

Figure 8: Performance of a sixth order bandpass SDM as a function of the filter parameter a *(OSR=100).*

The relation between the DT and CT loop filter transfer function can be found using the invariant impulse response transformation as discussed in section 3. As said, the CT loop filter structure should provide sufficient degrees of freedom to result in the designed DT loop filter. Here, the loop filter structure of the sixth-order bandpass SDM is based on a cascade of resonators (see Fig. 9). Three resonators realize the complex conjugate poles for the passband of the filter. Six feed-forward paths determine the location of the five required CT (and DT) transfer function zeroes. The resonators should have current inputs to simplify the summing nodes within the filter and voltage outputs such that the coefficients in the feed-forward paths can be realized by resistors (R_1 to R_6). The coupling resistors R_{c1} and R_{c2} are used for scaling the voltages within the

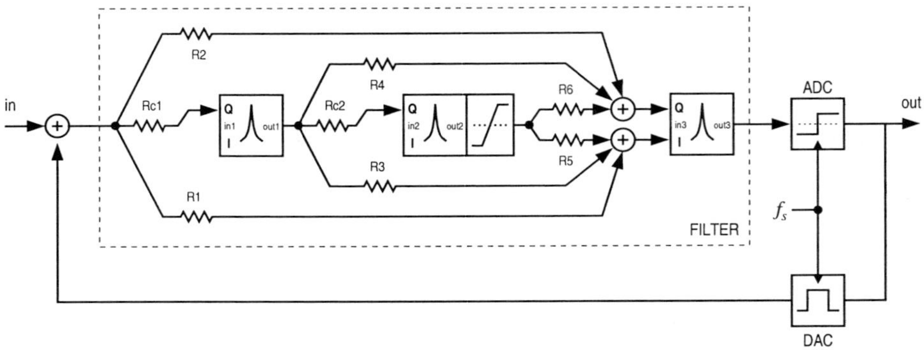

Figure 9: Diagram of the sixth order bandpass SDM.

filter to the same level. As the loop filter coefficients depend on the ratio of resistive values, an accuracy of $\pm 1\%$ can be achieved for the location of the transfer function zeroes. Tuning is only required for the resonator poles that determine the passband of the loop filter and modulator.

In order to ensure stable operation at large input signal amplitudes, the output of the second resonator is limited, as indicated in Fig. 9. When the limiter is active, the effective order of the loop filter is reduced to four and stable operation is ensured while degrading the performance of the SDM gracefully [14].

6 Implementation

The resonators of the loop filter use balanced integrators as shown in Fig. 10. The balanced integrator has a large linear output range and parasitic capacitances to the substrate have little influence as the integrating capacitors are placed in feedback. The tuning frequency equals $\omega_0 = 2\pi/RC$. As the tuning frequency is fixed, a choice for R automatically determines C. The choice for the value for R allows a trade-off between power consumption and noise as it determines the value of all resistors in the filter.

Two transistors are placed in parallel to the resistors to provide the required tuning mechanism. In order to assess the amount of nonlinear distortion introduced by these transistors, two versions of the SDM were made: with and without the tuning transistors. The capacitors C_Q limit the

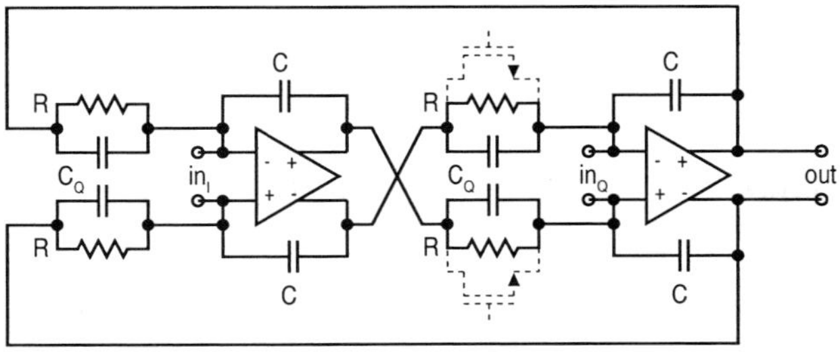

Figure 10: Balanced integrator resonator. Tuning transistors are dashed.

quality factor of the resonator. The resonators are fully balanced to allow easy implementation of negative coefficients by simply reversing positive and negative terminals. The ADC and DAC of the quantizer are also fully differential, improving the noise and distortion immunity. The supply voltage is 3.3V for the digital part and 5V for the analog sections. The common-mode DC level of the signals is 3V and the amplitude is $3V_{pp}$ maximally.

The key aspects of the design of the amplifiers of the resonator are the gain and the delay. A reasonable gain is required to prevent deterioration of the overall filter transfer characteristic. The delay of the amplifier should be small to prevent the resonator from oscillating. For these reasons, the single-stage transconductance amplifier of Fig. 11 is used [15]. The (negative) impedance of the cross-coupled transistors M1 and M2 together with the input pair M5 and M6 increase the voltage swing on the gates of the differential pair formed by M3 and M4. The overall transconductance is increased without seriously affecting the delay. The transistors M7 and M8 provide the common mode feedback. The amplifier consumes 1mA and has a gain-bandwidth product of 1.1GHz. The (simulated) equivalent input noise power density of the filter at the tuning frequency equals $8.4 \cdot 10^{-8}$ V$/\sqrt{\text{Hz}}$. Combined with a quantizer output power of -16.5dBV (see below), the noise of the filter results in a maximum achievable SNR of 72dB in 200kHz. Due to the high Q of the resonators, the noise caused by the filter will exceed the theoretical quantization noise and serve as a dither signal for the SDM.

The quantizer consists of a single bit ADC and DAC. The DAC uses a $T_s/4$-delayed return-to-zero (RTZ) pulse shape with a duration of $T_s/4$ and an amplitude of $1.2V_{pp}$ which gives a DAC power of -16.5dBV and

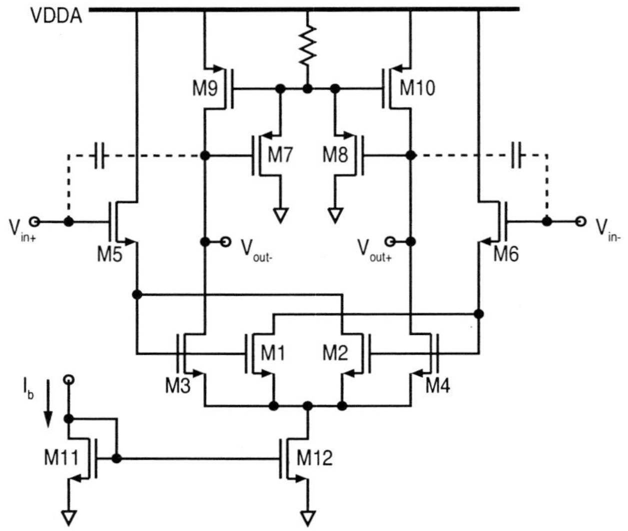

Figure 11: Transconductance amplifier.

a maximum differential input voltage of 200 mV_{pp}. The delay allows a longer decision time for the ADC, thereby decreasing the bit error rate and signal-dependent jitter. The RTZ shape of the DAC minimizes inter-symbol distortion. The quantizer ADC (see Fig. 12) consists of two flip-flops in a master slave configuration that reduces the bit error rate further without increasing the parasitic load on the filter. The master flip-flop consists of two cross-coupled transistors (M8,M9) and contains two diodes (M6,M7) for limiting the voltage swing and increasing the recovery speed of the comparator. Two non-overlapping clock signals are generated internally by dividing the external clock. A low offset voltage of the comparator is achieved by a separate input gain stage (M1,M3). The input stage is disabled by transistor M3 during latching to prevent input-signal induced switching of the master flip-flop.

The DAC of the quantizer is shown in Fig. 13. It consists of a logic block to create RTZ signals, a cascoded differential stage (M1-M4) and two source followers (M7,M8). Two dummy transistors (M5-M6) are added to the differential stage to reduce glitches caused by charge injection.

The loop of the SDM was closed off-chip to allow detailed measurements. The summing node at the input of the SDM (see Fig. 9) operates in

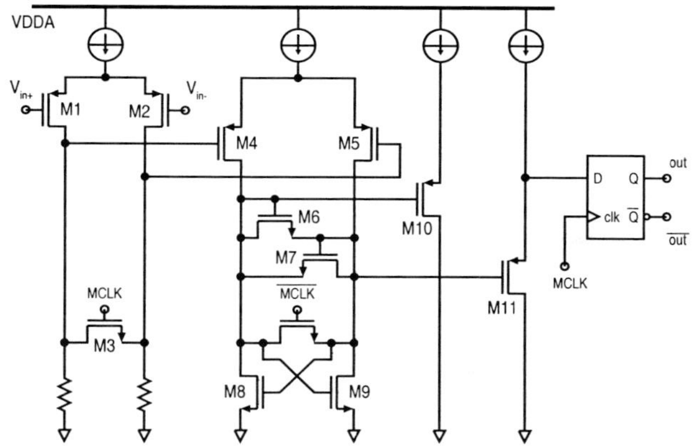

Figure 12: One bit ADC.

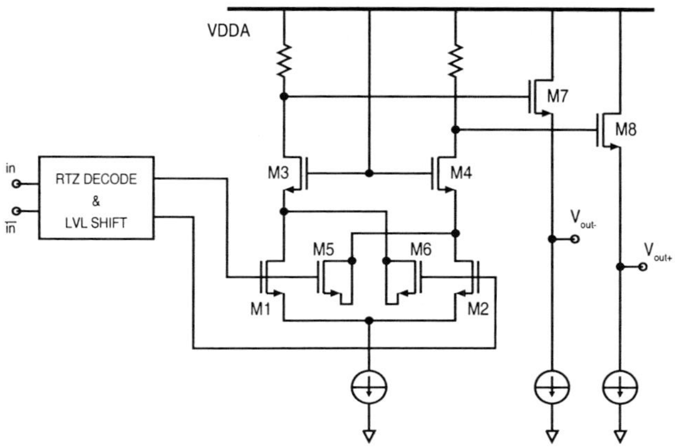

Figure 13: One bit DAC.

voltage mode and is implemented by a single transconductance amplifier (similar to Fig. 11) with an external feedback resistor and two input resistors (also external) connecting the SDM input signal and DAC feedback signal.

In order to monitor the individual outputs of the resonators in the loop filter, three analog buffers are also included in the design of the SDM. Both the tunable and non-tunable version of the SDM are realized in 0.5μm double-poly CMOS. A die photograph is shown in Fig. 14. The total chip area including bond pads measures 1.8x1.0mm^2. The core circuit (analog filter, ADC and DAC) measures 0.9x0.4mm^2 and consumes

60mW at a sample rate of 40MHz. The digital buffers consume 9.5mW at 40MHz with a load capacitance of 8pF. The analog buffers that are required for testing purposes only, consume 117mW. The total power consumed by the chip equals 186.5mW at 40MHz. As most of the power is consumed by the analog filter and the output buffers, the sample frequency is insignificant for the total power consumption.

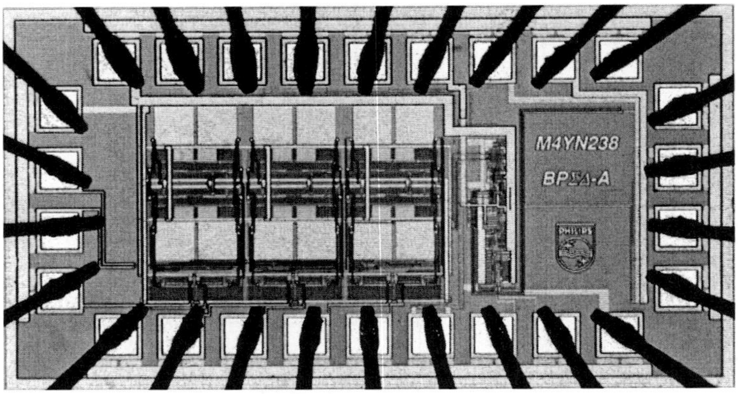

Figure 14: Die photo of the sixth order SDM IC.

7 Measurements

As the tuning frequency and sample frequency are not coupled, the sample frequency is variable and can range from 30MHz to 80MHz. For this sample frequency range, the SDM was stable and the performance (SNR and distortion) remained constant (within 0.5dB). This observation confirms the assumption that the SNR performance is limited by thermal noise. Unless stated otherwise, a sample frequency of 40MHz was used for the measurements.

Without the tuning transistors, the typical tuning frequency of the filter is 9.15MHz. The designed and measured transfer characteristic is shown in Fig. 15 after tuning to 10.7MHz. The quality factor of the resonators increased from $Q = 100$ (designed) to $Q = 180$ due to the parasitic delay of the amplifiers. The amplitude response has a maximum of 60dB and shows a loss of gain due mismatch in the tuning frequencies of the resonators (two in-band maxima). The phase response agrees well with

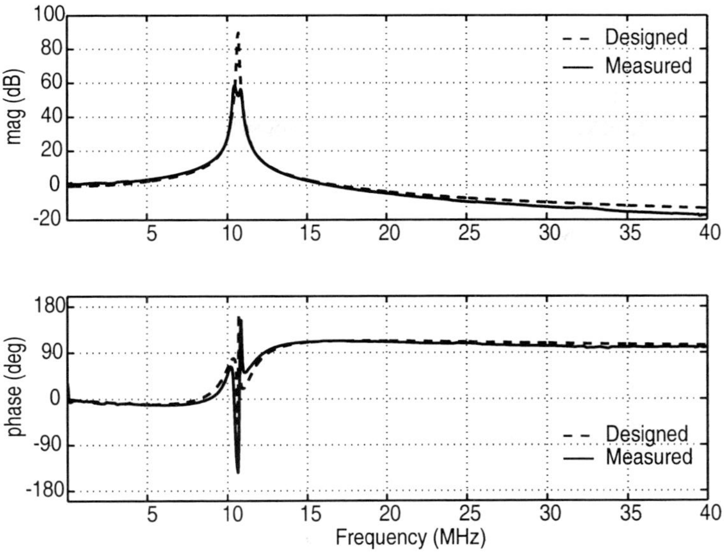

Figure 15: Measured filter transfer characteristic.

the desired characteristic which is important for stability. The phase shift of 360 degrees at the tuning frequency and the phase reduction of the feed-forward compensations (zeroes) are clearly visible.

A typical output spectrum of the tunable SDM is shown in Fig. 16. The distortion components near 1MHz and 19MHz are caused by the tuning transistors and are not present in the case of the non-tunable SDM. The non-tunable version has an idle channel noise of -91.5dBV in 200kHz and -105.5dBV in 9kHz, resulting in a dynamic range (DR) of 72dB and 86dB respectively. Figure 17 shows the signal to noise and distortion ratio (SNDR) vs. input power characteristic. At an input of -3dBFS, the maximum SNDR is 67dB in 200kHz and 80dB in 9kHz, giving an effective number of bits (ENOB) of 10.8 bit and 13 bit respectively. The performance of the tunable version is slightly less due to the nonlinearity of the tuning transistors, resulting in a DR of 67dB (200kHz) and 81dB (9kHz), and an SNDR of 63.5dB and 76dB respectively.

A two-tone measurement is performed to determine the third order intermodulation (IM3) distortion. Two input signals ('carriers') in adjacent channels (both below or above the tuning frequency) are applied to the modulator and the in-band response is measured. The measurements are

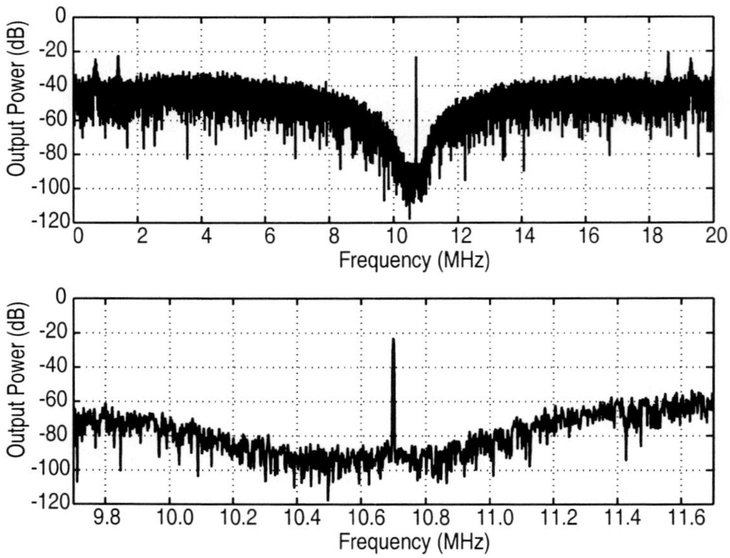

Figure 16: Measured output spectrum of tunable version (f_s=40MHz, f_{in}=10.7MHz, 32k FFT, RBW=1.22kHz).

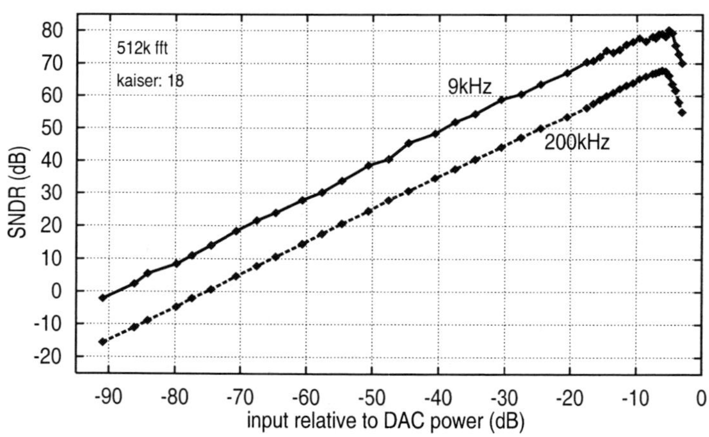

Figure 17: SNDR vs. input power characteristic of the untuned SDM.

performed for a 9kHz, 100kHz and 200kHz channel spacing (see Fig. 18). For the non-tunable version, the IM3 is -82dBc at a -10dBFS single carrier power, corresponding to an IP3 of +11.5dBV. The tunable version achieves an IM3 of -75dBc at the same input level. For an input carrier

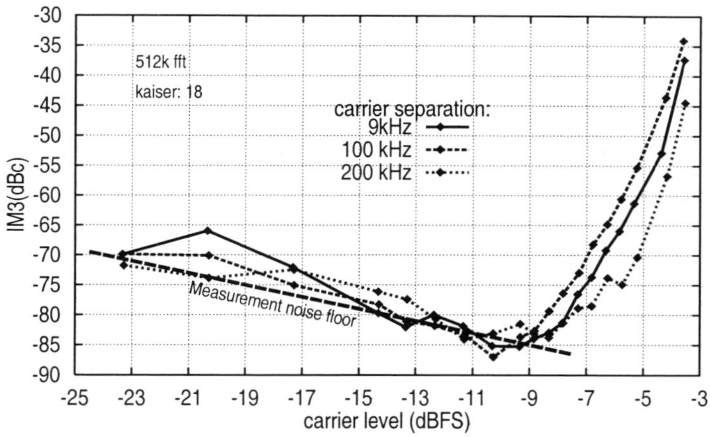

Figure 18: IM3 vs. carrier power of the non tunable SDM.

power below -13dBFS, the IM3 is masked by the noise floor. For carrier powers exceeding -8dBFS the IM3 distortion increases rapidly. At such input powers, the signals in the loop filter become large. The limiter in the second resonator becomes active and increases the distortion. Figure 19 shows a two-tone test of the tunable SDM for a carrier spacing of 200kHz. Note that the in-band IM3 component at 10.7MHz is more than

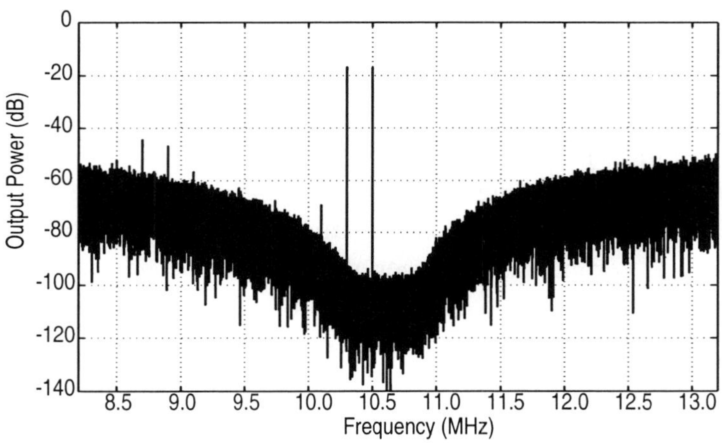

Figure 19: Output spectrum of the two-tone IM3 measurement of the tunable SDM (f_s=40MHz, f_1 =10.3MHz, f_2=10.5MHz, Δf =200kHz, 512k FFT, RBW=76.3Hz).

20dB lower than the out-of-band component at 10.1MHz. This shows that the IM3 distortion is partly suppressed by the feedback loop and suggests that the distortion is mainly caused within the filter. In case the distortion would have been caused in the DAC or summing node, the two distortion components would have been equal. Table 1 shows a summary of the performance of the sixth order bandpass SDM. The power consumption does not include the consumption by the analog output buffers which are not required for the operation of the SDM.

Table 1: 6th Order Bandpass SDM Performance

	non tunable		*tunable*	
technology	0.5μm CMOS			
supply voltage	5.0V analog, 3.3V digital			
power consumption	60 mW @ f_s=40MHz			
sample frequency	30 - 80 MHz			
tuning frequency	9.15 MHz		10.7 MHz	
	Bandwidth		Bandwidth	
	200kHz	9kHz	200kHz	9kHz
idle channel noise	-91.5dBV	-105.5dBV	-86.5dBV	-100.5dBV
DR	72dB	86dB	67dB	81dB
peak SNDR	67dB	80dB	63.5dB	76dB
ENOB	10.8	13	10.2	12.3
IM3 rel. to carriers	-82dBc		-75dBc	
max. diff. input	200mV$_{pp}$			

The DR and SNDR performance of the SDM is limited by the thermal noise of the filter. The performance can be improved by reducing the impedance level in the filter at the penalty of a higher power consumption. In particular, the noise contribution of the first resonator should be reduced. A comparison between the tunable and non-tunable version shows that the tuning transistors have a slight impact on the performance of the SDM. The tuning mechanism may be improved by adding two cross-coupled transistors to cancel the nonlinear part in the transconductance [16].

8 Conclusion

A sixth order single-loop one-bit continuous-time bandpass sigma delta modulator (SDM) for digitizing IF signals in a combined AM/FM radio receiver has been designed and tested. The SDM achieves a DR of 72dB in a 200kHz bandwidth centered at 10.7MHz. An important aspect in the design of the SDM is the stability analysis using the describing function method. Key to the analysis is the modeling of the phase uncertainty of a sampled quantizer. Together with the SDM presented in [7], this SDM shows that SDMs are a viable solution for high-performance A/D conversion of IF signals.

References

[1] K. Kianush and C. Vaucher, "A global car radio IC with inaudible signal quality checks," in *IEEE Int. Solid-State Circuits Conf. (ISSCC) digest of technical papers*, San Fransisco (CA), Feb. 1998, pp. 130–131.

[2] Armond Hairapetian, "Bandpass delta-sigma converters in IF receivers," in *Analog Circuit Design, MOST RF circuits, Sigma-Delta Converters and Translinear Circuits*, Willy Sansen, Johan H. Huijsing, and Rudy J. van de Plassche, Eds., pp. 193–209. Kluwer Academic Publishers, Dordrecht, NL, Apr. 1996, ISBN 0-7923-9776-2.

[3] Steven R. Norsworthy, Richard Schreier, and Gabor C. Temes, Eds., *Delta-Sigma Data Converters; Theory, Design and Simulation*, IEEE press, New York, 1997, ISBN 0-7803-1045-4.

[4] R. Schreier and M. Snelgrove, "Bandpass sigma-delta modulation," *Electronic Letters*, vol. 25, pp. 1560–1561, Nov. 1989.

[5] F.W. Singor and M. Snelgrove, "10.7MHz bandpass sigma-delta A/D modulators," in *Proc. of the IEEE 1994 Custom Integrated Circuits Conference (CICC)*, San Diego, CA, May 1994, pp. 163–166.

[6] Stephen A. Jantzi, Kenneth W. Martin, and Adel S. Sedra, "Quadrature bandpass $\Delta\Sigma$ modulation for digital radio," *IEEE Journal of Solid-State Circuits*, vol. 32, no. 12, pp. 1935–1950, Dec. 1997.

[7] Erik van der Zwan, Kathleen Philips, and Corné Bastiaansen, "A 10.7MHz IF-to-Baseband $\Sigma\Delta$ A/D conversion system for AM/FM radio receivers," in *IEEE Int. Solid-State Circuits Conf. (ISSCC) digest of technical papers*, San Fransisco (CA), Feb. 2000, pp. 340–341.

[8] Ichiro Fujimori, Lorenzo Longo, Armond Hairapetian, Kazushi Seiyama, Steve Kosic, Jun Cao, and Shu-Lap Chan, "A 90dB SNR, 2.5MHz output rate ADC using cascaded multibit ΔΣ modulation at 8x oversampling ratio," in *IEEE Int. Solid-State Circuits Conf. (ISSCC) digest of technical papers*, San Fransisco (CA), Feb. 2000, pp. 338–339.

[9] Toshio Hayashi, Yasunobu Inabe, Kuniharu Uchimura, and Tadakatsu Kimura, "A multistage delta-sigma modulator without double integration loop," in *IEEE Int. Solid-State Circuits Conf. (ISSCC) digest of Technical Papers*, Feb. 1986, pp. 182–183.

[10] Jurgen van Engelen and Rudy van de Plassche, *Bandpass Sigma Delta Modulators: Stability Analysis, Performance and Design Aspects*, Kluwer Academic Publishers, Dordrecht, the Netherlands, Oct. 1999, ISBN 0-7923-8698-1.

[11] Richard Schreier and Bo Zhang, "Delta-sigma modulators employing continuous-time circuitry," *IEEE Trans. on Circuits and Systems I: Fundamental Theory and Applications*, vol. 43, no. 4, pp. 324–332, Apr. 1996.

[12] J. van Engelen and R. van de Plassche, "A combined LC/gmC 4th order continuous-time bandpass ΣΔ modulator," in *Proc. of the European Solid-State Circuits Conference (ESSCIRC)*, The Hague (NL), 1998, pp. 156–159.

[13] J. van Engelen, R. van de Plassche, E. Stikvoort, and A. Venes, "A 6th order continuous time bandpass sigma delta modulator for digital radio IF," *IEEE Journal of Solid-State Circuits*, vol. 34, no. 12, pp. 1753–1764, Dec. 1999.

[14] E.J. van de Zwan, "A 2.3mW CMOS ΣΔ modulator for audio applications," in *IEEE Int. Solid-State Circuits Conf. (ISSCC) digest of technical papers*, San Fransisco, 1997, pp. 220–221.

[15] A. van Bezooijen, N. Ramalho, and J.O. Voorman, "Balanced integrator filters at video frequencies," in *Integrated Continuous-Time Filters*, Y.P. Tsividis and J.O. Voorman, Eds., pp. 129–132. IEEE Press, New York, 1993, ISBN 0-7803-0425-X, (reprint from Digest ESSCIRC'91, pp.1-4, 1991).

[16] Zdzislaw Czarnul, "Modification of the Banu-Tsividis continuous-time integrator structure," *IEEE Trans. on Circuits and Systems*, vol. 33, no. 7, pp. 714–716, July 1986.

Mixed Signal Design

Johan H. Huijsing

Today, the design of integrated circuits shifts more and more to integrated system design. Front end and back end analog circuits are mixed with digital circuitry on a single chip that is usually fabricated in a process dedicated to compact digital VLSI design. A first challenge to the designer is to optimize the partitioning between analog and digital circuits in a good top-down design. A second challenge is to minimize crosstalk from the digital part into the analog part. These challenges have been covered in six papers presented in this book.

The first two papers touch on general issues, George Gielen discusses System Level Design issues and telecom frontends, while Herman Casier touches on some specific mixed signal design issues and CAD tools.

The third and fourth papers talk about CAD tools: Ken Kundert talks about top down design of mixed signal circuits and Frederic Stubbe discusses computer aided design for integrated systems.

Finally the fifth and the sixth paper provide design examples. Giovanni Cesura shows a mixed-mode sigma-delta converter for high-quality audio and Domine Leenaerts presents mixed-mode telecom design issues.

R. J. van de Plassche et al. (eds.), Analog Circuit Design, 139.

SYSTEM-LEVEL DESIGN ISSUES FOR MIXED-SIGNAL ICs AND TELECOM FRONTENDS

Georges GIELEN*
Katholieke Universiteit Leuven, ESAT-MICAS
Kardinaal Mercierlaan 94, B-3001 Leuven, Belgium
e-mail : gielen@esat.kuleuven.ac.be

* also research associate of the Belgian National Fund of Scientific Research (FWO-V)

ABSTRACT

An overview is presented of the challenges and design issues in the system-level design of mixed analog-digital ICs and telecom frontends. The progress in VLSI technology allows the integration of complex systems on a chip, containing both analog and digital parts. In order to boost the design productivity and guarantee the optimality of such systems while meeting the time to market constraints, a systematic top-down design approach has to be followed with sufficient time and attention paid to system-level architectural design before proceeding to the detailed block design. High-level system exploration tools will be presented that enable analog-digital co-design and allow to analyze architectural alternatives and explore trade-offs such as finding the optimal analog-digital partitioning. This will be illustrated with several experimental results from experimental tools. Finally, the crucial underlying technology for such high-level design will be described in detail: analog behavioral modeling, efficient high-level simulation methods and analog power/area estimation.

R. J. van de Plassche et al. (eds.), Analog Circuit Design, 141–165.

1. INTRODUCTION

Advances in deep submicron VLSI integrated circuit processing technologies offer designers the possibility to integrate more and more functionality on one and the same die, enabling in the near future the integration of complete systems that before occupied one or more printed circuit boards onto a single piece of silicon [1]. An increasing part of these integrated systems contain digital as well as analog circuits, and this in application areas like telecommunications, automotive and multimedia among others.

The drive towards integrated mixed-mode analog/digital ICs is however posing some big problems. The complexity of the systems that can be integrated on a single IC can only be mastered by using advanced computer-aided design (CAD) tools and by shifting to a higher level of design abstraction. A typical design flow for mixed-signal integrated systems may look as shown in Fig. 1, where the following distinct phases can be identified: system specification, architectural design, cell design, cell layout and system layout [2].

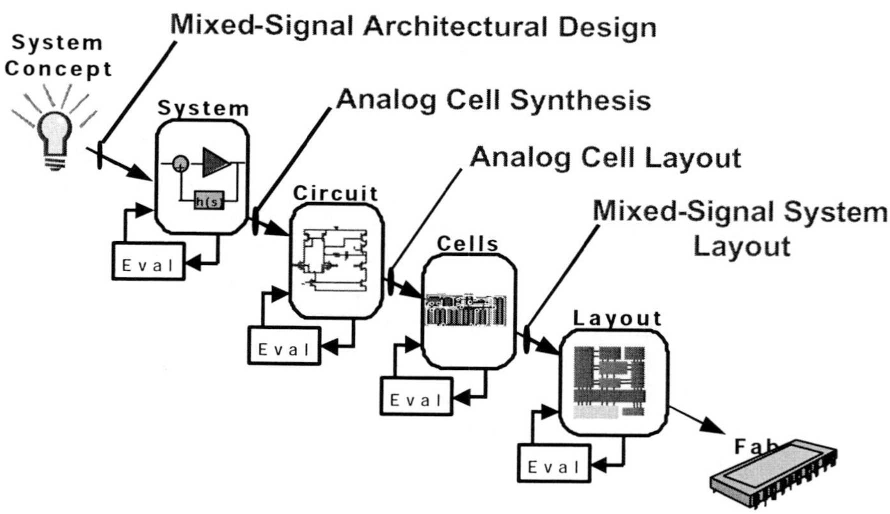

Fig. 1. View of the mixed-signal IC design process.

For the digital circuits, commercial simulation and synthesis tools, especially at the logic and layout level, have been around for some years now and a considerable part of the digital design flow has been automated, starting the design from an RTL description. In recent years the design entry level is shifting more and more towards an object-oriented system description. Most of the current system design environments however cannot handle the analog circuits, thereby excluding the possibility to explore system-level trade-offs and architectural decisions across the analog-digital boundary. For instance in telecom applications like GSM and xDSL, the analog frontend circuits are limiting the overall performance of the system and proper system architectural decisions can substantially relax the requirements on the analog interface circuits.

In addition, for complex integrated systems-on-a-chip (SoC) the use of soft or hard IP (intellectual property) blocks or so-called "virtual components" is expected to provide the extra productivity boost needed to realize these systems in the required time to market constraints [3]. The use of analog IP however is accompanied by a lot of scepticism and nightmare stories, not at least because of the strong technology dependence and stray sensitivity of analog circuit behavior.

The growing interest in mixed-signal ICs is also exposing the lack of mature analog CAD tools that can boost the productivity of analog designers. Therefore, although the analog circuits typically occupy only a small part of the area in mixed-signal ICs, they require a disproportionally large part of the overall design time. In times of increased integration with uniform deadlines for the analog and digital parts, this poses serious challenges to the analog designers and their productivity, prompting them to adopt some form of computer-aided design support or even synthesis. It has been reported that the use of analog CAD tools can drastically reduce the design time, without sacrificing on the performance [4].

In addition to all this, there are specific problems posed by the integration of analog and digital circuits on the same chip, mainly related to signal integrity issues [5]. Parasitic crosstalk due to capacitive or inductive wiring coupling, power supply coupling and especially substrate coupling introduces extra noise and limits the performance of the analog circuitry. These effects are today not yet sufficiently handled by the existing digital-oriented tools. Some specific backend verification tools exist, but no full closed-loop verification of complex mixed-signal systems is possible today, and the problem should also fully be incorporated in the frontend flow to anticipate rather than remedy problems.

In summary, some of the major challenges in mixed-signal integrated system design today are :

- need for a mixed-signal architectural exploration environment allowing to analyze and compare different architectural solutions in terms of performance, power and area;
- such an environment inevitably also requires higher-level analog modeling and power/area estimation to quantify and compare the different system solutions;
- need for analog CAD and synthesis tools to boost the analog design productivity up to a level compatible with the present time-to-market constraints;
- need for a full set of mixed-signal signal integrity analysis and design tools that can cope with the complexity of fully integrated systems.

This paper will address the first of these problems : the architectural exploration of mixed-signal telecom systems, and the underlying enabling techniques. The objectives and problem formulation will be presented in detail in section 2. Examples of architectural-level exploration and design of telecom frontends will be provided in section 3, including also the problem of analog-digital partioning. The underlying enabling techniques will then be described in section 4 : analog behavioral modeling, efficient high-level simulation methods and analog power/area estimation. Conclusions will be drawn in section 5.

2. MIXED-SIGNAL ARCHITECTURAL SYSTEM EXPLORATION

The general objective of analog architectural system exploration is twofold. First of all, a proper architecture for the system has to be decided upon. Secondly, the required specifications for each of the blocks in the chosen architecture must be determined, so that the overal system meets its requirements at a minimum implementation cost. The aim of a system exploration environment is to provide the system designer with the platform and the supporting tool set to fast explore alternatives and to take the above decisions based on quantified rather than heuristic information.

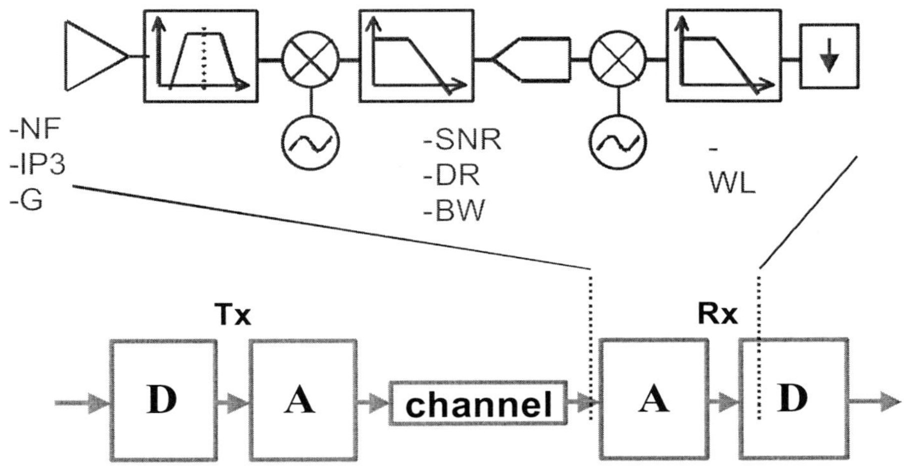

Fig. 2. Digital telecommunication link, indicating a possible receiver frontend architecture with some block specifications.

Consider for instance the digital telecommunication link of Fig. 2. It is clear that digital bits are going into the link to be transmitted over the channel, and that the received signals are being converted again in digital bits. One of the main considerations in digital telecom design is the bit error rate, which characterizes the reliability of the link. This bit error rate is impacted by the characteristics of the transmission channel itself, but also by the architecture chosen for the transmitter and receiver frontend and by the performances achieved by the subblocks. For example, the noise figure and nonlinear distortion of the input low-noise amplifier are key parameters. Similarly, the resolution and sampling speed of the used analog-to-digital converter (ADC) has a large influence on the bit error rate, but it also determines the requirements for the other analog subblocks. A higher ADC resolution may reduce the filtering requirements in the transceiver, resulting in simpler filter structures, but it will also consume more power and chip area than a lower-resolution converter. The minimum required ADC resolution however and therefore also the minimum power and area depends on the architecture chosen for the transceiver frontend. Clearly, there is a large interaction between system-level architectural decisions and the performance requirements for the different subblocks, which on their turn are bounded by technological limits that shift with every new process being taken up.

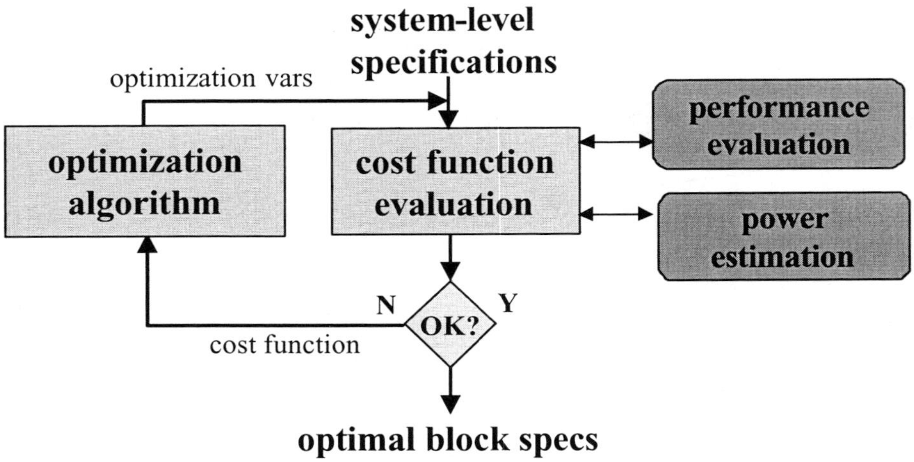

Fig. 3. General flow of analog high-level design, with behavioral performance evaluation and power estimation as tools to quantify performance and implementation cost.

In general, as shown in Fig. 3, analog high-level design is the translation of system-level specifications into a proper architecture of subblocks, in which the individual specifications of all subblocks are completely decided so that the overall system meets its specifications [2]. During the top-down design phase the subblocks however are not yet device-level circuit implementations. Hence they have to be represented as "behavioral models" representing their functional input-output behavior. The performance can be evaluated by carrying out a behavioral simulation of the architecture, and/or by evaluating (if available) a set of equations that describe the architecture's performances in terms of the subblock specifications. If equations can be obtained, this approach is in general faster, but the simulation-based approach with high-level model representations for the subblocks is more generally applicable. In this way the performance of an architecture with given building block specifications can be analyzed by the system designer and checked against the system specifications.

A difficult step however is the translation of the system-level specifications into individual specifications for the subblocks. This may require detailed experience in analog circuit design, and even then there is no guarantee that the designer will take optimal decisions. It is therefore better to exploit the power of computers and base the decision on quantitative rather than heuristic information. The most general way for a high-level synthesis program to do this is by using optimization techniques to find the optimal set of subblock specifications, such that the overall system meets the target specifications Ψ_j :

$$P_j(specs_i) \geq \Psi_j \text{ or } P_j(specs_i) \leq \Psi_j \tag{1}$$

and at the same time a user-defined design objective (e.g. the overall power consumption) is minimized. This implies that at each iteration of the optimization both the performance of the system and the optimization targets are evaluated. The performance can be analyzed as described above. The design targets can be estimated using power and area estimators (see Fig. 3) that predict the power or area required to implement each subblock i in the architecture as a function of the subblock's specifications :

$$power_i = f(specs_i) \tag{2}$$

$$area_i = f(specs_i) \tag{3}$$

If this high-level optimization is repeated for different architectures, then a comparison between these architectures becomes possible before they are implemented down to the transistor level, by assessing and quantifying their suitability to meet the system specifications at a minimum implementation cost. Bad or less optimal architectures can be eliminated in this way, and only the most promising one can be selected for further implementation down to the device level. Today the alternative architectures that are considered are still to be provided by the system designer, but future tools might also derive these architectures automatically from a high-level language description [6].

This approach of analog system-level architectural exploration will now be illustrated with some examples. The important underlying techniques - simulation algorithms, behavioral models and power/area estimators - will then be described afterwards.

3. EXAMPLES OF ARCHITECTURAL-LEVEL EXPLORATION AND DESIGN OF TELECOM FRONTENDS

3.1. Transceiver frontend exploration and optimization

A first experimental prototype tool towards the architectural-level exploration and optimization of receiver frontends was the ORCA tool [7]. The tool focused on receivers as their design is more difficult due to the uncertainty and wide range of possible input signals (desired channel, noise, neighboring channels and blocking signals, etc.). The tool contains a library with behavioral models and power estimation functions for the typical blocks encountered in receivers. The behavioral model of each block includes both its nominal behavior (amplification, filtering, mixing) as well as its most important nonidealities (e.g. noise, distortion, aliasing, phase noise, etc). In ORCA the simulations were performed with a dedicated frequency-domain simulation algorithm that processes stochastic input power spectra typical for the targeted application (e.g. GSM, DECT). The power spectral distributions at every node in the circuit were calculated in an efficient though approximate way to enable a short response time of ORCA to the designer during the architectural exploration phase of the receiver. The tool allowed the designer to interactively explore alternative receiver architectures and to investigate design trade-offs within each architecture at the architectural level, before designing each individual subblock. As the performance analysis routine was also integrated within an optimization loop, the tool could also perform an optimal high-level synthesis of a given architecture towards a specific application such that the complete receiver meets the required signal quality while the overall estimated power consumption of the entire architecture is minimized [7].

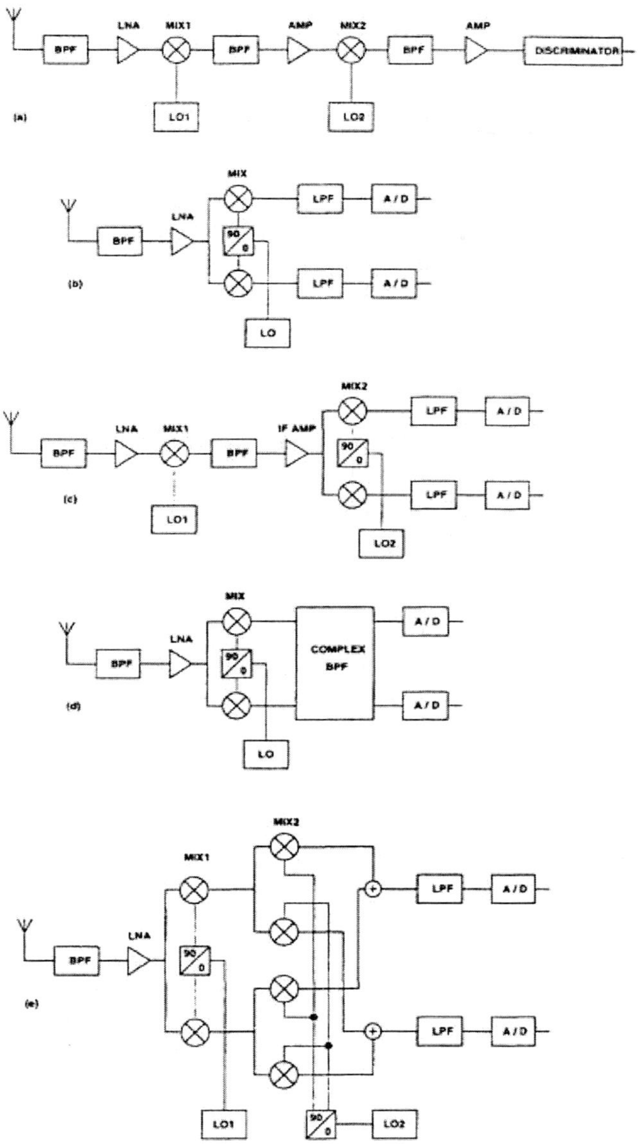

Fig. 4. Different RF receiver architectures : (a) classical double-IF receiver, (b) zero-IF receiver, (c) combined IF zero-IF receiver, (d) low-IF receiver and (e) quasi-IF receiver architecture.

The recent boom of wireless communication applications has resulted in a demand for very small, highly integrated RF transceivers dissipating low power. To achieve this goal, several new transceiver architectures different from the classical superheterodyne receiver have been proposed in recent years. Fig. 4 shows some of these architectures : the direct conversion or zero-IF architecture (b), the combined IF zero-IF architecture (c), the low-IF architecture with complex signal processing (d), and the quasi-IF architecture (e). These architectures can then be analyzed by ORCA for input power spectra typical for their application, for instance the spectra shown in Fig. 5 for cellular phone applications (e.g. GSM). For the architecture of Fig. 4c and for the input spectrum of Fig. 5a, the resulting spectrum at the receiver output as calculated by ORCA is shown in Fig. 6 [8]. The signal band of interest in this example is 100 kHz. The different channels have been approximated as rectangular spectra. The power spectrum of the wanted signal and the different signal-degrading effects (thermal and phase noise contributions from the subblocks, aliased signals and distortion introduced by the subblocks) are calculated and displayed separately by ORCA, allowing to analyze the different effects separately. The resulting SNDR in the signal band due to the different effects are summarized in Table 1. The reliability of these results of course depends on the accuracy of the behavioral models, the used signal representations and the simulation algorithms.

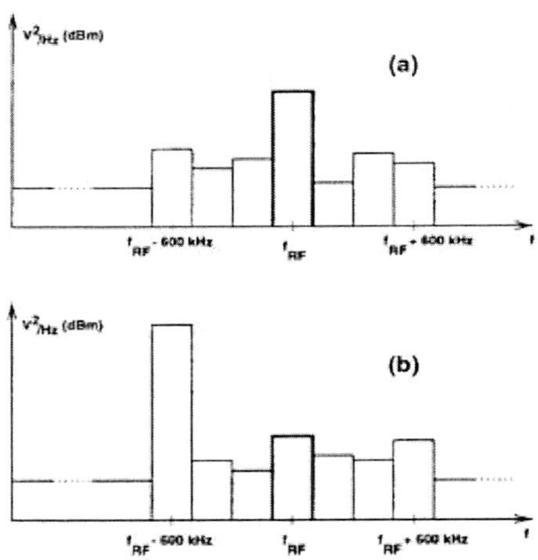

Fig. 5. Typical input power spectra for cellular phone applications.

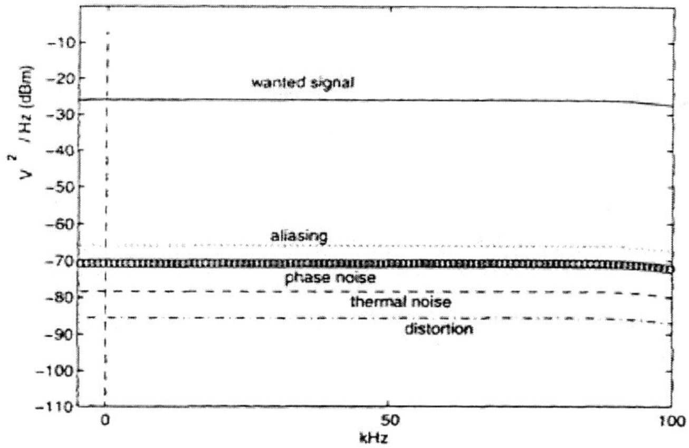

Fig. 6. Plot of an ORCA behavioral simulation result showing the receiver output spectrum with desired signal and major signal-degrading effects.

	SNDR (dB)
thermal noise only	52.5
phase noise only	44.8
aliasing only	39.9
distortion only	59.9
DC peak	25.8
total (without DC peak)	38.8

Table 1. Contribution of the different signal degrading effects to the overall SNDR.

As a second example, consider a frontend for a cable TV modem receiver, based on the MCNS standard [8]. The MCNS frequency band for upstream communication on the CATV network is from 5 to 42 MHz (extended subsplit band). Two architectures are shown in Fig. 7 : an all-digital architecture where both the channel selection and the downconversion are done in the digital domain, and the classical architecture where the channel selection is performed in the analog domain. A typical input spectrum is shown in Fig. 8. For this example we have used 12 QAM-16 channels with a 3 MHz bandwidth. We assume a signal variation of the different channels of maximally ±5 dB around the average level. The average channel noise is 30 dB below this level. Fig. 9 and 10 show the spectrum simulated by ORCA for the all-digital architecture of Fig. 7a. Fig. 9 shows the spectrum after the analog-to-digital converter, whereas Fig. 10 shows the spectrum at the output after digital channel selection and quadrature downconversion. The wanted signal and the effects of the channel noise, the ADC quantization noise, and the second- and third-order distortion are shown separately. The resulting SNDR is equal to 22.7 dB in this case, which corresponds to a symbol error rate of less than 10^{-10} for QAM-16.

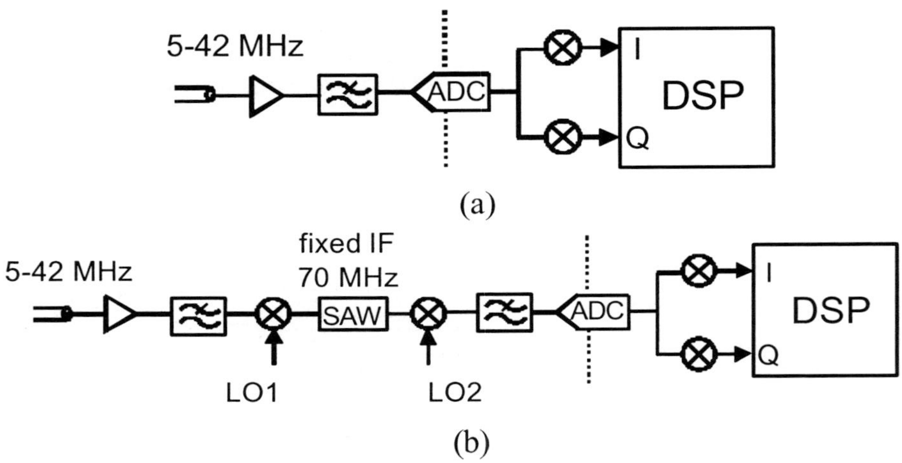

Fig. 7. Two architectures for CATV application : (a) all digital architecture, (b) classical architecture.

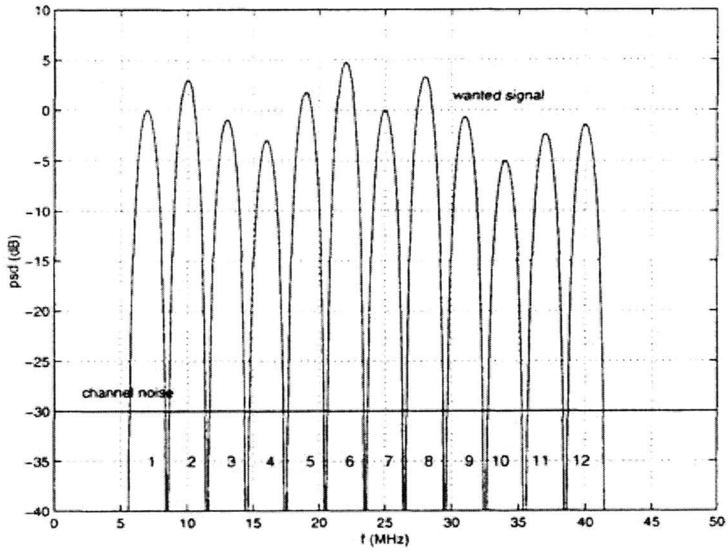

Fig. 8. Typical input spectrum for a CATV frontend architecture using 12 QAM-16 channels.

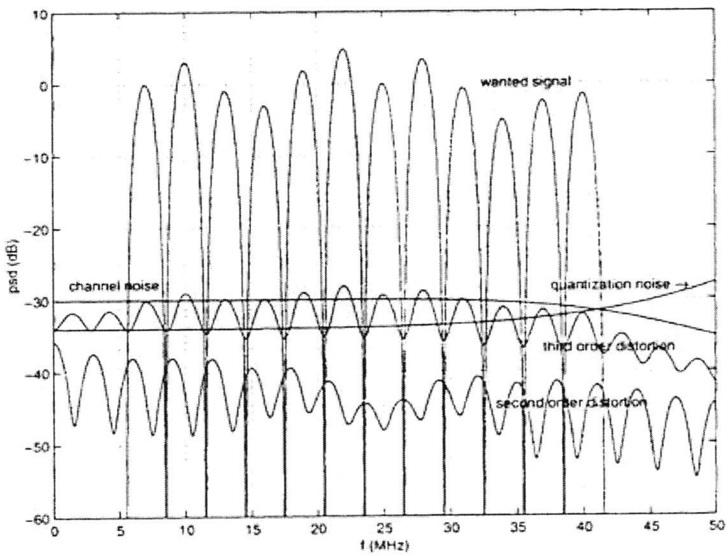

Fig. 9. Simulated spectrum of the all-digital CATV architecture after the ADC.

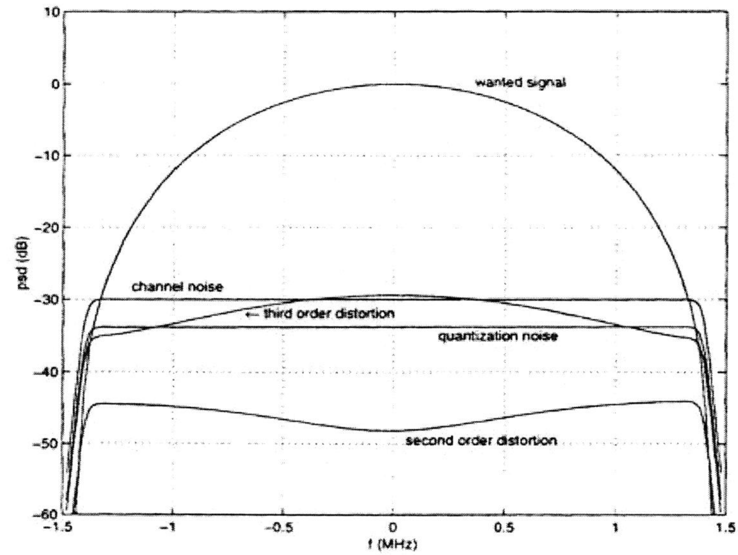

Fig. 10. Simulated spectrum of the all-digital CATV architecture at the receiver output.

3.2. Architectural exploration and trade-off analysis

By performing the same analysis for different architectures and by linking the required subblock specifications to the power and/or chip area required to implement the subblocks, a quantitative comparison of different alternative architectures becomes possible with respect to 1) their suitability to implement the system specifications, and 2) the corresponding implementation cost in power consumption and/or silicon real estate. To assess the latter, high-level power and/or area estimators must be used to quantify the implementation cost. In this way the system designer can choose the most promising architecture.

Fig. 11 shows a comparison between the estimated total power consumption required by the all-digital and by the classical CATV receiver architectures of Fig. 7 as a function of the required SNR [9]. These results were obtained with the simulator FAST [10]. Clearly, with the current state of the technology, the classical architecture still requires much less power than the all-digital solution. Fig. 12 gives the distribution of the power consumption over the different subblocks for the CATV architectures.

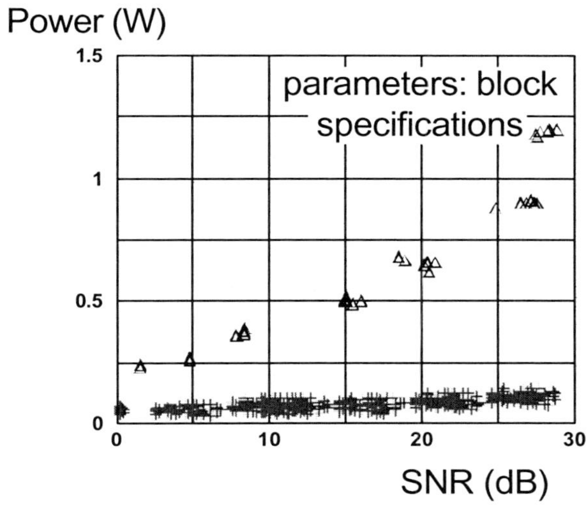

Fig. 11. Power consumption comparison between the all-digital CATV architecture (blue triangles) and the classical architecture (red crosses) as a function of the required SNR.

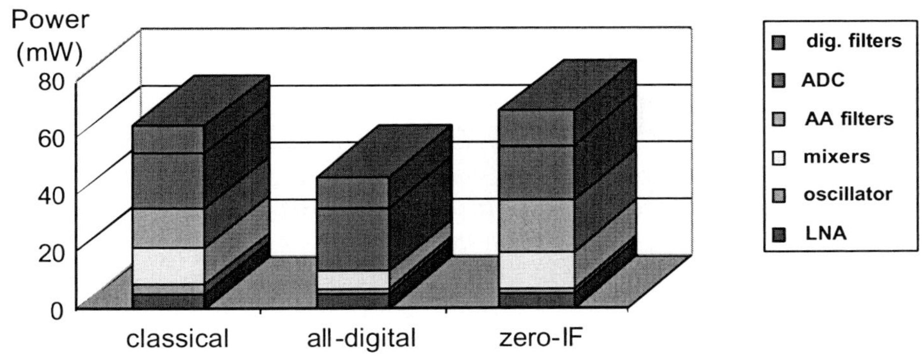

Fig. 12. Power consumption of the different subblocks in different CATV receiver architectures.

3.3. Analog-digital system-level partitioning

Another important problem in mixed-signal system design is finding the optimal partitioning between analog and digital [11]. Today this partitioning is often performed ad hoc, with some crude calculations, by an experienced system designer. It is however often not feasible to investigate many alternative solutions in the design space. And although there might be good reasons for implementing as many functions in the digital domain as possible, two problems might result. Firstly, the specification constraints on the few remaining analog subblocks might become too stringent to realize in the given technology. For instance, the required analog-to-digital converter specifications in terms of resolution and bandwidth might become unrealizable. Secondly, the digital solution might require more power than the solution with more analog functionality. Therefore, high-level exploration can quantitatively find the optimal analog-digital partitioning in mixed-signal systems. Optimal here means for instance lowest overall power consumption for the complete system. Tools like ORCA can find the optimal partitioning by also incorporating architectural variations within the optimization variables. The method is intended to assist system designers with quantitative data in their comparing different alternative system architectures.

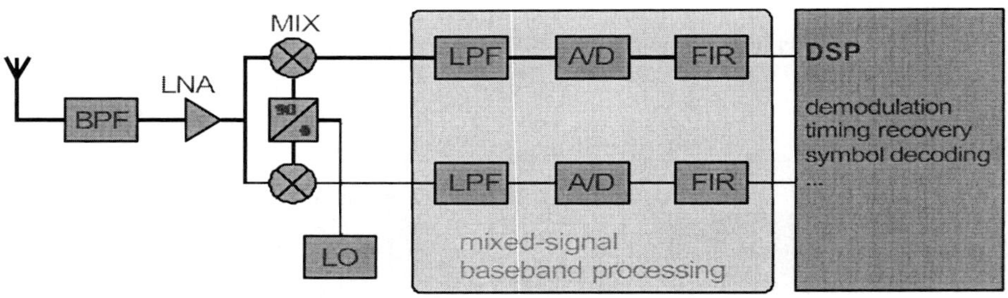

Fig. 13. Baseband signal processing channel in a direct conversion receiver.

This is now illustrated for the baseband signal processing part of a direct conversion receiver as shown in Fig. 13. The partitioning trade-off is in the position of the analog-to-digital converter and the amount of filtering in the analog versus the digital domain. Hence, the order of the analog low-pass filter and of the digital FIR filter are the architectural optimization variables. Subblock specifications include for instance the resolution and the speed of the analog-to-digital converter. The main system-level specification to be met is a 10-dB signal-to-noise requirement at the output of the digital multirate filter for an input signal spectrum to the baseband filtering blocks shown in Fig. 14. The system is simulated using the ORCA tool [7], extended with behavioral simulation models for digital (multirate) filters.

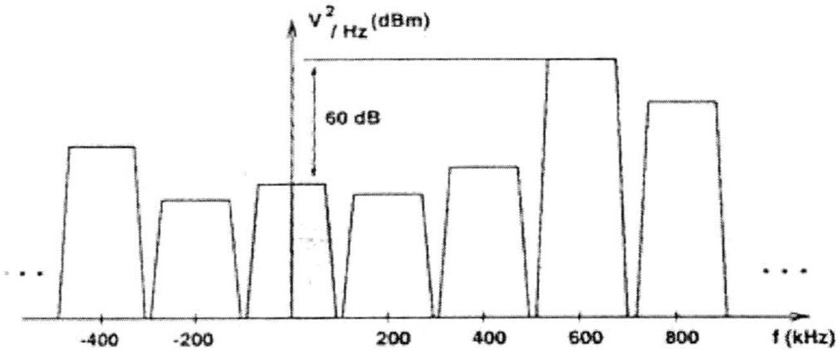

Fig. 14. Input spectrum of the baseband (channel) filtering blocks.

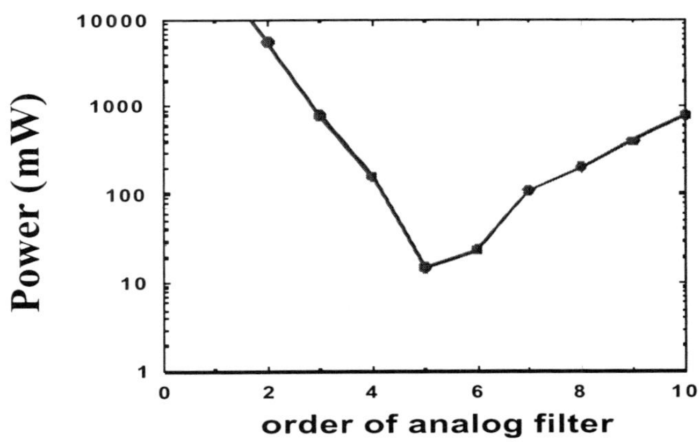

Fig. 15. Power consumption of the baseband signal processing channel as a function of the amount of analog filtering (and inversely the amount of digital filtering).

The resulting (estimated) power consumption varies as a function of the order of the analog filter (and inversely also of the digital filter), as shown in Fig. 15. The smallest overall power consumption is obtained with a 5th-order analog filter and a 28-tap digital filter. Hence this also determines the optimal insertion point of the analog-to-digital converter in the given technology. Clearly a lot of power can be gained by carefully making the trade-off between analog and digital filtering in this example. The reliability of these results of course depends on the accuracy of the behavioral models, the signal representations, the simulation algorithms and the power estimators used.

4. UNDERLYING TECHNOLOGY : BEHAVIORAL MODELING, SIMULATION ALGORITHMS AND POWER ESTIMATION

This section will now discuss in more detail the underlying technology needed for this mixed-signal system-level exploration : analog behavioral models, simulation algorithms and signal representations, and power estimators.

4.1. Analog behavioral modeling

A crucial element for system-level top-down design to succeed is the availability of analog behavioral models that describe the behavior of an analog block without knowing the details of the underlying circuit implementation. These models must describe the desired behavior of the block, while still including the major nonidealities of real implementations with sufficient accuracy. For this purpose a library of generic behavioral models can be developed for analog blocks that are frequently encountered in the targeted application domain.

For telecom frontends for instance, the following behavior can be included in every block's model (see Fig. 16) [8] :

- intended behavior, e.g. amplification for an LNA, filtering for a filter, mixing for a mixer, etc.
- noise behavior introduced by the block – this could be specified by characteristics like the noise figure
- phase noise introduced by some blocks like the local oscillator
- distortion behavior introduced by the block – this could be specified by characteristics like the harmonic distortion or intercept point
- undesired aliasing or mirroring of signals by some blocks like aa mixer or an analog-to-digital converter

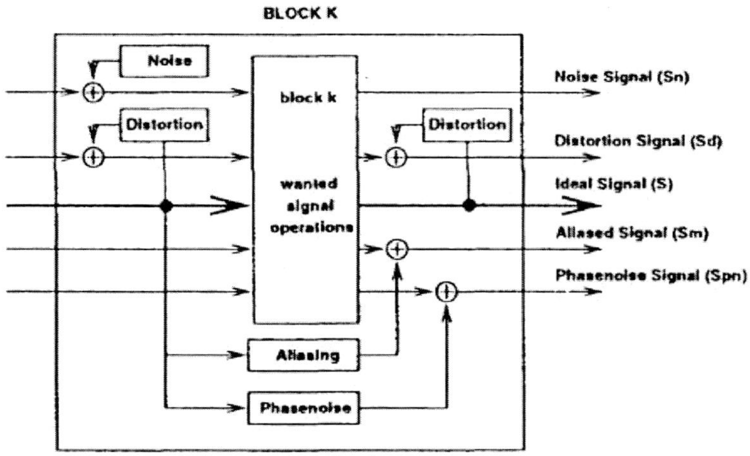

Fig. 16. General behavioral model template for an analog telecom frontend block.

A generalization for the compact modeling of the frequency-dependent nonlinear distortion of telecom frontend circuits using Volterra kernels has been presented recently [12]. The key problem however remains the development of a good mathematical description for the block behavior, for which no good systematic methods exist yet. For efficiency reasons, preference is given to explicit models without iteration. As an illustration, consider for instance the model that was derived for the class of current-steering digital-to-analog converters [13]. The model includes both the static and dynamic behavior (settling time and glitch energy). As shown in Fig. 17, the latter can accurately be described by superposition of an exponentially damped sine and a shifted hyperbolic tangent :

$$\begin{aligned} i_{out} = A_{gl} \sin\left(\frac{2\pi}{t_{gl}}(t - t_0)\right) \exp\left(-sign(t - t_0)\frac{2\pi}{t_{gl}}(t - t_0)\right) + \\ \frac{level_{i+1} - level_i}{2} \tanh\left(\frac{2\pi}{t_{gl}}(t - t_0)\right) + \frac{level_{i+1} + level_i}{2} \end{aligned} \tag{4}$$

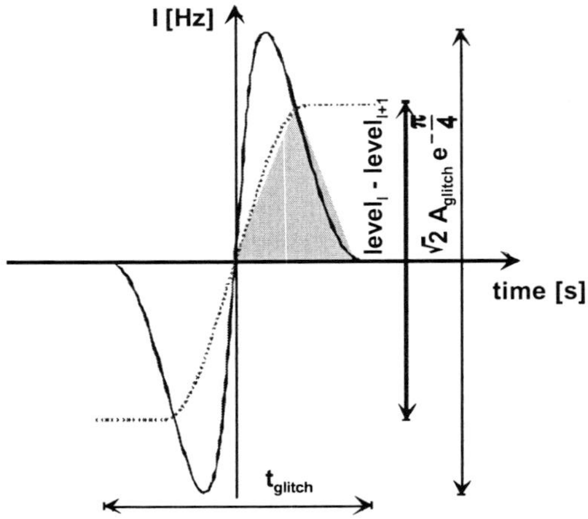

Fig. 17. Dynamic behavior of a current-steering digital-to-analog converter.

4.2. Simulation algorithms and signal representations

The second crucial element for system-level exploration as well as for analog high-level synthesis is an efficient way to evaluate the performance of the circuit. Analytic equations that explicitly characterize an architecture's behavior as a function of the subblock parameters are a fast alternative if they are are available, but the most general solution is to use some sort of simulation. Yet, different simulation techniques can be used, and they vary widely in CPU time consumption for systems that contain both high-frequency signals (e.g. antenna signals in an RF application) and downconverted baseband signals.

A standard integration-based simulation like in SPICE would be too time consuming due to the tight time-step control dictated by the high-frequency signals and also would loose a large time simulation the start-up transients before the steady-state situation is reached. Special RF simulation techniques like harmonic balance or shooting methods on the other hand directly calculate the steady-state solution in an efficient way, but they are interesting for a small number of sinusoidal input tones only which is a poor approximation for complex digital telecom signals. Circuit envelope methods can handle digital telecom signals more efficiently as they separate the calculation of the carrier and the modulating signal, but as current commercial implementations treat all nodes in the system in the same global way they are suboptimal.

To make system-level exploration really fast and interactive, dedicated algorithms have to be developed that speed up the calculations by maximally exploiting the properties of the systems under investigation and using proper approximations where possible. The ORCA tool for example represents the signals as a sum of band-limited piecewise rational approximations. This formulation requires less memory than a sampled data-point representation, and most of the signal operations can be performed symbolically. For example, if we assume that the input signals X to a block are a random process with zero-mean Gaussian distribution, which is a good approximation for multi-channel applications, then the power spectral densities of the second-order and third-order distortion at the output Y can efficiently be calculated using :

$$S_{Y_2}(f) = a_2^2\,(R_X^2(0)\,\delta(f) + 2\,(S_X(f) \otimes S_X(f))) \tag{5}$$

$$S_{Y_3}(f) = a_3^2\,(9\,R_X^2(0)\,S_X(f) + 6\,(S_X(f) \otimes S_X(f) \otimes S_X(f))) \tag{6}$$

where a_2 and a_3 are the second-order and third-order nonlinearity coefficients and R and S represent the autocorrelation and power spectral density functions respectively. Fig. 18 shows a comparison between the results obtained using this analytic approach and the results from the time-domain simulation of a large number of independent QPSK symbols modulated on 5 different carriers. The figure shows a good agreement between the power spectral content of both cases for the third-order distortion output.

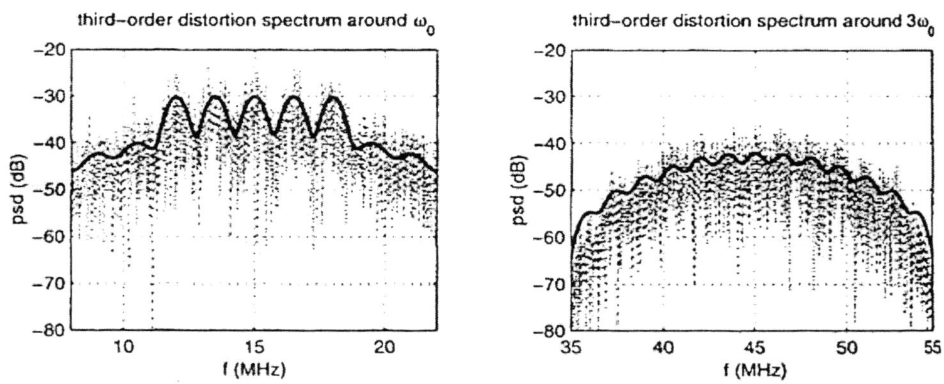

Fig. 18. Simulated results versus analytic expression approach (thick black line) of the third-order distortion spectrum for 5 QPSK-modulated channels.

A more recent development is the FAST tool which performs a dataflow type of simulation without iterations [10] and which therefore easily allows the co-simulation with digital blocks. Compared to commercial simulators like COSSAP, PTOLEMY or SPW, this simulator is more efficient by using block processing instead of point-by-point calculations for the different time points in circuits without feedback. In addition, the signals are represented as complex equivalent baseband signals with multiple carriers that each may have a different timestep. Fig. 11 was obtained with this simulator.

Finally, the FONZIE simulation environment uses the most efficient signal representation depending on the block type, ranging from damped complex exponentiels which are suited for the time-domain simulation of weakly nonlinear circuits to sampled-data representation for strongly nonlinear circuits, and automatically converts from one representation to the other during the simulation [14].

4.3. Analog power estimation

The last crucial element to compare different architectural alternatives and to explore trade-offs in system-level exploration as well as for analog high-level synthesis, are accurate and efficient power and area estimators needed to assess the optimality of different alternatives. Such estimators are functions that predict the power or area that is going to be consumed by a circuit implementation of an analog block (e.g. an analog-to-digital converter) with given specification values (e.g. resolution and speed). Since the implementation of the block is not yet know during high-level system design and considering the large number of different possible implementations for a block, it is very difficult to generate these estimators with high absolute accuracy. This however is not so important since the most crucial aspect is the relative accuracy needed to reliably compare different alternatives.

An example of a general yet relatively accurate power estimator that was derived for the whole class of CMOS high-speed Nyquist-rate analog-to-digital converters (such as flash, two-step, pipelined... architectures) was presented in [15] and is given by :

$$power = \frac{V_{dd}^2 \cdot L_{\min} \cdot (F_{sample} + F_{signal})}{10\hat{\ }(-0.1525 \cdot ENOB + 4.8381)} \tag{7}$$

The estimator is technology scalable, has been fitted with published data of real converters, and has an accuracy of 2.2 for a confidence factor of 85%. Similar functions are now being developed for other blocks.

5. CONCLUSIONS

An overview has been given of the challenges and design issues in the system-level design of mixed analog-digital ICs and telecom frontends. In order to establish the design productivity and design optimality needed to design future systems on a chip within the tightening time-to-market constraints, a systematic top-down design approach has to be followed with sufficient time and attention paid to system-level architectural design before proceeding to the detailed block design. High-level system exploration tools have been presented that enable analog-digital co-design and that allow to analyze architectural alternatives and explore trade-offs such as finding the optimal analog-digital partitioning. Examples of system-level architectural exploration for telecom frontends, including analog-digital partitioning, have been given. Techniques to derive the crucial underlying techniques for such high-level mixed-signal system design, i.e. analog behavioral modeling, efficient high-level simulation methods and analog power/area estimation, have then been described in detail.

It can be concluded that a systematic top-down approach will be unavoidable for designing future complex systems, but that major challenges remain in the development of accurate yet efficient analog behavioral models and power/area estimators. Also the mapping from system-level specifications (an "algorithm") into an appropriate architecture remains an open problem.

ACKNOWLEDGEMENTS

The author acknowledges all Ph.D. researchers who have contributed to the reported results, including Stéphane Donnay, Erik Lauwers and Piet Vanassche. This work was supported in part by the ESPRIT project SALOMON under the Low-Power Design Initiative.

REFERENCES

[1] Semiconductor Industry Association (SIA), "The national technology roadmap for semiconductors," 1994.

[2] Georges Gielen, "Top-down design of mixed-mode systems: challenges and solutions," chapter 18 in Advances in Analog Circuit Design (edited by Huijsing, Sansen, Van der Plassche), Kluwer Academic Publishers, 1999.

[3] "Virtual Socket Interface Architecture Document," version 1.0, VSI Alliance, 1997.

[4] L.R. Carley, G. Gielen, R. Rutenbar, W. Sansen, "Synthesis tools for mixed-signal ICs: progress on frontend and backend strategies," proceedings ACM/IEEE Design Automation Conference (DAC), pp. 298-303, 1996.

[5] H. Huijsing, W. Sansen, R. Van der Plassche, "Advances in analog circuit design," different chapters by different authors, Kluwer Academic Publishers, 1999.

[6] A. Doboli, A. Nunez-Aldana, N. Dhanwada, S. Ganesan, R. Vemuri, "Behavioral synthesis of analog systems using two-layered design space exploration," proceedings ACM/IEEE Design Automation Conference (DAC), pp. 951-957, June 1999.

[7] J. Crols, S. Donnay, M. Steyaert, G. Gielen, "A high-level design and optimization tool for analog RF receiver front-ends," proceedings International Conference on Computer-Aided Design (ICCAD), pp. 550-553, 1995.

[8] S. Donnay, "Analog high-level design automation in mixed-signal ASICs," Ph.D. dissertation, Katholieke Universiteit Leuven, December 1998.

[9] P. Wambacq, G. Vandersteen, S. Donnay, M. Engels, I. Bolsens, E. Lauwers, P. Vanassche, G. Gielen, "High-level simulation and power modeling of mixed-signal front-ends for digital telecommunications," proceedings International Conference on Electronics, Circuits and Systems (ICECS), pp. 525-528, September 1999.

[10] G. Vandersteen, P. Wambacq, "FAST," proceedings ACM/IEEE Design Automation Conference (DAC), June 2000.

[11] S. Donnay, G. Gielen, W. Sansen, "High-level analog/digital partitioning in low-power signal processing applications,"

proceedings 7th international workshop on Power and Timing Modeling, Optimization and Simulation (PATMOS), pp. 47-56, 1997.

[12] P. Wambacq, "Compact modeling of nonlinear distortion in analog communication circuits," proceedings Design And Test in Europe conference (DATE), March 2000.

[13] J. Vandenbussche, G. Van der Plas, G. Gielen, W. Sansen, "Behavioral model of reusable D/A converters," IEEE Transactions on Circuits and Systems, Part II, Vol. 46, No. 11, pp. , November 1999.

[14] P. Vanassche, G. Gielen, W. Sansen, "System-level time-domain simulation of telecom frontends using damped complex exponentials," internal report, Lathlike Universiteit Leuven, October 1999.

[15] E. Lauwers, G. Gielen, "A power estimation model for high-speed CMOS A/D converters," proceedings Design, Automation and Test in Europe (DATE) conference, pp. 401-405, March 1999.

Mixed Signal Design
- Design Issues -

H. J. Casier

Alcatel Microelectronics

Abstract

The continuously increasing chip density makes it now possible to integrate more functions and even a complete system on a single chip. Since a complete system contains analog and digital signal processing and is often controlled and supported by an embedded microcore, this evolution has favoured mixed signal chips, that combine these functions on a single chip. At the same time, mixed signal design itself has changed. In the eighties, the analog circuits dominated the area of the mixed signal chips. Nowadays, the situation is reversed and the digital circuits dominate the area. As a consequence, the design issues for mixed signal chips have changed accordingly. Three of these design issues are discussed below. The strong demand to use a digital technology for the mixed signal chip, the increased interference between digital and analog circuits and the concurrent design of analog, digital and embedded software for the mixed system-on-a-chip.

1 Introduction

Mixed signal design has evolved from the addition of digital circuits to an analog function in order to improve the analog performance to a true mixed signal system-on-a-chip that combines analog functions with digital and DSP/software functions and controls. The analog circuits are mainly used as interface to the outside world and for high speed or lower resolution signal processing on-chip. A small part of the digital circuits is still used as before, to improve the analog

R. J. van de Plassche et al. (eds.), Analog Circuit Design, 167–183.

functions. They are tightly linked to the analog circuits e.g. A/D and D/A converters, digital AGC control, PLL ... The largest part of the digital circuits replaces analog for high resolution and lower speed signal processing. Furthermore, the integration of DSPs and microcontrollers adds more intelligence and flexibility to the mixed signal chip and makes it a true system-on-a-chip. The result of this evolution is that the proportion of digital circuits in the mixed signal chip has steadily grown and it now dominates the total chip area in most cases. In the domains where the analog content is still dominant, such as RF, High Voltage and High Power circuits, the same evolution can be seen.

This evolution has changed the technology frame for Mixed Signal Design. When the digital circuits were an add-on to the analog circuits, they followed the analog technology and requirements. Now, the situation is reversed and the digital circuits dominate the area and the chip cost. This explains the pressure to implement the mixed signal system-on-a-chip in the latest, mainstream digital technology.

At the same time, it remains true that everything, lost in the analog interface never can be recovered. The overall system performance is hence strongly dependent on the analog performance. The steady increase of system performance (Figure 1) created a second, parallel push towards higher analog performance. This makes the key analog circuits more vulnerable to interference from the on-chip digital circuits and from off-chip interference (EMC). It also increases the design effort for the key analog circuits.

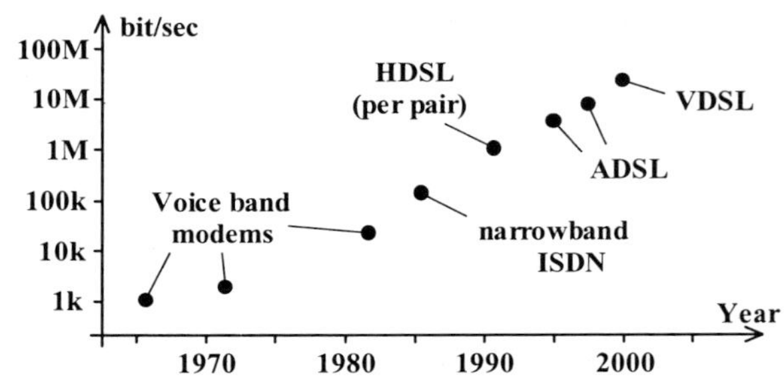

Figure 1: Evolution of Transmission Speed over a Twisted Pair Line

In this paper, three design issues, which result from this evolution, are discussed. The next section deals with the digital technology roadmap issues: lower supply voltages and a push to the mainstream digital technology. Section 3 discusses the on-chip interference between the digital and the analog circuits. The concurrent system design and trailing analog design efficiency are the subject of the last section.

2 Technology Issues

There is a conflict in the technology-requirements between analog and digital circuits. Analog circuits are performance critical and require a stable and well-characterised technology. They do not benefit from the reduced dimensions of the next generation technology but they benefit from extra, analog processing steps. The preferred analog technology is hence a stable, analog enhanced technology. Digital circuits at the contrary are mainly density critical and benefit from the next generation technology for both area and power-delay product. The preferred digital technology is hence the most advanced mainstream technology.

As explained before, digital circuits dominate the mixed mode chip area (typically 70...95% of the active area) and consequently also the silicon cost. As a result, the mainstream digital technology is the preferred choice for the mixed system-on-a-chip, as long as the analog performance remains adequate. If the analog performance has to be increased, special analog enhanced technologies can be used. The design issues, resulting from the implementation of analog circuits in a pure digital technology are discussed in the first point below. Analog enhanced technologies are discussed in the second point and the last point shows two case studies.

2.1 Design Issues for analog circuits in a digital technology

The most important issue for analog circuits in mainstream digital technology is the decreasing supply voltage, that reduces dynamic range in noise dominated circuits and decreases resolution in matching dominated circuits. Provided the same analog circuit topology can be used, the dynamic range and resolution can only be kept constant at the expense of a higher power dissipation [1]. At the same time, the area of critical analog circuits remains roughly constant for the same performance level [2]. The push towards higher analog performance increases both power and area of the analog circuits and the power and area increase further if the supply voltage decrease imposes circuit topology changes towards multi-stage circuits. Performance critical analog circuits do not benefit from the technology roadmap.

The next important issue for analog circuits in mainstream digital technology is lower analog performance of some devices and lack of analog characterisation of all devices. In a digital technology, PMOS and NMOS transistors are only optimised for speed/power and not for matching and noise. Their analog performance is not necessarily worse but they have to be characterised and guaranteed for analog design. Capacitors and resistors are parasitic devices in a digital technology. Some of them are linear and can be used in analog design. Compared to the special capacitors and resistors of an analog enhanced technology, they have a similar analog performance but require much larger area (5…10 times [3]) and exhibit more parasitic effects. The area increase becomes very important for circuits, which contain large capacitors e.g. charge redistribution type converters, low frequency filters and high accuracy sampled data systems. It is shown below (Table 1) that a special analog technology is usually more economical for these circuits, unless new, innovative circuit techniques are used [4].

2.2 Analog enhanced technologies

The cheapest analog enhanced technology is the "Analog CMOS" technology. It is based on the mainstream digital technology to which special steps are added, that do not impact basic CMOS characteristics. The most common options are a very linear thin oxide capacitor, a linear high value resistor and low Vt transistors. These optional devices improve analog performance, lower analog area and counterbalance supply voltage decrease.

The required processing steps are not critical and do not add a lot of production cost to the technology [3]. Roughly 5% (1 mask) is added to the silicon cost for each option. The charged silicon cost however used to be considerably higher since the analog enhanced CMOS technology deviates from the mainstream, has a more limited availability and runs in smaller production volumes. Today, with the increasing production volumes of mixed signal systems, an increasing number of foundries include analog CMOS options in their standard offering.

A second important cost factor in the comparison between analog CMOS and digital CMOS is the time lag, because this impacts optimal integration of digital circuits. The analog CMOS options do not delay the mainstream technology development and are easily added to a new technology. They typically lag the mainstream digital technology by less than half a generation (Figure 2). The real technology choice for mixed system-on-a-chip is between analog enhanced CMOS and digital CMOS, on which all analog components have been characterised. The time lag between these two technologies is even less than half a generation and digital circuit density is normally not penalised.

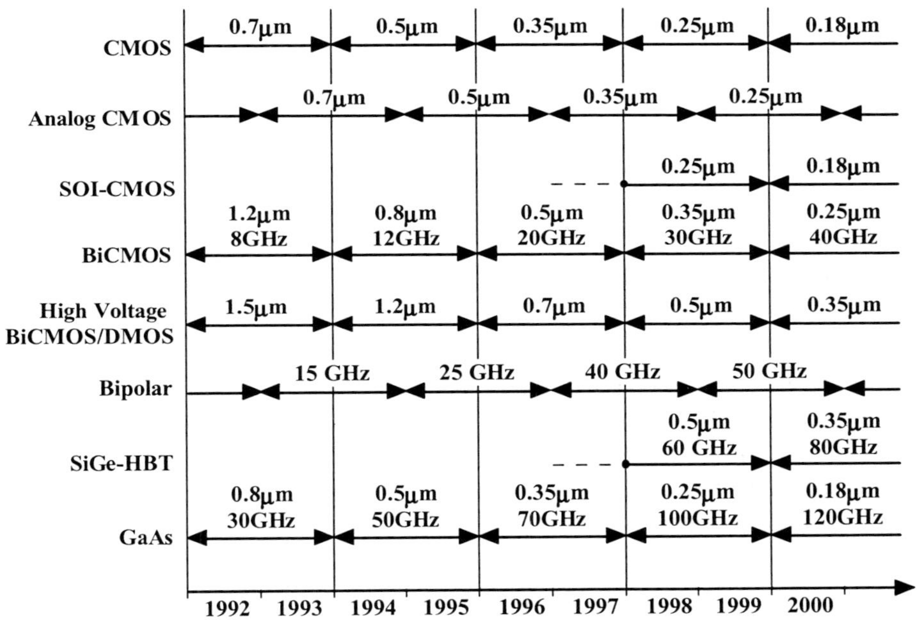

Figure 2: Relative Roadmaps for various Technology Options

Other high performance analog technologies (BiCMOS, High Voltage DMOS, EEPROM…) are optimised for analog performance of one or more devices. They reach the best analog performance for these devices but also have a high cost penalty. These technologies are quite different from the mainstream technology and include several, critical processing steps. These increase production cost and lower silicon yield. Their limited availability and lower production volume further increases the charged silicon cost. Finally, they show a considerable time lag to mainstream technology (Figure 2), which penalises the digital circuits.

As a conclusion, in many cases, there is a real choice between the analog characterised digital CMOS and the analog enhanced CMOS technology for the integration of a mixed signal system. Other high performance analog technologies are fairly expensive and are only selected if there is no other choice or if the design risks would be to high in an analog CMOS. e.g. high voltage, on-chip programmability, RF requirements.

2.3 Examples

The first example is a single chip ISDN circuit that was developed in a 0.5μm analog CMOS technology (Figure 3). The circuit contains a U-interface to the telephone line and a S0-interface to the user equipment. It also contains a micro-controller, RAM, ROM and glue logic. The on-chip analog circuitry consists of a 2nd order ΣΔ converter, a sampled data and a continuous time smoothing filter and a line driver. The analog area is 19.5% of the total chip area.

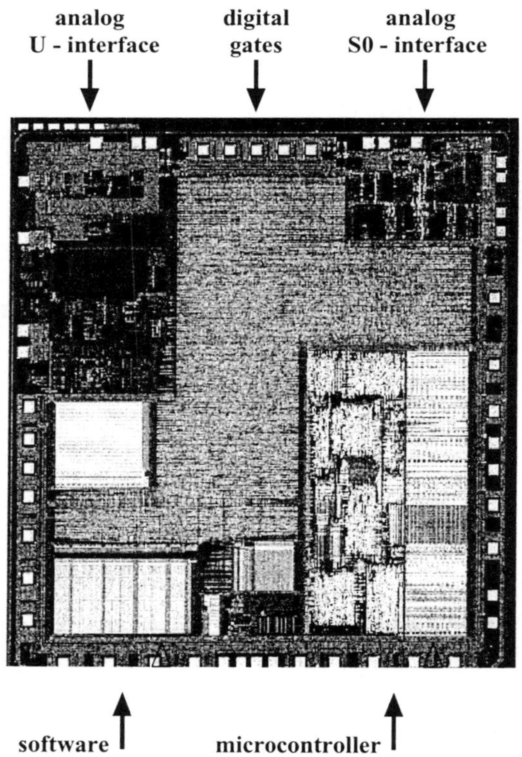

Figure 3: ISDN NT chip

In Table 1 a comparison is shown for an implementation of this circuit in a 0.5µm digital CMOS, a 0.35µm analog CMOS and a 0.35µm digital CMOS technology [3]. The different scaling factors per device and per technology are easily seen. Analog characterised digital CMOS is obviously not a good choice since the chip is 13...14% bigger than the analog enhanced CMOS chip of the same generation while silicon cost decreases less than 10%. A shrink of the circuit in a 0.35µm analog enhanced CMOS technology is the best option.

ISDN - NT chip (all areas in % of the reference area: Total area of first column)	0.5µ CMOS		0.35µ CMOS		
	Analog CMOS	Digital CMOS	Analog CMOS	Digital CMOS	
Analog Circuit areas					
M_L Capacitors	1.26	12.60	1.26	8.82	%
NM_L Capacitors	0.00	0.00	0.00	0.00	%
M_L Resistors	0.15	0.23	0.15	0.23	%
NM_L Resistors	0.23	1.60	0.23	1.60	%
NL Trans., C & R	18.02	18.02	18.02	18.02	%
Total Analog area	19.67	32.46	19.67	28.68	%
Digital Circuit area	64.92	64.92	32.46	32.46	%
Pad + Scribe area	15.42	15.67	14.69	14.90	%
Total area	100.00	113.05	66.81	76.04	%

(M_L: Matching-Linear, NM_L: Non-Matching-Linear, NL: Non-Linear)

Table 1: Next generation evaluation for ISDN-NT chip

The second example is a cable modem chip. This chip was originally developed in 0.35µm digital CMOS technology. Since the DDFS DAC and the smoothing filter capacitor are placed outside, the on-chip analog circuitry is limited to a 4-stage pipeline converter and a smoothing filter resistor. The analog area is only 9.4% of the total chip area.

In Table 2 a comparison is shown for the next generation of this circuit in 0.25µm analog enhanced CMOS and 0.25µm analog characterised, digital CMOS technology. The 0.35µm analog CMOS technology is also added [3]. Assuming that the DAC and the capacitor remain outside in the next generation, digital CMOS is obviously the best choice since the chip area increases only 1.5% wit respect to the analog CMOS chip area of the same generation while the silicon cost decreases more than 5%. A shrink of the circuit in 0.25μm digital CMOS technology is the best option.

Cable Modem (all areas in % of the reference area: Total area of 2nd column)	0.35μ CMOS		0.25μ CMOS		
	Analog CMOS	Digital CMOS	Analog CMOS	Digital CMOS	
Analog Circuit areas					
M_L Capacitors	0.14	1.01	0.14	1.01	%
NM_L Capacitors	0.00	0.00	0.00	0.00	%
M_L Resistors	0.00	0.00	0.00	0.00	%
NM_L Resistors	0.11	0.76	0.11	0.76	%
NL Trans., C & R	7.61	7.61	7.61	7.61	%
Total Analog area	7.86	9.38	7.86	9.38	%
Digital Circuit area	76.51	76.51	38.26	38.26	%
Pad + Scribe area	14.07	14.11	13.19	13.23	%
Total area	98.45	100.00	59.31	60.87	%

(M_L: Matching & Linear; NM_L: Non-Matching & Linear; NL: Non-Linear)

Table 2: Next generation evaluation for Cable modem chip

3 On-Chip Interference between analog and digital circuits

Interference between analog and digital circuits can go both ways but since digital circuits have a high noise margin, large analog disturbances are required to interfere with the digital functions e.g. high energy transients in a high voltage, high power technology or external EMC pulses. Analog circuitry on the other hand has a low noise margin and is much more vulnerable to interference through resistive, capacitive, inductive or thermal coupling between on-chip circuits. Although the noise source can also be an analog circuit e.g. a sensor actuator, a line driver, RF circuits… the main noise source in mixed signal designs is switching noise of the synchronous CMOS logic. Large switching current peaks generate ground and supply bounce through the resistive and inductive behaviour of on-chip supply lines and the bondwires.

Since, in most digital libraries, the ground supply line is connected to the substrate to prevent latch-up, digital noise is transferred to the substrate. In modern CMOS technologies the substrate consists of a thin, lowly doped epilayer on top of a thick highly doped bulk wafer. This highly doped bulk-substrate acts as an equipotential node and distributes the injected noise over the whole chip [5].

This chip wide, common-mode substrate noise couples in a different way to NMOS and PMOS transistors, to capacitors and to resistors. Except for special technologies, such as BiCMOS, triple well CMOS or SOI, the NMOS transistor bulk is the substrate and substrate noise is hence directly coupled into the active transistor. The PMOS transistor bulk is the n-well, that is capacitively coupled to the substrate. A properly contacted n-well shields the PMOS device from substrate noise. Capacitors and Resistors can also be shielded from the substrate by a properly contacted n-well underneath. The remaining substrate noise affects the analog circuit by direct incoupling, demodulation and sampling [6].

It is difficult and costly to reduce the substrate noise. A good method is to connect the equipotential substrate to a clean PCB Ground via a low-ohmic contact at the backside of the chip. This method is limited to frequencies below the skin effect e.g. 500MHz and it requires non-standard processing. Therefore, it is not commonly done. On highly doped substrates, guard rings only provide shielding in a very small area (< 4•Tepi [5]), but they are very effective for lowly doped substrates in e.g. High voltage CMOS and BiCMOS technologies. High frequency on-chip capacitors between the supply lines also limit the substrate noise but they require area and care must be taken with the LC-resonance of this capacitor with the bondwire inductance [7].

The above mentioned methods only reduce substrate noise but do not eliminate it. Analog circuits must hence be able to deal with substrate noise. A good strategy to limit the effect of substrate noise on analog performance is demonstrated in [6]. For best results, on-chip signals should be differential and the signal transistors should be balanced PMOS structures. Both differential signals and PMOS transistors should be referred to a clean analog supply voltage. NMOS transistors should not be used in the signal path and should be referred to the substrate and not to the clean analog supply voltage. In many cases, input signals at the chip to PCB interface are neither differential nor referenced to the analog supply but single ended and referenced to the analog PCB ground. In this case, the best solution is to bring the analog PCB ground into the chip as a signal input and to reference the input signal to this analog signal ground. The single ended signal should be converted to a differential signal in the first analog stages on the chip (Figure 4).

Another common noise path in mixed signal chips is coupling of the digital supply noise into the analog signal path via the clock lines [9]. This path can be suppressed by inserting at least one level of clean logic between the digital and analog circuits (Figure 4). The clean logic is connected to a clean analog supply and a clean analog ground line. The analog ground line is not connected to the chip-substrate.

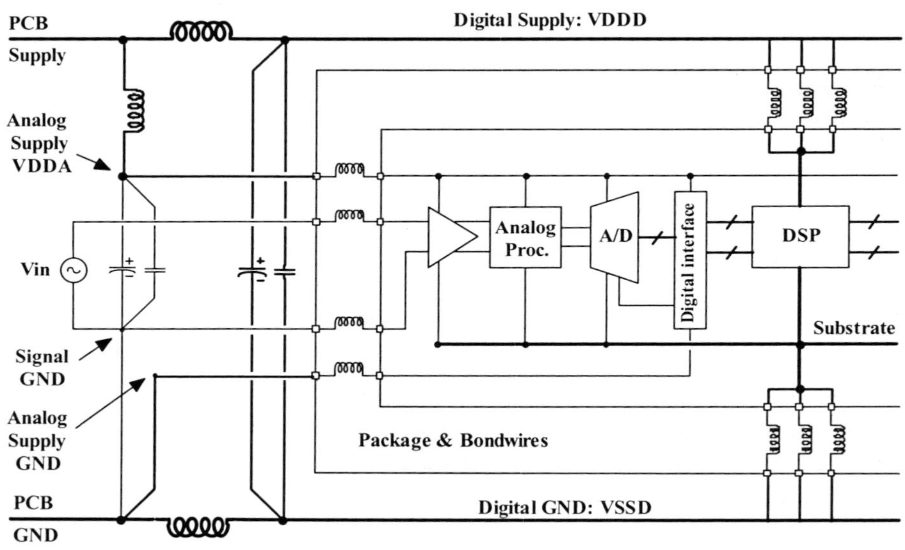

Figure 4: recommended supply and reference routing for a mixed signal IC (after Figure 4 of reference [6])

Other measures to suppress capacitive and inductive coupling are to place sensitive inputs and digital outputs far from each other on the chip [8], to put a shielding bondwire in-between a sensitive input and a noisy neighbouring bondwire and to surround a sensitive signal line with shielding metal layers.

If the application permits, one of the best measures is to halt the digital circuitry during an analog acquisition. This is not always possible but gives good results in e.g. high accuracy metering applications. If the digital circuits can not be halted, their clock can be delayed with respect to the analog sampling clock. This gives good results if the digital substrate bounce is damped before the next analog sample time and if it has no impact on analog clock jitter.

Digital circuitry can never be completely halted since relatively large digital blocks inter-operate with the analog circuitry e.g. clock generators and drivers, comparators, data latches ... The digital complexity of these blocks is limited and they generate less substrate noise than the full digital circuitry. If required, this noise can be further suppressed by using a low noise library e.g. gates with separate ground and substrate contacts or gates with low peak currents. The area penalty for these special libraries is small (10%) and the latch-up problem is manageable.

Substrate noise, induced by digital logic contains a wide range of harmonic frequencies, resulting from the clock period and from clock signal slopes. Differential circuits reject the lower frequency components but this is limited by matching and by functional asymmetry in the circuit e.g. bandgap, gm-C stage ... [6, 10]. The high frequency components are difficult to reject. A very careful, symmetrical layout for the transistors, the routing and their parasitics is required.

4 System Design issues

State-of-the-art mixed signal design requires the concurrent development of analog and digital circuits and of software. Besides the planning challenges, these projects also push the design tools and methodology. Indeed, high level architectural exploration, partitioning and optimisation is required for the proper specification of each part and their interfaces. Top-down, bottom-up design methodology requires the derivation of analog behavioural models and multi-level, mixed signal simulations are needed to cope with the widely spaced time constants of the different levels. These CAD issues are the subject of the next paper and will not be discussed here.

Although the area of state-of-the-art mixed signal designs is dominated by the digital circuitry, the analog circuits dominate the design effort due to their much lower design efficiency. In Figure 5, the mean design effort and complexity for high-voltage mixed mode products is shown. The complexity rises in accordance with the law of Moore, which means that the chip area over the different technology generations remains roughly constant. The design efficiency clearly does not improve at the same speed and the total design effort per project rises. Analog circuit design takes the largest share of the design effort increase due to analog performance increases and low performance tools. Besides the inherent analog circuit design risks, this low analog design efficiency is a key problem for mixed system-on-a-chip projects to follow the digital technology roadmap. As a result, first generation mixed chips use state-of-the-art analog technology while subsequent generations usually follow much slower the technology roadmap than digital chips. An example is shown in the digital and analog technology roadmaps of the DynaMiTe-ADSL chip set at the bottom of Figure 6.

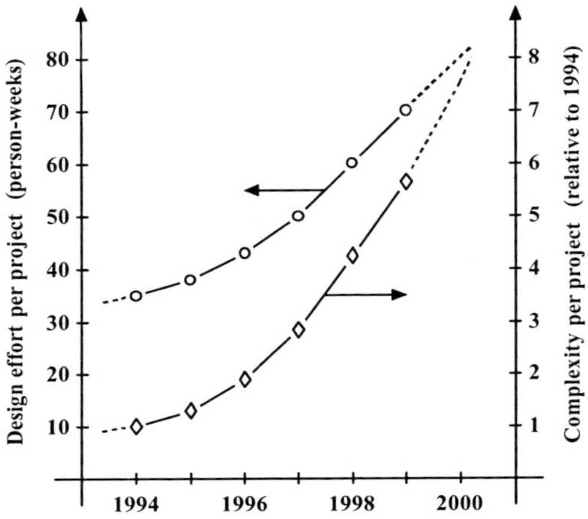

Figure 5: Mean Design-Effort and Project Complexity for High-Voltage Automotive and Industrial products.

Digital design efficiency has been increased by designing at a higher level of abstraction, and by increasing the re-use of functional blocks. In the next section, it is discussed how this methodology has to be adapted for analog design.

The first higher digital design level for is at gate level. A gate is a, general usable, not application specific basic function. Differential pairs, current mirrors, cascode stages and complete OTAs are functionally equivalent to digital gates and flip-flops but their

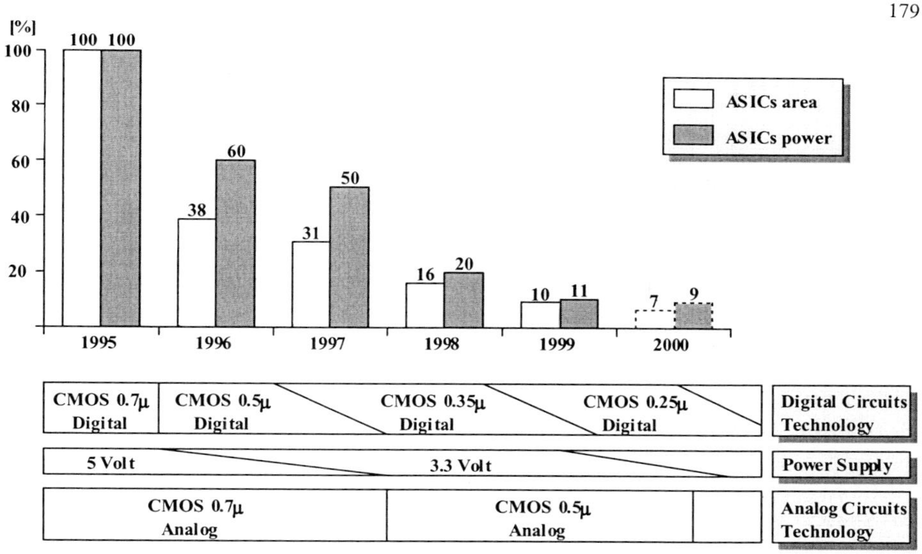

Figure 6: Dynamite - ADSL Technology Roadmap

performance is very dependent on the preceding and following circuits and on analog feedback over the complete function. A digital gate, at the contrary, has a weak and predictable dependence on the source and load impedance and there is no analog feedback in digital circuits. There is hence no direct equivalent to the digital gate level in analog design.

The next higher design level in analog is building blocks e.g. OpAmps with included feedback, buffer OpAmps, PLLs, A/Ds and D/As ... These building blocks have a weak and predictable source and load dependence and can be adequately modelled at a higher level. They are however more application specific, requiring large block libraries to cover all applications.

These analog building blocks are made as fixed GDSII blocks or as compiled blocks from a library of leaf-cells [11, 12]. They are highly reusable within the same technology and within the specification capabilities of the block, but a large effort is required to port these blocks to another technology or redesign them to other specifications. In most cases, GDSII blocks are confined to one application area and have to be redesigned for other product areas. Compiled blocks cover a larger performance space but they also require a larger development and porting effort. They are developed for a limited number of functions with a high reuse capability e.g. D/As, A/Ds, SC-filters… Improving analog design efficiency by using a library of fixed or compiled analog building blocks requires a large library development effort, continuous porting to other technologies and redesign to new application areas.

One of the problems with reuse of GDSII or compiled blocks is their technology link. It is a better approach to limit the reuse part of the design to the symbolic design level, which is much less linked to a technology and to simplify or eventually automate the lower level design tasks and the GDSII generation.

In fact, a good analog designer starts with a symbolic evaluation of the proposed topology in order to understand the circuit and find the best way to optimise it. Only after this step, SPICE simulations with real device models are used to validate the topology and to optimise the circuit for worst case processing tolerances and voltage and temperature ranges. Reuse of this symbolic design level essentially means to keep track of the design plan of the analog building block. Topology history is an add-on, not required for reuse but facilitating the development of new, improved analog blocks and also providing some encapsulation of the expert designer know-how, without consuming his time. The design plan is not limited to analog building block level but can be extended to the analog system design level and include other circuit types, mathematical models, DSP algorithms… Analog designers nowadays use hand calculations or symbolic equation handling tools (Maple, Mathematica, ISAAC [13]…). Several dedicated CAD tools are currently in development [14, 15].

Complementary to this design plan are CAD tools that simplify and/or automate the lower level design tasks and the GDSII generation. The two, most time consuming lower level design tasks are circuit characterisation and layout. Circuit and block characterisation is application specific and can not be automated except for standard circuit classes e.g. OpAmps, comparators… Many fixed and flexible characterisation tools exist on the market and many companies have developed their own tools [16]. Analog layout can be generated automatically from the pre-characterised block if the proper constraints are passed to the layout tools. The generation of these constraints is simplified due to a better understanding of the circuit during the design plan evaluation.

5 Conclusion

The evolution from analog dominated to digital dominated mixed signal chips and the increasing analog performance and digital complexity of the mixed signal circuits, has given rise to a technology issue and has aggravated the interference, CAD and design efficiency issues. None of these issues however, will impede further performance and complexity rise of mixed signal system-on-a-chip integration.

It has been shown that an upgrade of deep-submicron mainstream technology with simple, cheap analog process steps is an efficient means to decrease technology issues at the expense of a slightly higher chip cost. A design know-how increase is required to cope with the growing interference problem. Reuse of symbolic level design and the resulting design plan, together with further automation of characterisation and layout generation is proposed to increase analog design efficiency. This requires new CAD tools and a stricter analog design discipline.

Acknowledgement: The content of this paper is the result of many discussions within Alcatel Microelectronics and within the ESPRIT 29261-Mixmodest and MEDEA A409-SADE projects. In special I would like to thank E. Huyskens, M. De Mey and W. Dehaene of Alcatel Microelectronics for their contributions in the preparation of this paper.

6 References

[1] K. Bult, “Analog Broadband Communication Circuits in Pure Digital Deep Sub-Micron CMOS”, ”, ISSCC 1999, Digest of Technical Papers, Slide supplement, Paper MP 4.6.

[2] H. Samueli, “Broadband Communications ICs: enabling High-Bandwidth Connectivity in the Home and Office”, ISSCC 1999, Digest of Technical Papers, Slide supplement, Paper MA 1.3.

[3] H. Casier, "Mixed Design in a Digital CMOS Technology, Trade-offs and Case Study", Proceedings of the IEEE CAS Workshop on Mixed-Signal Integrated Circuit Design MSICD, Long Beach 1999.

[4] H. Yoshizawa, Y. Huang, P.F. Ferguson Jr. and G.C. Temes, "MOSFET-Only Switched-Capacitor circuits in Digital CMOS Technology", IEEE Journal of Solid-State Circuits, vol. 34 no. 6, June 1999.

[5] D. Su, M. Loinaz, S. Masui and B. Wooley, "Experimental Results and Modeling Techniques for Substrate Noise in Mixed-Signal Integrated Circuits", IEEE Journal of Solid-State Circuits, vol. 28 no. 4, April 1993.

[6] B. Nauta and G. Hoogzaad, "Substrate Bounce in Mixed-Mode CMOS ICs", Digest of Technical Papers, workshop on Advances in Analog Circuit Design, Copenhagen 1998.

[7] D.W.J. Groeneveld, "Groundbounce in CMOS", Digest of Technical Papers, workshop on Advances in Analog Circuit Design AACD, Leuven 1993.

[8] M. Ingels, "CMOS Interface Circuits for Optical Communication", Chapter 5, Electrical interferences in mixed-mode integrated circuits, Ph.D. dissertation, Katholieke Universiteit Leuven, March 2000.

[9] T. Blalack and B. Wooley, "The effects of Switching Noise on an Oversampling A/D Converter", ISSCC 1995 Digest of Technical Papers, pp 200-201.

[10] W. Dehaene, "CMOS Integrated Circuits for Analog Signal Processing in Hard Disk Systems", Paragraph 5.5, Results for the filter and tuning system, Ph.D. dissertation, Katholieke Universiteit Leuven, November 1996.

[11] J. Hendrickx, "MIDAS and Analog CAD at Philips", Proceedings of the 2nd international Summer Course on Systematic Analog Design, Heverlee, Belgium, May 31, 1995.

[12] ADMIR-x, Family of ADCs (10 to 16 bits) for measurement/sensor applications, Dolphin Integration, Grenoble France.

[13] G. Gielen, H. Walscharts and W. Sansen, "ISAAC: a symbolic simulator for analog integrated circuits", IEEE Journal of Solid-State Circuits, vol. 24 no. 6, December 1989.

[14] J-O. Fischer-Binder, "Analog circuit sizing using CISE," MEDEA/ESPRIT Conference on System-level Design, Antwerp, Belgium, September 1999.

[15] R. Sommer, E. Hennig, M. Thole, T. Halfmann and T. Wichmann, "Symbolic Modeling and Analysis of Analog Integrated Circuits," Proceedings of the European Conference on Circuit Theory and Design ECCTD, Stresa, Italy, September 1999.

[16] CHARACTER, Automated Cells characterization, MIDI Ingenierie, Labege Innopole, France

Top-Down Design of Mixed-Signal Circuits

Ken Kundert

Cadence Design Systems
San Jose, California, USA

Abstract

With mixed-signal designs becoming more complex and time-to-market windows shrinking, designers cannot hope to keep up unless they change the way they design. They must adopt a more formal process for design and verification: top-down design. It involves more than simply a cursory design of the circuit block diagram before designing the blocks. Rather, it also involves developing and following a formal verification plan and an incremental and methodical approach for transforming the design from a abstract block diagram to a detailed transistor-level implementation.

1. Introduction

At the Design Automation Conference in 1998, Ron Collett of Collett International presented findings from a 1997 productivity study in which his firm analyzed 21 chip designs from 14 leading semiconductor firms. The study revealed a productivity gap of 14× between the most and least productive design teams. The study also revealed that developing analog and mixed-signal circuitry requires three to seven time more effort per transistor than designing digital control logic, though this factor was normalized out of the 14× ratio.

In my experience, the primary culprits behind the poor productivity of those at the bottom of the scale are increasingly complex designs combined with a continued preference for bottom-up (i.e., transistor-level) design methodology and the occurrence of simulation late in the design cycle, which leads to errors and re-spins. There's a huge disparity in productivity between those mixed-signal designers who have transi-

R. J. van de Plassche et al. (eds.), Analog Circuit Design, 185–202.

tioned to a "top-down" design methodology and use mixed-signal hardware description languages (MS-HDLs), and those who practice "bottom-up" design and rely solely on SPICE.

1.1. Getting to Market First

With the internet and wireless technology as the latest market drivers, the pace of the electronic marketplace continues to quicken. New products and new product categories are being created faster than ever before. In order to keep up with the rapid pace of the market, designers must get their products to market more quickly than ever. Those that are successful at bringing significant new capabilities to the market first are usually rewarded with higher profit margins and greater market share. Conversely, those that are late must face an uphill battle against entrenched competition. To understand this, consider three scenarios for developing a product, illustrated in Figure 1. First consider

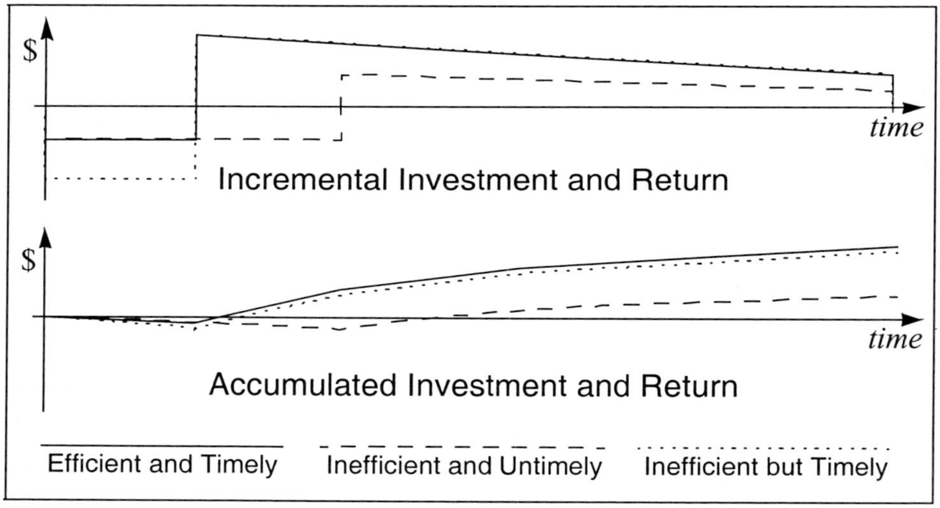

Figure 1: The expected investment and return for the same product developed using three different approaches.

employing an efficient product development process and being first to market. Then consider using the same number of developers with an inefficient development process. Furthermore, assume that this causes the product to be late to market, which results in a much lower return because the product enters a market where a competitor has already established leadership position and because there are fewer available customers left. Finally, consider using an inefficient development process but increasing the number of developers in order to reach the market first. If this were possible, the development costs would be higher, but the total return is almost the same as in the first case. This is because the returns are expected to be much greater than the initial development costs.

This example shows that it is more important to get a product to the market first than it is to control development costs. Of course this assumes that the product is the right product in that it satisfies the customers needs, and that it has some new and very valuable capability. With follow on products, the situation is somewhat different. Here, the market leadership position is largely determined and the need to develop the product in a timely manner is balanced by the need to control development costs.

1.2. Design Productivity Gap

Moore's observation that the number of transistors available on an integrated circuit doubles every 18 to 24 months continues to hold. Competitive pressures compel designers to use these transistors to either provide additional functionality and to increase the integration level and thereby decreasing the size, weight, power and cost of the product. As a result, designers are confronted with larger and more complex designs. The increasing size and complexity of these designs combines with the shrinking time available to develop and get them to market; making the job of the circuit designer today much more difficult than in the past.

Circuits are getting more complex in two different ways at the same time. First, circuits are becoming larger. Consider wireless products. 30 years ago a typical receiver contained between 5 and 10 transistors whereas it is common for a modern cell phone to contain 10M transistors. Second, the operation of the circuits are becoming more complex. 30 years ago integrated circuits generally consisted of simple functional blocks such as opamps and gates. Verification typically required simulating the block for two or three cycles. Today, mixed-signal chips implement complex algorithms that require designers to examine their operation over thousands of cycles. Examples include PLLs, ΣΔ converters, magnetic storage PRML channels, and CDMA transceivers. The result of these two effects together is that complexity is increasing at a blistering pace, and is outstripping the designers ability to keep up. For example, 30 years ago you could build a radio from a rock and a wire, whereas to build a modern radio requires more compute power than existed in the largest supercomputer available back then.

The CAD tools and computers employed by designers continually improve, which serves to increase the productivity of designers. However, the rate of productivity increase is not sufficient to allow the designers to keep up with the increasing complexity of designs and decreasing time-to-market requirements. The growing difference between the improvement in productivity needed to satisfy the demands of the market and the productivity available simply by using the latest CAD tools and computers is referred to as the *Design Productivity Gap*. To close this gap, one must change the way design is done. A design style that reduces the number of serial steps, increases the likelihood of first time working silicon, and increases the number of designers that can work together effectively is needed. If a design group fails to move to such a design style, it will become increasingly ineffective. It eventually will be unable to get products to market in a time of relevance and so will be forced out of the market.

2. Bottom-Up Design

The traditional approach to design is referred to as bottom-up design. In it, the design process starts with the design of the individual blocks, which are then combined to form the system. The design of the blocks starts with a set of specifications and ends with a transistor level implementation. At this point, each block is verified as a stand-alone unit against specifications and not in the context of the overall system. Once verified individually, the blocks are then combined and verified together, but at this point the entire system is represented at the transistor level.

While the bottom-up design style continues to be effective for small designs, large designs expose several important problems in this approach.

- Once the blocks are combined, simulation takes a long time and verification becomes difficult and perhaps impossible. The amount of verification must be reduced to meet time and compute constraints. Inadequate verification may cause projects to be delayed because of the need for extra silicon prototypes.
- For complex designs, the greatest impact on the performance, cost and functionality is typically found at the architectural level. With a bottom-up design style, little if any architectural exploration is performed, and so these types of improvements are often missed.
- Any errors or problems found when assembling the system are expensive to fix because they involve redesign of the blocks.
- Communication between designers is critical, yet an informal and error prone approach to communication is employed. In order to assure the whole design works properly when the blocks are combined, the designers must be in close proximity and must communicate often. With the limited ability to verify the system, any failure in communication could result in the need for additional silicon prototypes.
- Several important and expensive steps in the bottom-up design process must be performed serially, which stretches the time required to complete the design. Examples include system-level verification and test development.

The number of designers than can be used effectively in a bottom-up design process is limited by the need for intensive communication between the designers and the inherently serial nature of several of the steps. The communication requirements also tend to require that designers be co-located.

3. Top-Down Design

In order to address these challenges, many design teams are either looking to, or else have already implemented, a top-down design methodology. In a basic top-down approach, the architecture of the chip is defined as a block diagram and simulated and optimized using either a MS-HDL simulator or a system simulator. From the high-level simulation, requirements for the individual circuit blocks are derived. Circuits are then designed individually to meet these specifications. Finally, the entire chip is laid out and verified against the original requirements.

This represents the widely held view of what top-down design in. And while this is a step towards top-down design, it only addresses one of the issues with bottom-up design and there is much more that can be done. To overcome the remaining issues, one must go further. Also needed is a formal verification planning procedure and a mixed-level verification strategy. Both act to reduce risk and are added with the understanding that with complex mixed-signal circuits, complete final transistor-level verification is at a minimum very expensive and is often impractical. With careful planning one can use mixed-level simulation to move the verification up in the design process where it is less expensive and so can be more comprehensive. It also tends to find errors earlier in the design process when recovery is easier and less expensive.

A well designed top-down design process methodically proceeds from architecture- to transistor-level design. Each level is fully designed before proceeding to next and each level is fully leveraged in design of next. It acts to partition the design into smaller, well defined blocks, and so allows more designers to work together productively. This tends to reduce the total time required to complete the design. A top-down design process also formalizes and improves communications between designers. This reduces the number of flaws that creep into a design because of miscommunication. The formal nature of the communication also allows designers to located at different sites and still be effective.

Following a top-down design methodology also reduces the impact of changes that come late in the design cycle. If, for whatever reason, the circuit needs to be partially redesigned, the infrastructure put in place as part of the methodology allows the change to be made quickly. The models can be updated and impact on the rest of system can be quickly evaluated. The simulation plan and the infrastructure for mixed-level simulations would already be available and can be quickly applied to verify any changes.

3.1. System Architect

The system architect is a new member of the design team. He or she is the leader of the top-down design process and is expected to develop the simulation and modeling plans and to coordinate with the other designers to assure that the plans are followed. The primary responsibility of the system architect is to assure that the system operates as expected when finally implemented. This must be a designer that has experience in the type of system being designed so that he or she can anticipate and plan for issues that are likely to occur. Preferably, the experience covers aspects of both system and block design. The system architect may be the system engineer, the one that actually designs the block diagram, but it need not be. However, the system architect should not also have block design responsibilities. Block design has a tendency of consuming an engineer.

The system architect should be comfortable with modeling and MS-HDLs to the point where he or she can write the descriptions of both the system and the individual blocks. Since most designers are not skilled at modeling and not familiar with MS-HDLs, the system architect should also be able to train the other engineers on the project on the use of MS-HDLs. However, it is important to recognize that modeling is a skill that is distinct from design. Designer often have neither the skill nor the inclination to write sophisticated models. If sophisticated models are required, they generally must be developed by the system architect.

The system architect must have a good understanding of simulation. In particular, where and how simulation can be used to verify areas of concern in the design, and where it cannot. This knowledge is leveraged heavily during the development of the simulation plan.

The system architect owns the top-level schematic for the design. This schematic must be captured before any block design begins, even though it is likely to change before the design is complete. The top-level schematic specifies the partitioning of the design into blocks and the interface for each block. So each block should be "pin-accurate". By this is it meant that in the top-level schematic, each block, and each pin on each block, is represented, and the type of each pin is carefully defined and documented. For example, an enable line on a block may be denoted "3V CMOS active high" or a trigger line may be described with "5V TTL positive edge triggered". In this way, the top-level schematic provides "clarity of intention" to the design team.

Once the top-level schematic is captured, the top-level models are written, usually by the system architect, and the system completely verified according the simulation plan. The top-level schematic and models are then distributed to everyone on the design team. As the design progresses, the system architect would approve and coordinate any changes to the block interfaces, and then distribute updated models of the system or the blocks to the team. As the block designers work, they would provide transistor-level schematics (pre- and post-layout) to the system architect, who would then verify them with mixed-level simulation, again according to the simulation plan, before accepting them.

During the design process, the system architect would work with the test engineers to develop the test plan and test programs. The availability of a working model of the system early in the design process allows test engineers to begin the development and testing of test programs early. Moving this activity, which used to occur exclusively after the design was complete, so that it starts at the same time the block design begins significantly reduces the time-to-production [1,2,3,6].

3.2. Simulation and Modeling Plans

An important focus in a good top-down design methodology is the development of a comprehensive simulation plan, which in turn leads to a modeling plan. This is done by the system architect with input from the whole design team. The process begins by identifying particular areas of concern in the design. Plans are then developed for how each area of concern will be verified. The plans would specify how the test would be preformed, and which blocks would be at the transistor level during the test. For example, if an area of concern is the loading of one block on another, the plan might specify that one test should include both blocks represented at the transistor level together. For those blocks for which models are used, the effects required to be included in the model are identified for each test. This is the beginning the modeling plan. Typically, many different models will be created for each block. These models may be written either by the system architect or the block designer.

It is important to resist the temptation to specify and write models that are more complicated than necessary. Start with simple models and only model additional effects as needed (and as spelled out in the modeling plan). Also, the emphasis when writing models should be to model the behavior of the block, not its structure. A simple equation that relates the signals on the terminals is preferred to a more complicated model that tries to mimic the internal working of the block. This is counter to the inclination of most designers, whose intimate knowledge of the internal operation of the block usually causes them to write models that are faithful to the architecture of the block, but more complicated than necessary. Following these general rules will result in faster simulations and less time spent writing models.

A formal planning process generally results in more efficient and more comprehensive verification, meaning that more flaws are caught early and so there are fewer design iterations. The simulation and test plans would initially be applied to the high-level description of the system, where they can be quickly debugged. Once available, they can be applied during the mixed-level simulations of the blocks, reducing the chance that errors will be found late in the design cycle.

3.3. System-Level Verification

System-level design is generally performed by system engineers. Their goal is to find an algorithm and architecture that implement the required functionality while providing adequate performance at minimum cost. They typically use system-level simulators, such as Matlab, Simulink or SPW [4,5], that allow them to explore various algorithms and evaluate trade-offs early in the design process. These tools are preferred because they represent the design as a block diagram, they run quickly, they support the abstract data types used at the system level, and have large libraries of predefined blocks for common application areas.

This phase of the design provides a greater understanding of system early in the design process. It also allows a rapid optimization of the algorithm and moves trades to the front of design process where changes are inexpensive and easy to make. Unworkable approaches are discarded early. Simulation is also moved further up in the design process where it is much faster and can also be used to help partition the system into blocks and budget their performance requirements.

Once the algorithm is chosen, it must be mapped to a particular architecture. Thus, it must be refined to the point where the blocks used at the system level accurately reflect the way the circuit is partitioned for implementation. The blocks must represent sections of the circuit that are to be designed and verified as a unit. Furthermore, the interfaces must be chosen carefully to avoid interaction between the blocks that are hard to predict and model, such as loading or coupling. The primary goal at this phase is the accurate modeling of the blocks and their interfaces. This contrasts with the goal during algorithm design, which is to quickly predict the output behavior of the entire circuit with little concern about matching the architectural structure of the chip as implemented. As such, mixed-signal hardware description languages (MS-HDLs) such as Verilog-AMS [6] or VHDL-AMS [8] become preferred during this phase of the design because they allow accurate modeling of the interfaces and support mixed-level simulation.

The transition between algorithm and architecture design currently represents a discontinuity in the design flow. The tools used during algorithm design are different from the ones used during architecture design, and they generally operate off of different design representations. Thus, the design must be re-entered, which is a source of inefficiencies and errors. It also prevents the test benches and constraints used during the algorithm design phase from being used during the rest of the design.

On the digital side, tools such as SPW do provide paths to implementation via Verilog and VHDL generation. Similar capabilities do not yet exist for the analog or mixed-signal portions of the design. An alternative is to use Verilog-AMS or VHDL-AMS for both algorithm and architecture design. This has not been done to date because simulators that support these languages are just now becoming available. It will probably take a while for this approach to become established because of the absence of application specific libraries.

3.4. Mixed-Level Verification

Digital synthesis maps digital behavior onto digital gates that are arranged in a rather constrained topology. The simple nature of gates combined with the constrained topology makes synthesis feasible. With analog circuitry, the fundamental building blocks are much more complex and varied and the topology is completely unconstrained. These two factors make analog synthesis a fundamentally much more difficult problem than digital synthesis. Analog synthesis so far has resisted all attempts at automation except in limited cases, such as analog filters. Work continues, but we are still far from having universal analog synthesis.

Without analog synthesis, analog design is done the old fashioned way, with designers manually converting specifications to circuits. While this allows for more creativity, it also results in more errors, particularly those that stem from miscommunication. These miscommunications result in errors that prevent the system from operating properly when the blocks are assembled even though the blocks were thought to be correct when tested individually.

To overcome this problem, mixed-level simulation is employed in a top-down design methodology for analog and mixed-signal circuits (this represents a significant but essential departure from the digital design methodology). Mixed-level simulation is required to establish that the blocks will function as designed in the overall system.

To verify a block with mixed-level simulation, the model of the block in the top-level schematic is replaced with the transistor level schematic of the block before running the simulation. In mixed-level simulation, the system, described at a high level, acts as a test-bench for the block, which is described at the transistor level. Thus, the block is verified in the context of the system, and it is easy to see the effect of imperfections in the block on the performance of the system. Mixed-level simulation requires that both the system and the block designers use the same simulator and that it be well suited for both system- and transistor-level simulation.

Mixed-level simulation allows a natural sharing of information between the system and block designers. When the system level model is passed to the block designer, the behavioral model of a block becomes an executable specification and the description of the system becomes an executable test bench for the block. When the transistor level design of the block is complete, it is easily included in the system level simulation by the system architect.

Mixed-level simulation is the only feasible approach currently available for verifying large complex mixed-signal systems. Some propose to use either timing simulators (sometimes referred to as fast or reduced accuracy circuit simulators) or circuit simulators running on parallel processors. However, both approaches defer system-level verification until the whole system is available at transistor level, and neither provide the performance nor the generality needed to verify most mixed-signal systems.

3.5. Bottom-Up Verification

Once a block is implemented, one could update the models that represent it to more closely mimic its actual behavior. This would improve the effectiveness of mixed-level and system-level simulation. This process is referred to as bottom-up verification. To reduce the chance of errors, it is best done during the mixed-level simulation procedure. In this way, the verification of a block by mixed-level simulation becomes a three step process. First the proposed block functionality is verified by including an idealized model of the block in system-level simulations. Then, the functionality of the block as implemented is verified by replacing the idealized model with the netlist of the block. This also allows the effect of the block's imperfections on the system performance to be observed. Finally, the netlist of the block is replaced by an extracted model. By comparing the results achieved from simulations that involved the netlist and extracted models, the functionality and accuracy of the extracted model can be verified. From then on, mixed-level simulations of other blocks are made more representative by using the extracted model of the block just verified rather than the idealized model.

When done properly, bottom-up verification allows the detailed verification of very large systems. The behavioral simulation runs quickly because the details of the implementation are discarded while keeping the details of the behavior. Because the details of the implementation are discarded, the detailed behavioral models generated in a bottom-up verification process are useful for third-party IP evaluation and reuse.

Though bottom-up verification is helpful when verifying the performance of large systems, it is rarely done today. Generating behavioral models that include the detailed behavior of even simple blocks is quite difficult and requires a specialized skill not commonly found in the design team. This situation is not expected to change until automated tools and methodologies develop to generate detailed behavioral models.

Mixed-level simulation is currently the best approach to verifying large mixed-signal systems that are designed with a top-down methodology. However, eventually systems will be too large to completely verify with mixed-level simulation, in which case a bottom-up verification approach will become necessary.

3.6. Final Verification

In a top-down design process, SPICE-level simulation is used judiciously in order to get the benefits without incurring the costs. All blocks are simulated at the transistor level in the context of the system (mixed-level simulation) in order to verify their functionality and interface. Areas of special concern, such as critical paths, are identified up front and simulated at the transistor level. The performance of the circuit is verified by simulating just the signal path or key pieces of it at the transistor level. Finally, if start-up behavior is a concern, it is also simulated at the transistor level. The idea is not to eliminate SPICE simulation, but to reduce the time spent in SPICE simulation while increasing the effectiveness of simulation in general by careful planning.

4. MS-HDLs

Both Verilog-AMS and VHDL-AMS have been defined and commercial implementations are emerging. These languages are expected to have a big impact on the design of mixed-signal systems because they provide a single language and a single simulator that are shared between analog and digital designers. It will be much easier to provide a single design flow that naturally supports analog, digital and mixed-signal blocks, making it simpler for these designers to work together. It also becomes substantially more straight-forward to write behavioral models for mixed-signal blocks. Finally, the AMS languages bring strong event-driven capabilities to analog simulation, allowing analog event-driven models to be written that perform with the speed and capacity inherited from the digital engines.

It is important to recognize that the AMS languages are primarily used for verification. Unlike the digital languages, the AMS languages will not be used for synthesis because the only synthesis that is available for analog circuits is very narrowly focused.

4.1. Verilog-AMS

Verilog-A is an analog hardware description language patterned after Verilog-HDL. Verilog-AMS combines Verilog-HDL and Verilog-A into a MS-HDL that is a super-set of both seed languages [7]. Verilog-HDL provides event-driven modeling constructs, and Verilog-A provides continuous-time modeling constructs. By combining Verilog-HDL and Verilog-A it becomes possible to easily write efficient mixed-signal behavioral models. Verilog-AMS also provides automatic interface element insertion so that analog and digital models can be directly interconnected even if their terminal / port types do not match. It also provides support for back annotating interconnect parasitics.

A commercial version of Verilog-AMS that also supports VHDL is expected soon from Cadence Design Systems.

4.2. VHDL-AMS

VHDL-AMS adds continuous time modeling constructs to the VHDL event-driven modeling language [8]. Like Verilog-AMS, mixed-signal behavioral models can be directly written in VHDL-AMS. Unlike with Verilog, there is no analog-only subset.

VHDL-AMS inherits both the good and the bad aspects of VHDL. For example, VHDL-AMS inherently supports configurations and abstract data types. However, VHDL is also strongly typed, which is a serious problem for mixed-signal designs. You are not allowed to interconnect digital and analog ports, and there is no support for automatic interface element insertion. In fact, you are not even allowed to connect ports from an abstract analog model (a signal flow port) to a port from a low-level analog model (a conservative port). This makes it difficult to support mixed-level simulation. VHDL-AMS also does not provide support for back-annotation of RC interconnect. These represent fundamental flaws that will have to be overcome by a simulation environment, making VHDL-AMS much more dependent on its environment. This should slow deployment of effective VHDL-AMS-based flows.

A commercial version of VHDL-AMS that also supports Verilog is available from Mentor Graphics [9]. A VHDL-AMS simulator is also expected soon from Analogy [9].

5. Example

Though this example is several years old, it is representative of the type of circuit complexity that is becoming mainstream today. It is a PRML channel chip that it difficult to simulate for two reasons. First, it is a relatively large circuit that involves both analog and digital sections that are closely coupled. Second, the architecture involves complex feedback loops and adaptive circuits that take many cycles to settle. The combination of many transistors and many cycles combines with the result being a simulation that is so expensive as to be impractical. In this case, the expected simulation time was predicted to be greater than a month.

The traditional approach to simulating a complex circuit like this would be to simulate the blocks individually. Of course this verifies that the blocks work individually, but not together. In addition, for this circuit it is difficult to verify the blocks when operating outside the system, and it is difficult to predict the performance of the system just knowing the performance of the individual blocks.

When the architecture was simulated at a high level with each block represented by a pin-accurate behavioral model, the simulation time was less than 10 minutes. Then, when a single block was run at the transistor level, the simulation ran overnight. Even though the full system was never simulated at the transistor level, it worked the first time because this methodology verified the blocks in the context of the system and it verified the interfaces between the blocks.

6. Development of System Architects

The primary barrier to widespread adoption of a top-down design style for complex mixed-signal circuits is a lack of engineers with the skills and training to be system architects. A system architect must

- Be fluent in an AMS language and skilled in the art of modeling
- Be an experienced designer
- Have a good understanding of the top-down design process
- Be proficient in the use of circuit and AMS simulation
- Have the ability to lead and manage complex projects

Given the high pressure world that most designers live in, it is unlikely that they will be able to acquire such a broad and deep set of skills while on the job, even if they are motivated to do so. Rather, it is important for their employers to look for engineers that have the interest and the relevant background and invest the time and training to develop them into system architects. In addition, it is essential that appropriate training becomes available from universities and continuing education centers.

7. Conclusion

Top-down design is a formal design process that requires a serious commitment throughout the entire design process. It is not a piece of software or something you do in your spare time. It is not a way to reduce headcount or something you can try after the design is complete. It is considerably more than simply doing the initial design of the block diagram with Simulink and it is not something you can be successful at without a significant investment in time and training. However, it is much easier the second time around and once mastered provides a dramatic return. Using top-down design usually results in needing fewer design iterations, which provides a more predictable design process. It also results in more optimal designs in a shorter time. Finally, it allows design teams to be larger and more dispersed, giving the option of trading a higher initial investment for a shorter time-to-production.

Acknowledgements

Much of the material presented is based on discussions with Dan Jefferies and Henry Chang of Cadence Design Systems and Jim Holmes of Texas Instruments.

Bibliography

[1] C. Force, T. Austin. Testing the design: the evolution of test simulation. *International Test Conference*, Washington 1998.

[2] C. Force. Integrating design and test using new tools and techniques. *Integrated System Design*, February 1999.

[3] Dantes virtual test environment, *www.virtualtest.com*.

[4] Matlab and Simulink, *www.mathworks.com*.

[5] *Signal-Processing Worksystem User's Guide.* Cadence Design Systems, San Jose, CA.

[6] SpectreVX and SaberVX virtual test environments, *www.teradyne.com*.

[7] *Verilog-AMS Language Reference Manual: Analog & Mixed-Signal Extensions to Verilog HDL*, version 2.0. Open Verilog International, 2000. Available from *www.ovi.org*.

[8] VHDL-AMS, *www.vhdl.org/analog*.

[9] VHDL-AMS simulators, *www.vhdl-ams.com*.

COMPUTER AIDED DESIGN FOR INTEGRATED SYSTEMS

Frederic M. Stubbe
Conexant Systems, Inc., Newport Beach, California, USA

ABSTRACT

This paper describes a design framework for integrated system design. It discusses the design methodology, the design system, the CAD tools, the models and their interaction that enable optimized integrated system design. Examples of this design approach are discussed in the context of an integrated GSM chip.

1. INTRODUCTION

The continued downscaling of CMOS technologies over the last decades made CMOS not only the technology of choice for pure digital and analog design, but made it also feasible for RF circuit design, which enabled the economical integration of full systems on 1 CMOS die. Prototypes of full-integrated systems in CMOS have already been demonstrated [1,2]. In the coming years we can therefor expect to witness the introduction of numerous system on chip products in the marketplace.

The rational behind the integration is that it allows to combine more functionality on the same substrate, and have circuits benefit from the existence of each other (functionality reuse). It reduces the system overhead, and can reduce system test time. It also avoids unnecessary signal degradation (from package parasitics), power consumption (to drive external loads) and increased area (for bond pads with ESD protection) related to crossing the pad boundaries and interfacing with other dies and or external components.

On the other hand integration does not come for free. If the integrated system is not core limited, the die size will increase by integrating

R. J. van de Plassche et al. (eds.), Analog Circuit Design, 203–227.

more functionality, which reduces the yield compared to the combined yield of two smaller dies. Since all circuits share the same substrate and same technology, signal integrity can deteriorate and you might be paying a premium to have simple digital circuitry integrated together with RF functionality, which sometimes requires special process steps. You also give up on visibility because less signals are accessible, which can translate into longer debugging times and increased system test cost, unless testability and observability is taken into account during the design phase.

Clearly just combining functionality in the same technology will not result in a successful system on chip product. In order to benefit from integration, the system needs to be re-engineered with integration in mind rather than combining existing dies on one substrate [5]. A clear example is the trend in wireless transceivers. Just integrating all the active circuitry involved with dual heterodyne architectures, is not optimal, because they were architected to minimize the (at that time) expensive active circuitry, and exploit the relative cheap passive components like bandpass filters. The increased interest in low and zero IF receivers as ideal candidates for full integrated systems, stems from the fact that they are re-engineered with integration in mind, and exploit the cheap active circuitry to minimize external component cost and interfaces related to it.

So, what does it take to enable efficient re-engineering for system integration ? To answer that question it pays to realize that an integrated system encompasses not only the integrated circuit, but also a package, external components, a PCB and software. In order to optimize the overall system performance, all these components and their interaction need to be understood and their behavior with respect to each other needs to be scrutinized. This process used to be managed by system engineering. Based on their experience, knowledge and discussions with the different engineering disciplines involved, they come up with an interface specification that the individual engineering groups design to. This divide and conquer process is inherently limited by the shared knowledge and complexity that both engineering teams can handle and can therefor severely restrict the solution space and obscure an optimized system solution.

Since nowadays computers and CAD tools can handle larger and larger databases more and more efficiently, they are ideal candidates

to help manage the complexity that integrated system design requires. But in order for them to do a good job at efficient database management, all this data needs to be entered in a disciplined way, meaning that all engineering teams need to enter data that is usable by all the other engineering teams in the same database. Using a unified design methodology can enforce this.
To handle the increased complexity that is related to system on a chip design and still meet the stringent time to market constraints, it is required to have a common database (design system) in which resides the design knowledge that is optimized by designers through the interaction with CAD tools, guided by a unified design methodology.

This paper is structured as follows : in the second section the design methodology is discussed that enables integrated system design. The third section discusses the type of simulations and models that integrated system design mandates. In the fourth section the type of layout tools that allow the fast realization in silicon of these integrated systems are described and a conclusion ends this paper.

2. DESIGN METHODOLOGY FOR INTEGRATED SYSTEM DESIGN

To facilitate system integration the design system should allow interaction between all different functions that will be integrated.
This interaction can only be achieved if the data is entered in one design database that supports the needs of these different disciplines (figure 1). Therefor the main task for the design system is to allow efficient capturing of all the necessary data for all involved disciplines (PCB, external components, package, analog and digital circuits, software, production testing, documentation, ...) to optimize their interaction. It is the main task of the designer to attach all this data to his functional block to support this interaction. This process is guided through a design methodology. It is a set of guidelines that ensures capturing all necessary information in a standard way to improve interactivity and reuse, and minimize error and overhead.
Typical examples of design methodology guidelines are signal naming conventions for power supplies, bias lines, control signals and sensitive analog signals. The design methodology also requires the

designer to describe the function's behavior to efficiently simulate the behavior and interface of a design. This model depends on the level of interactivity with other blocks or other hierarchies and has different representations. The highest level behavioral model of the block involves simple first order input output relations. A lower level model describes more real world non-idealities that influence system behavior. The lower level is the transistor schematic, which models all non-ideal behavior that is taken into account in the models of the primitive components. The design methodology requires also capturing all important layout information into the schematic as properties attached to components (symmetry, currents and styles). During the design review these guidelines are checked for completeness.

The real advantages of the design system become visible once initial behavioral models are captured according to these guidelines. A top-down approach can be used to translate high-level system specs into lower level block specification. This can be used to quantify the sensitivity of block level specs to higher level system and the appropriate trade offs become visible. Also early on system feasibility can be assessed.

One difficulty is that second-order effects of components at the lowest level govern some of the high-level system specifications. Examples are the dynamic range that depends on offsets, noise figure that depends on accurate high frequency noise models, and offset that can depend on isolation between LO and RF ports of a direct down conversion mixer. Therefor it is crucial for the design system to handle the complexity that is related to bridging the gaps from lower level physical behavior to system level performance. Behavioral modeling is key to bridging this gap. It allows to focus on those mechanisms that are impacting the high level behavior, and simplifying non essential contributors, such that the overall complexity can still be handled by the design system.

Besides the performance complexity there is another type of complexity related to system on chip design. Since after all the performance of the whole chip needs to be guaranteed, it is important to simulate the overall behavior (e.g. simulating the overall receive chain filter characteristics). Therefor, the design system needs to handle a big number of interacting blocks that might be described in

different formats (transistor level, gate level, behavioral level, algorithms, ...) Again behavioral modeling can help the design system to handle the latter complexity. The design system needs to be designed for both to be an efficient mean in the design verification of a system on a chip.
It becomes clear that more than one model of the functionality needs to be created to handle the two different types of complexity : transistor schematics with very detailed transistor models that allow simulation of all known non idealities, and on the other extreme, first order behavioral model to allow top level simulations. These different behavioral models are all linked to the same symbol, such that each designers can pick and choose the model for each block that allows him to simulate most efficiently the behavior he is verifying.
The design system needs to manage this complexity, guided by the designer's knowledge of its relations embedded in behavioral models and allow relevant results in a reasonable amount of time.

These blocks must tie into the common design database to allow the different design groups to examine the relation of their block to the others. If interface discrepancies are found, the designer will update the database. This mechanism provides visibility of the updated data to the whole team. It is important to stress the importance of the common database. Matlab and Mathcad are ideal behavioral modeling tools. But unless they are integrated (meaning co-simulation is possible) in a common database, their results can not be validated with other blocks, and there exists a good chance that its behavior is not be optimized for the complete system. In other words it cannot benefit from the knowledge of the other designers that interface with that particular block. Moreover, its behavioral model cannot be verified against transistor level schematics nor can it be used as a part of a higher level simulation.

If the designers provide all the information according to a design methodology, which guarantees the design system can handle all the functionality, the complexity related to system on a chip design is in fact distributed. This allows each system spec to be designed and verified with another optimal combination of models. This approach optimizes interaction between different building bocks which widens the solution space and enables a more optimal system solution. In this respect, it is crucial to minimize the levels of hierarchy, not to obscure any of the interactions.

3. SIMULATION TOOLS AND MODELS FOR INTEGRATED SYSTEM DESIGN

Although most of the simulations are still based on the well known Spice like time based simulators for analog design and cycle based simulators for digital design, some significant improvements in the simulation tools used specifically for integrated system design are going mainstream. Mixed mode simulators are increasingly used for co-simulation of analog, digital and behavioral. Other simulators based on harmonic balance and shooting methods that address specifically the unique characteristics of RF, but also some mixed signal circuits (like switched capacitor circuits) are finding their way in mixed signal design environment [6,7]. Their advancements in CAD have resulted in significant reduction of simulation times without sacrificing too much accuracy, if implemented smart.
They both address the issue of handling the complexity and are therefore important steps towards efficient simulation of integrated systems.

Simulation results can only be as accurate as the models used to describe the physical behavior of the devices. Therefore, significant effort has been put in place to obtain accurate models for all devices.
Especially in the RF CMOS arena where the technology is pushed to its extremes to get the necessary system performance, accurate transistor models are even more important to avoid over design.
The approach that is taken to obtain the best trade off between accuracy and efficiency, is to split up the transistor in an intrinsic part and an extrinsic part [3,4] (figure 2). The intrinsic part is an efficient BSIM3v3 MOS transistor compact model, with all extrinsic components disabled. The actual RF transistor is then implemented as a subcircuit (or composite model), calling the intrinsic compact model and adding the extrinsic components being the source and drain diodes, the terminal resistances, the terminal capacitances and a

substrate model. Advantages of the subcircuit approach are manifold. It allows accurate modeling of physical parasitics well into the GHz frequencies, to provide accurate results for non-linear, transient, non-quasi-static, coupling and even noise behavior, all characteristics that play a fundamental role for RF circuits. It does not compromise simulation time, since only few extra nodes are added in the composite model versus the intrinsic compact model. The extrinsic parameters can easily be extracted using a direct extraction method based on S parameter measurements and finally it can accurately take the layout style into account (e.g. handle odd / even number of fingers and substrate contact configuration)
For those circuits where high frequency effects have less impact on the performance, or where simulation time is a bigger concern, the designer can choose not to use the composite model, but the compact BSIM3v3 MOS model in which the extrinsic elements are not disabled. In other words, the same way a designer can choose between a behavioral model and a transistor schematic for a block, he can choose between a simple and fast or a somewhat slower accurate transistor model, depending on the behavior that is verified.
In addition passive devices (resistors, capacitors, inductors and varactors) have been modeled as either simple two terminal devices with lumped impedance or 4 terminal devices, relating input to output through distributed and de-Q'd impedances including the parasitics to the substrate as well as a model for the lossy substrate.

Another part that requires careful modeling is the package. Especially for systems involving RF interfaces (e.g. to and from antenna), impedance matching is crucial for optimum power transfer, and can be seriously degraded by package parasitics. Signal coupling through the package parasitics can degrade the system performance. Both these characteristics can be taken into account during the design process and their impact can be minimized if the proper package model exist. Proper package interconnect modeling consists of modeling the bondwires that connect the bondpad with the pad lead tip and the leadframe or substrate that connects this lead tip with the package pin. The actual modeling is most of the time based on electromagnetic simulations using the package information and the padring layout as input to generate a netlist of capacitively and inductively coupled

lossy inductors on all pins and their parasitic capacitance to the flag. Again for those simulations where these effects don't impact the behavior, the package is modeled as a short between the pin and the pad.
Once these package parasitics are accurately modeled, they can be used to the advantage of the designer. Bondwire inductances can be used to reduce the need for external inductors (e.g. inductive degeneration in LNA's), pads can be properly arranged to avoid crosstalk or can be properly positioned with respect to the leadframe to minimize the influence of the package parasitics.

In order to clarify some of the concepts mentioned above and understand how these elements fit together to realize optimized integrated systems, 2 areas of an integrated CMOS GSM transceiver chip will be discussed in more detail : the offset cancellation circuit in the receive path and the power amplifier control loop in the transmit path. Each of these areas will clearly show the need for multilevel simulation and mixed mode simulation and show how the overall design can be optimized by integration.

3.1 OFFSET CANCELLATION CIRCUIT

The offset cancellation circuit in the receive path of this GSM transceiver is needed to eliminate the DC component generated from the LO self-mixing or blocker self-mixing in this zero IF architecture as well as offset generated in the baseband. If these offsets are not eliminated, they can overload the subsequent stages under high gain conditions with loss of dynamic range and SNR as a consequence. One approach would be ac coupling to eliminate the DC content. But to minimize the signal loss at low frequencies, an uneconomically large capacitor needs to be integrated, which eliminates this approach for integrated systems. An alternative approach is offset cancellation. The straightforward way to cancel offsets is to short the inputs at RF and measure the offset at the ADC output and add or subtract an amount in the analog baseband section to compensate for the generated offset. This analog offset cancellation works fine, but has as disadvantage that still a large time constant needs to be realized to measure only the DC in the signal content. The chosen approach in this receive section uses a mostly analog loop to do the coarse offset cancellation just to avoid any overload in the analog baseband section, and relies on the already integrated sinc filter to get the residue of the DC content. This DC content is then subtracted from the each digital sample to do the fine offset cancellation (reduce the offset voltage below the noise floor).

Clearly this loop involves RF, analog circuitry and digital circuitry, both continuous time and discrete time signals, as well as a finite state machine (FSM) to control the timing of this offset cancellation. This is realized during the calibration cycle prior to the receive slot. But since there is already a FSM existent on this die, adding the extra states in there are a small overhead to pay for the extra functionality. Moreover, integration of the digital receive signal processing, allows us to shift the fine offset cancellation to the digital domain, and leave only the coarse offset cancellation to be done in the analog, reducing the area and power of the overall system. A system solution where the analog section of the receive strip is separate from the digital processing would make this approach not feasible. Or the fact that both the analog baseband section and the digital channel processing are integrated on the same die enabled a more optimal solution for the offset cancellation.

The first benefit of mixed mode simulation is that it enables verification of the correct connectivity of all circuits throughout the loop. This applies to connectivity of the power supply, connectivity of control signals and signal polarity check. A more interesting area where mixed mode simulation contributes is where it drives the design of the loop.

First, a very detailed simulation of the RF amplifier and mixer was performed based on the composite transistor model that includes substrate coupling effects, including actual layout parasitics and measurement of these functions, to come up with a realistic range of generated offsets in these blocks. Also the offset of baseband components in the loop were estimated based on process matching information and simulations. Then behavior models were created with the offset information embedded. These behavior models all tied into the common database, such that they can be swapped with the view that is most appropriate for the intended simulation and to verify the mutual correctness of model versus circuit. All the digital parts of this loop were modeled using a digital behavioral model (VHDL). Once all the models were available, a mixed mode simulation was setup with the created models to design and verify the loop performance The operation of the calibration cycle is explained in figure 3 and the timing is given in figure 4.

The calibration cycle consists of two phases, a coarse calibration and a fine calibration, governed by a finite state machine. During the complete calibration cycle, the receive inputs are shorted such that the LO to LO self-mixing product and all other offsets can be measured. During the coarse calibration, the gains of the different baseband gain stages are lowered to avoid any overload and the decimation filter is set to have a bigger decimation ratio compared to normal operation to get a more representative measure of the DC content. This digital representation of the DC offset is then converted into an analog voltage, which then can be subtracted from the signal to cancel the coarse offset. In figure 5 the remainder of the offset at the input of the ADC after coarse calibration for a wide range of forced input offsets at the output of the mixer is shown. During the fine calibration all gains are reset to their programmed values and the previously measured coarse offset is subtracted from the signal path. The remainder offset is then measured by the sinc filter with the bigger decimation ratio to get the best representation of the remaining DC offset. This remaining DC offset is then subtracted in the digital domain before the rest of the channel filtering is performed.

The waveforms for the coarse and fine calibration phases are shown in figure 6. The resulting offset after coarse and fine calibration is below 8 μV. These are results of the analog / digital co-simulation of the Rx offset cancellation loop.

3.2 POWER AMPLIFIER CONTROL (PAC) LOOP

The PAC is another good example of functionality that could be optimized by behavioral level simulations and integration.

The main function the PAC loop is to tightly control the output power of the power amplifier (PA) that is radiated through the antenna to meet the stringent GSM transmit mask under all conditions of temperature, process and power supply. Since the loop behavior is influenced by the power amplifier, the package, RF, analog and digital circuits, this PA control loop is a prime candidate for behavioral level simulations to design it properly. Straightforward simulation at the circuit level would translate in impractical long simulation times and are therefor useless for architecture exploration and trade off analysis.

The functionality of the PAC is described as follows. An external coupler taps off part of the PA output signal into the chip. A rectifier generates a DC voltage proportional to the power in the input signal, which is logarithmically amplified to get a linear relation between the rectifier output voltage and the power in dB. This signal is then compared to the ramp profile for the PA, and an error signal is generated, that needs to be predistorted to compensate for the non-linearity of the control voltage versus output power of the PA. This signal will then be driving the PA power control pin.

This loop could be implemented in the analog domain, but would be much less flexible compared to a part analog / part digital implementation as shown in figure 7. In this architecture, the output of the logamp is converted to a digital word, and compared to the filtered rampstore word, which is digital to begin with. The resulting digital error signal can then easily be predistorted in the digital domain with a so-called power shaper. Finally this digital sample is converted to an analog signal before it is driven out to control the PA. The predistortion characteristic can be programmed and adjusted to accommodate different PA's, therefore increasing the flexibility of the PA control loop. The latter would be very difficult to implement in an analog only loop.

Please note that to properly dimension this loop (A/D and D/A resolution, gains, programmability of pre-distortion network), external components (non-linear PA power versus control voltage) , packaging, RF, analog and digital components are involved. In order to get a handle on the implementation of this loop, the relevant behavior of each of these blocks with respect to the PAC loop was modeled, and all these models were plugged into the one database so that their interaction with each other can be simulated and optimized.

For the PA control loop, the behavioral simulations allowed us first of all to verify connectivity (power supply / signal polarity) and control logic, which is not trivial, since this loop involves RF, analog and digital building blocks, each with their own power supplies and control logic. But more important, it enabled to simulate the stability of this loop (avoid ringing during the constant output power versus time) and optimize both the digital low pass filter pole and power shaper characteristics such that the overall transmit mask was never violated. In figures 8,9 and 10 the transmit mask is shown and the influence of both the low pass filter pole and power shaper characteristics on the output power in the time domain. These results are obtained by behavioral simulation of the complete PA control loop with relatively simple behavioral models for the analog and RF functionality and VHDL models for the digital part. The advantage of this behavioral simulation is that it allowed digital designers to really tweak the behavior of their building blocks (the low pass filter and power shaper) based on behavioral models of the PA and the other analog building blocks in the loop. Normally digital designers are not involved in analog design, let alone RF aspects of PA's. This is made possible by sharing information in a common database, in which the actual behavioral model is written by the expert of the block, such that other designers could use it to make relevant trade offs, even if they are not 100% familiar with the complete loop.

4. LAYOUT TOOLS FOR INTEGRATED SYSTEM DESIGN

Mixed signal CAD has also undergone revolutionary changes in the layout area. Although we are not seeing similar levels of automatic layout generation of mixed signal circuits as currently exist for digital synthesized circuits, significant improvements to automatic analog layout generation at different levels of complexity are noticed [8]. Most major layout tools already provide device generators for primitive components being transistors, capacitors and resistors. But in more advanced design systems, not only more flexibility is added to drive the layout generators for primitives, but also higher level subblock generators and a template driven layout methodology are used to speed up the layout phase to minimize time to market. The timesaving that results from their implementation is manifold. On the one hand the generation is faster than can be done manually what translates directly in timesaving. On the other hand these primitives are created "correct by construction" and therefor also reduce the verification time. These generators basically reduce the layout work to the interconnection of generated devices. It also allows the designer to do initial floorplanning and relative placement without relying on layout engineers to give an initial estimate, therefor skipping the first turnaround between design and layout. This advantage does not come for free. It requires a more disciplined way of designing circuits. Designers are required to add layout information in the common database, typically the schematic from which the layout is generated (schematic driven layout).

The automatic layout approach is different for RF intensive layout, mixed signal intensive layout and digital intensive layout.

First of all in the RF world, where parasitics have a bigger impact on the performance, a limited set of RF standard cells is used. The main reasoning behind this approach is that this limited set of structures are precisely characterized at high frequencies and its behavior is accurately modeled in high frequency transistor models, taking all the layout parasitics into account exactly as they will be used on chip. Accuracy is more important than flexibility. So once the transistor layout has been optimized for RF, the device generator is programmed to generate this layout. This generator can than automatically generate the layout a wide range of RF transistors that span the different dimensions (W, L, NF,I) and fabricated on a test wafer surrounded with pads that allow S-parameter measurements for proper high frequency characterization. The transistor models are finally tweaked to fit the measurements. Not only the active devices are handled this way, also the passives : resistors, capacitors, inductors and baluns. Since its behavior is more predictable at high frequency, extra degrees of freedom are given to the designer to guide the primitive generators, like resistor width or capacitor aspect ratio. But again a wide range of passives are added on the test wafer for high frequency characterization and its measurement results are reflected in the high frequency models of the passives.

In the mixed signal layout area the situation is different. Although the primitive's parasitics are not negligible, their accuracy is traded off to provide extra flexibility and / or simplify the primitive's model to save simulation time. E.g. the transistors have different styles to accommodate the current (figure 11) or specify the substrate tie (figure 12). Only first order parasitics for the primitives are taken into account. But still there is a close link between the style and the model.

One level above the device generators are the so-called subblock generators. These subblocks are a combination of primitives that are used very frequently and have special layout constraints. Typical examples are current mirrors where matching is important, common centroid differential pairs where matching and symmetry are important or cascode transistors where the diffusion between the two gates is minimal to avoid signal loss (figure 13). Once the layout is optimized, the subblock generator is programmed accordingly for reuse in all current mirrors and differential pairs. In order to drive this subblock generator from the schematic (schematic driven layout), a unique symbol is required in the schematic. Again, this is a small price to pay for using this technology.

A great number of mixed signal circuits use operational amplifiers as critical building blocks. Therefor a tremendous time saving could be realized if they could be automatically generated. The template driven layout methodology specifically addresses this type of blocks. Initially the analog layout knowledge is captured once in a reference layout. Then the template driven layout system regenerates the layout for different designs based on the same reference amplifier while preserving the relative device configurations, wire trajectories and material types and symmetry as well as matching constraints. An example of this resizing is shown in figure 14. This technology is based on the above mentioned device generators and a specialized analog compactor that can handle analog constraints.

Then finally there is the digital intensive layout, where the trend is towards full synthesis through place and route tools.

In principle the same place and route tools can be used for the toplevel layout involving RF, mixed signal and digital parts. Unfortunately the constraints that current toplevel routers can handle do not easily support analog constraints like minimum power bus widths (to limit IR drops), shielding and minimum distance between sensitive analog wires and digital wires (to limit coupling).

The third area where mixed signal CAD adds significant amount of value to the design process is post layout simulation, where layout parasitics are back annotated into the design environment.

The main challenge for post layout simulation is to limit the parasitic information to the relevant sections to still be able to extract relevant information on performance in a reasonable amount of simulation time. Since the interconnects are only defined in layout, back annotation of electrical interconnect parameters (lumped capacitance to ground, coupled capacitance between nets, distributed rc, and coupled distributed rc) and post layout simulation are the only means to verify their impact on the performance. It is important also extract the coupled capacitance between nodes as opposed to the more common capacitance lumped to ground. Especially with the process that use multiple layers of interconnect (4, 5 or 6), the line to line capacitance can be significantly bigger than the lumped capacitance to ground. Parasitic line resistances allow the simulations of de-Q'd elements and ir drops in power supplies. Once these parasitics are extracted, they can either be back annotated to the schematic, allowing the designer to check their magnitudes, but more importantly a post layout netlist can be generated. This netlist can then be considered another model that the designer can choose to do final verification that the circuit meets the specification.

5. CONCLUSION

In this paper we discussed in detail the elements that are needed in a design environment to design integrated systems. The need for a design methodology that enforces correct data entry in a common database, that then can be used by all engineers to design and verify their specifications using a top down approach is explained. The importance of behavioral simulation tools is shown to manage the extra complexity that comes with integrated circuit design. But also the accuracy of the models can not be neglected to avoid over design. Finally the use of layout generators at different levels can significantly reduce the layout phase. The combination of all these elements allow the design to be verified more thoroughly during the design phase, reducing the time to market and increasing the chance for a first time success.

REFERENCES

[1] A. Rofougaran, G. Chang, J. Rael, et al. "A Single Chip 900 MHz Spread-Spectrum Wireless Transceiver in 1um CMOS", IEEE Journal of Solid State Circuits, April 1998

[2] J. Rael, A. Rofougaran, A. Abidi et al., "Design Methodology used in a Single-Chip CMOS 900MHz Spread-Spectrum Wireless Transceiver", *IEEE Design Automation Conference*, 1998.

[3] S. Jen, C. Enz, D. Pehlke, M. Shroter, "Accurate Modeling and parameter Extraction for MOS Transistor Valid up to 10 GHz", *IEEE Transactions on Electron Devices*, Vol. 46. No. 11, November 1999, p2217-2225.

[4] C. Enz and Y. Cheng, "MOS Transistor Modeling for RFIC Design", *IEEE Journal of Solid State Circuits*, Vol. 35, No 2, February 2000, p186-201.

[5] P. Miliozzi, K. Kundert, K. Lampaert, P. Good et al., " A Design System for RFIC : Challenges and Solutions", *to be published IEEE Proceedings*, Dec 2000.

[6] K. Kundert, J. White, A. Sangiovanni-Vincentelli, "Steady-State Methods for Simulating Analog and Microwave Circuits", *Kluwer Academic Publishers*, 1990.

[7] K. Kundert, "Introduction to RF Simulation and its Application", *IEEE Journal of Solid State Circuits*, Vol. 34, No. 9, September 1999.

[8] K. Lampaert, G. Gielen, W. Sansen, " Analog Layout Generation for Performance and Manufacturability", *Kluwer Academic Publishers*, 1999

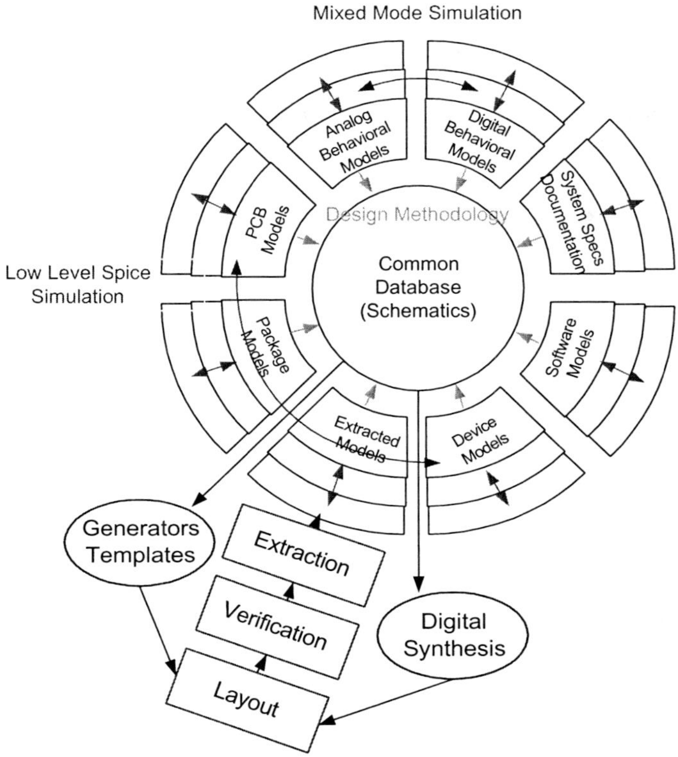

Figure 1 : Integrated System Design Methodology

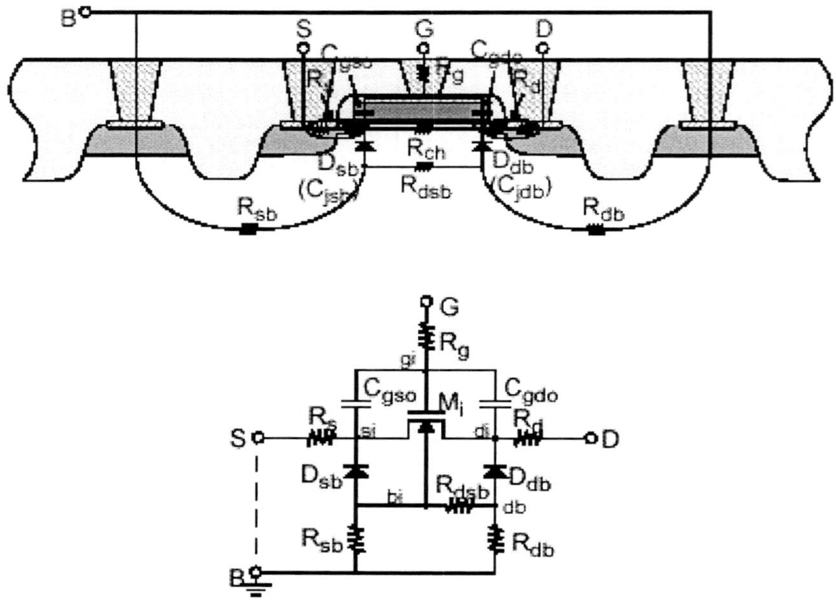

Figure 2 : Composite RF MOS Model

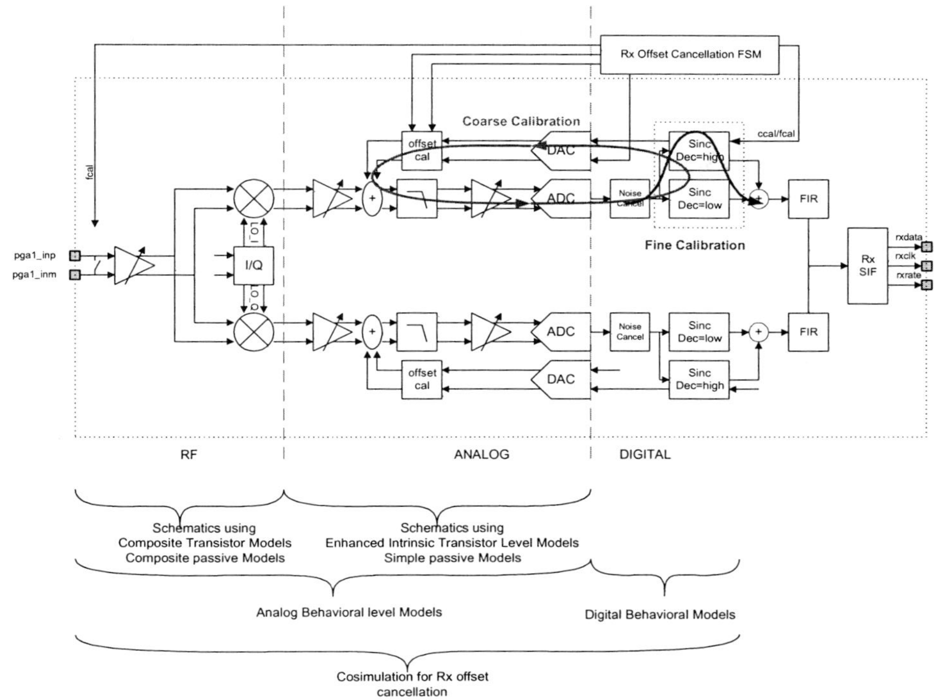

Figure 3 : Receive Offset Cancellation Circuit

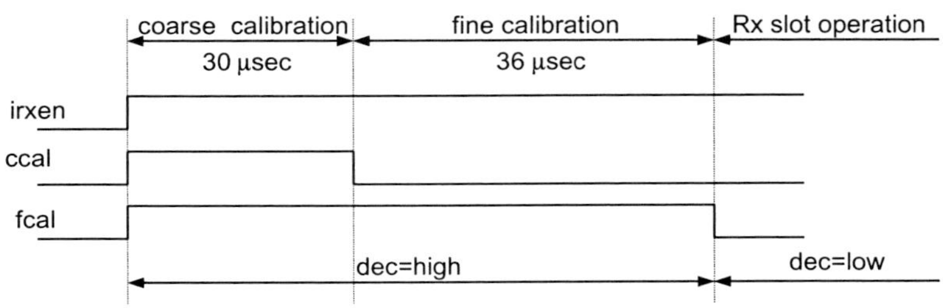

Figure 4 : Receive Offset Cancellation Loop Timing

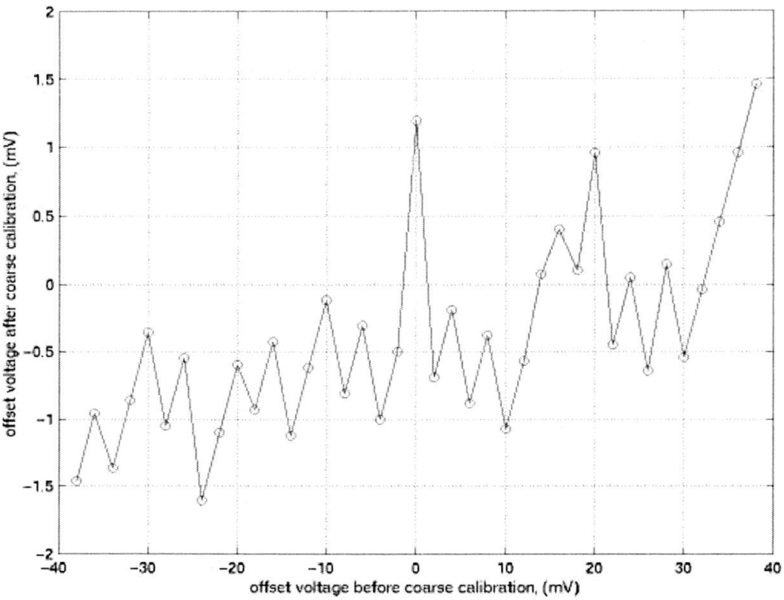

Figure 5 : Coarse Offset Calibration Characteristic

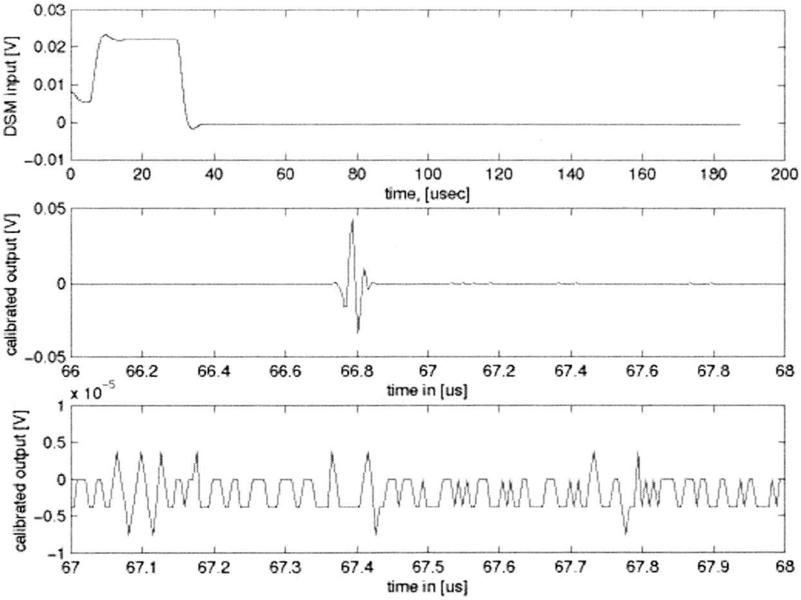

Figure 6 : Fine and Coarse Offset Cancellation

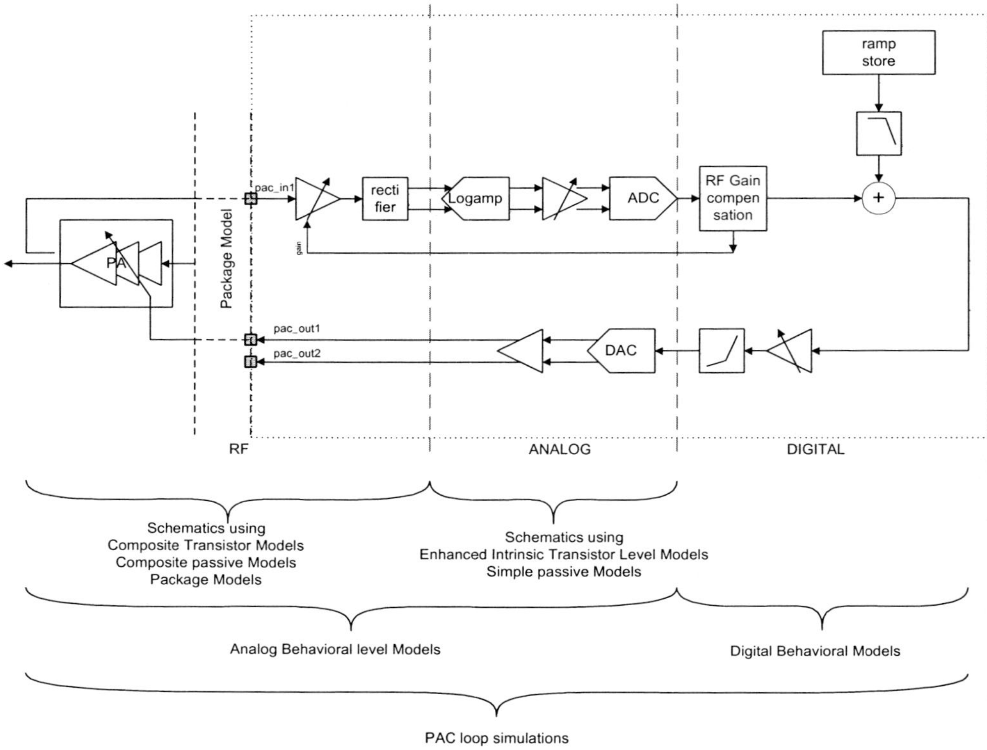

Figure 7 : PAC loop

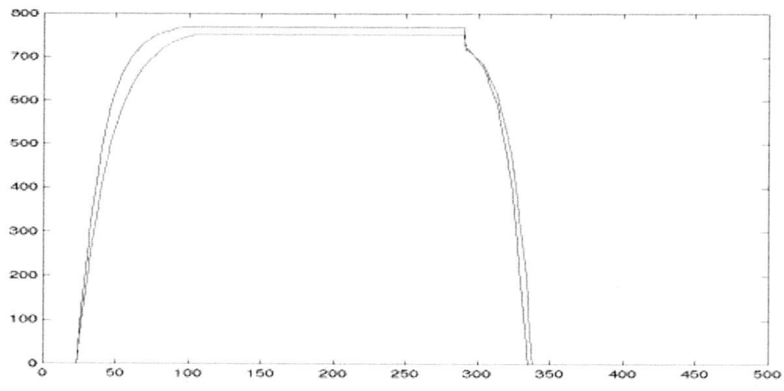

Figure 8 : Influence of Low Pass Filter Pole

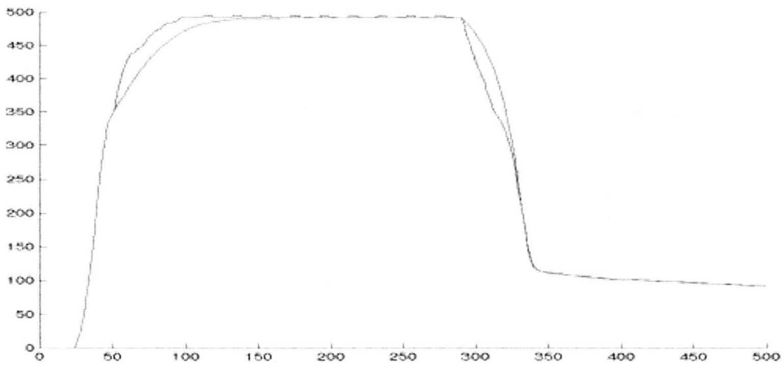

Figure 9 : Influence of Power Shaper

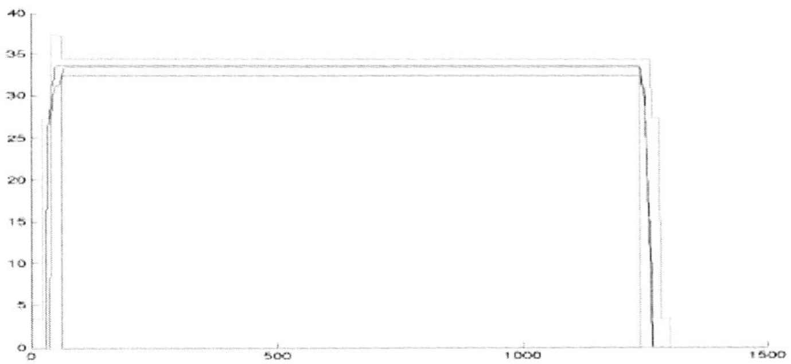

Figure 10 : PAC loop behavior versus Transmit Mask

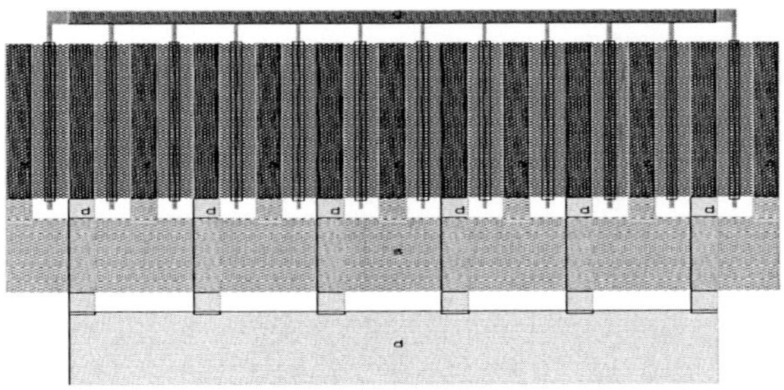

Figure 11 : Transistor drain and source are generated based on the designer specified current

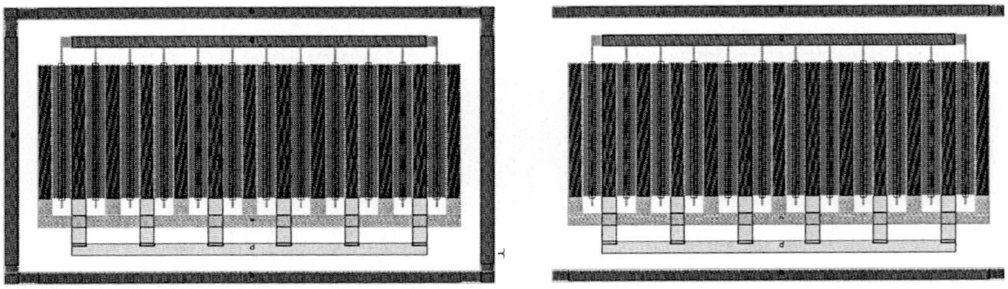

Figure 12 : Guard ring style specified by the designer

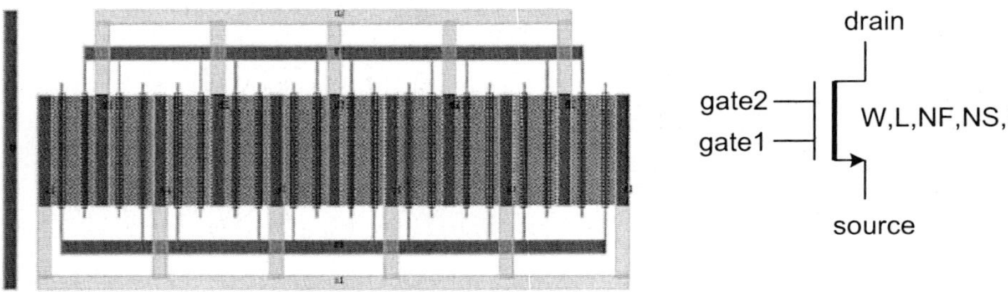

Figure 13 : Cascode or dual-gate subblock generator

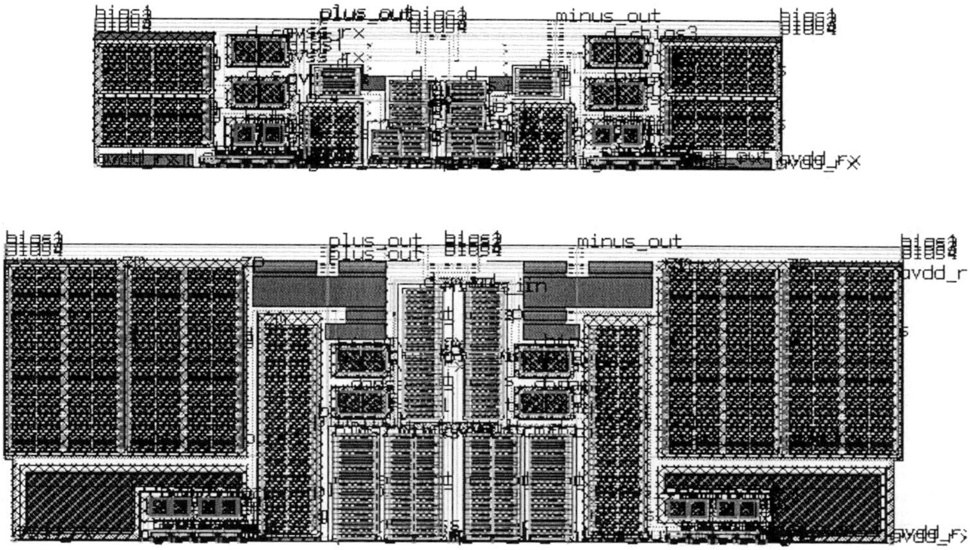

Figure 14 : Automatic amplifier layout generation based on templates

MIXED MODE SIGMA-DELTA ADC DESIGN FOR HIGH QUALITY AUDIO

Giovanni CESURA, Alessandro VENCA, Vittorio COLONNA, Gabriele GANDOLFI, Sandro DALLE FESTE and Rinaldo CASTELLO*

ST Microelectronics, Pavia, Italy

*Department of Electronics University of Pavia, Italy

ABSTRACT

The design of Sigma-Delta ADC embedded in large mixed analog and digital systems is critical especially when high quality audio performance must be retained. Idle tones reduction and noise modulation suppression can be achieved through proper architecture choice, building block design and careful layout.

1. INTRODUCTION

Modern scaled down VLSI technologies open up the possibility to implement on the same silicon many functional blocks in a cost and power efficient way. To fully exploit the potential of such systems on one chip approach the front-end blocks that connect analog and digital part (A/D and D/A) must be carefully designed taking into account that they will operate embedded in a harsh environment.

In high quality audio systems A/D converters based on the use of oversampled sigma-delta ($\Sigma\Delta$) modulators followed by decimators are widely used to achieve the requested performances. In these systems not only a very high dynamic range (in excess of 100dB for consumer and of 120dB for professional on 22.05kHz signal bandwidth) is required but also a distortion-free spectrum is mandatory.

R. J. van de Plassche et al. (eds.), Analog Circuit Design, 229–246.

To achieve a large dynamic range, many modulator topologies are possible. Among them, at a first sight, those that use a single bit quantizer are more promising because single bit quantizers have the undisputed advantage that they exhibit linear characteristic by having only two quantization levels. Having fixed the quantizer resolution to one bit, only two parameters are left in order to get the desired *DR*, namely the oversampling ratio (*OSR*) and the modulator order (*L*). Unrespectively to the actual architecture, the linear analysis of the ΣΔ modulators tell us that more than 100dB of dynamic range can be achieved, for example, with a second-order modulator oversampled 256 times or with a third-order modulator oversampled 128 times.
Therefore the architecture choice must be based on a different criterion.
Moreover the linear analysis is based on the assumption that the quantization error from ΣΔ modulators is white. This is never true especially for those modulators employing single bit quantizers. Indeed, for dc inputs, the quantization error is periodic, resulting in the so called idle channel tones [1-8]. The spectrum contains tones whose frequency and amplitude are a function of the input dc level. Also for ac inputs the quantization error is periodic and tones whose frequencies are related to the signal frequency and amplitude appear in the modulator output spectrum.
The presence of tones in high quality audio devices is detrimental since human ear is very sensitive to these periodicity in the sound.
Therefore, for these kind of applications, the modulator architecture less prone to the generation of tones in the output spectrum must be used. In large mixed-mode systems, like those in which the ADC will be used, not only the direct in band generation of tone is of concern. Possibly a more serious problem arises from the generation of tones near $f_s/2$. Those tones have very high power. Any small clock coupling near half the sampling frequency due to the presence of large digital circuitry can demodulate these tones down into base band resulting in an unacceptable spiky spectrum.
In the following sections three modulator topologies namely, a second-order single loop, a third order single loop and a third-order 2-1 cascaded, all capable of achieving 16+ resolution, will be analyzed

from the point of view of their tones generation properties. The spectrogram method will be used as a powerful tool to evidentiate the presence of idle channel tones. The 2-1 cascaded architecture will be shown to be the most suitable for high quality mixed-mode audio systems due to its lower inherent tonicity and smoother noise modulation. The limits of its good tonal properties will be highlighted. Measurements, that validate the choice, will be shown for a ΣΔ converter embedded in a large single chip digital radio.

2. IDLE TONES IN ΣΔ MODULATORS WITH DC INPUT

Three ΣΔ A/D modulators have been simulated to characterize them from the point of view of their tonal properties. The block diagrams of the architectures are shown in figure 1. The first one (figure 1a) is a well known second-order single loop single bit topology with integrator gains $k_{1a}=k_{1b}=0.5$; the second one (figure 1b) is a third-order single loop single bit in which the integrator gains have been set to $k_{1a}=0.2$ and $k_{1b}=k_{1c}=0.5$ respectively to assure the loop stability. The last one (figure 1c) is a third-order modulator in a 2-1 cascaded or MASH topology. In the latter architecture the two integrators with gains $k_{1a}=k_{1b}=0.5$ and the single bit quantizer in a loop constitute a second order modulator that is the first stage of the cascade; the integrator with gain $k_2=0.5$ and its single bit quantizer in feedback loop form a first-order modulator that is the second stage. The second integrator output of the first stage provides the input of the following stage as shown in figure 1c. An alternative approach would have been to use the quantization error of the first-stage as input of the second-stage; the reason why the first solution has been adopted will be clarified later.

In the cascaded architecture a digital network, also called digital noise canceller filter, combines the two bit-streams in order to cancel out the quantization error of the first-stage modulator.

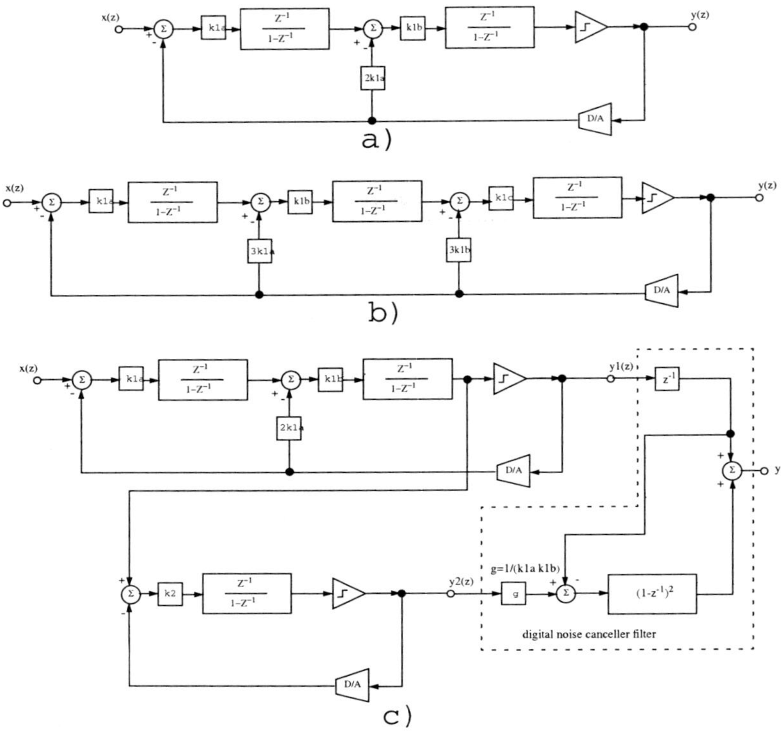

Figure 1 a)Second-order single loop modulator; b)third-order single loop modulator; c) third-order 2-1 cascaded modulator.

If a linear analysis is carried out, it is found that the dynamic range of the converter can be expressed as a function of the *OSR* and the modulator order *L* by the following equation:

$$DR = 7.78 - 10\log_{10}\left(\frac{\pi^{2L}}{2L+1}\right) + 3.01(2L+1)\log_2(OSR) \qquad (1)$$

Equation (1) predict a dynamic range well in excess of 100dB when the second-order modulator is oversampled 256 times and the third-order modulators are oversampled 128 times. However the hypothesis on which the linear analysis is based are, as previously said, never satisfied in single bit ΣΔ modulators. Therefore, is common practice to simulate the modulator architecture in the time domain with a variable amplitude sine wave as input and to plot the signal to noise ratio (*SNDR*) against the signal amplitude. The information of signal and noise power are usually recovered from a FFT of the output bit-stream. The *SNDR* vs. input amplitude of the three modulator topologies are sketched in figure 2.

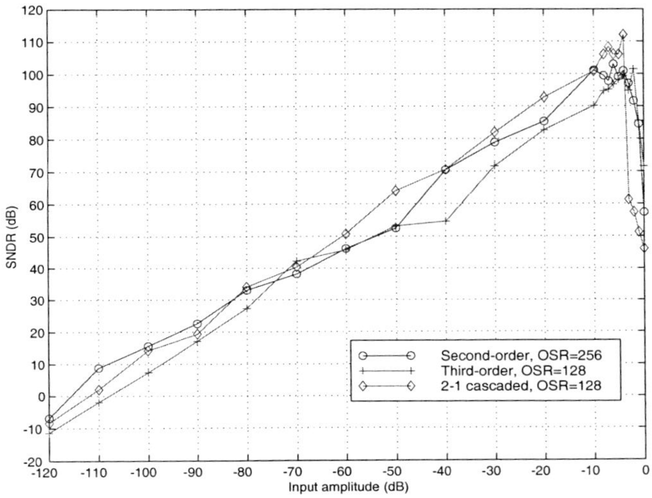

Figure 2 *SNDR* vs input amplitude for the three modulator architectures, the input frequency is 5kHz.

Since the information given by such a plot is of integral nature the knowledge of the presence of idle tones that can possibly carry most of the quantization noise power is completely lost. It is, therefore, evident that another type of simulation (and measurement procedure) must be used.

To reveal the presence of tones a slowly varying sawtooth signal is presented at the input of each modulator architecture. The resulting output bit-stream sequence has been analyzed using the spectrogram method. Using this method, FFT's made at different time intervals and therefore at different dc input are put side by side in a time-frequency-intensity plot. In this way, tones that will be difficult to perceive through a normal spectral analysis are visible thanks to the characteristic pattern they leave on the plot.

Figure 3a shows the spectrogram of in-band frequencies for the second-order single bit modulator of figure 1a. As can be seen the pattern left by the tones on the plot is clearly evident.

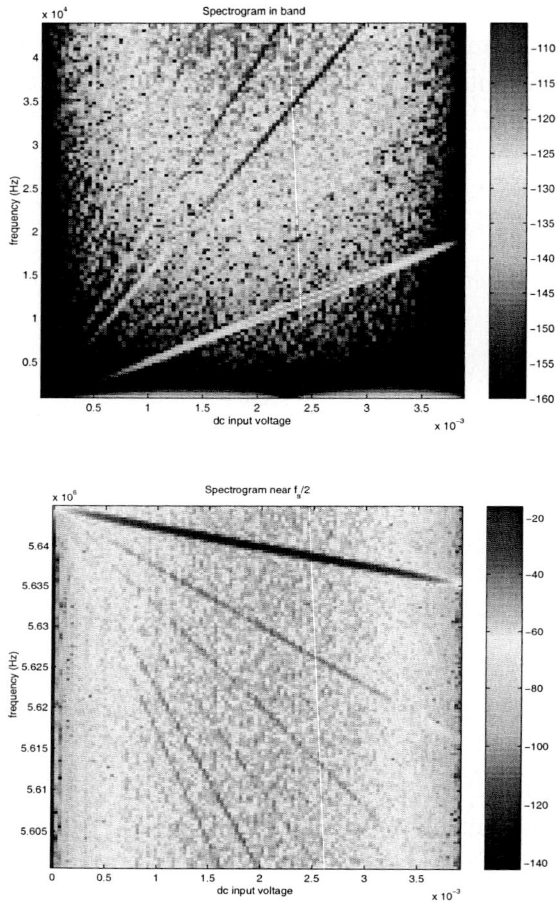

a)

b)

Figure 3 a)In-band and b)out of band spectrograms for a second-order single loop modulator.

From the analysis of pattern noise in delta modulation of Iwersen [2], the subsequent study of Candy and Benjamin [1] for continuous time first order modulator and the comprehensive work of R. M. Gray [3] it appears that, when the input is a dc, the bit-stream spectrum is discrete with tones at frequencies:

$$f_{tones} = f_s < k\left(\frac{1}{2} + \frac{|v_{dc}|}{\Delta}\right) > \quad , \quad k=1,2,... \qquad (2)$$

where v_{dc} is the input dc voltage, Δ is the quantizer step and <r> denotes the fractional part of r, i.e. r modulo 1.
From the spectrogram of figure 3 it can be shown that the tones follow the equation (2) with k=1,2,3.
Figure 4 shows a 3D plot of the in-band spectrum of the second-order modulator as a function of the dc offset at the input of the modulator; again the first tone travelling from dc to th signal bandwidth f_b is apparent.

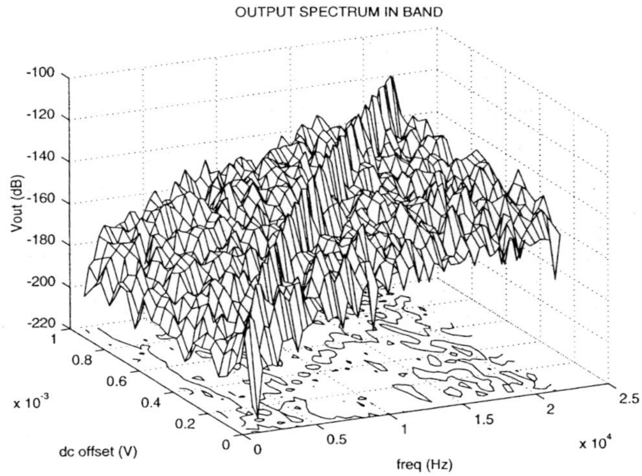

Figure 4 3D plot showing the tone travelling in the signal bandwidth for a second-order modulator.

As said in the introduction, in large mixed mode systems, the presence of tones near $f_s/2$ is an even more serious problem since those tones have very high power and their demodulation in the signal band due to digital clock coupling or intermodulation can, in principle, be much more destructive than direct in band tone generation. Figure 3b, shows the presence of high frequency tones in the spectrogram of the second order modulator.
For the third-order single loop modulator the same spectrogram are plotted in figure 5a and 5b.

Also in this modulator the pattern left by the tones are visible. The frequency of the highest in-band and out-of-band tones are well in agreement with the theoretical value predicted by equation (2).
The 2-1 cascaded architecture of figure 1c has been also simulated with a dc input signal and, as for the other cases, using ideal integrator blocks in order to achieve a complete cancellation of the first stage quantization errors. The results are shown in figure 6a and 6b. It is clear that the spectrograms are much more smooth than for the others single loop architectures.

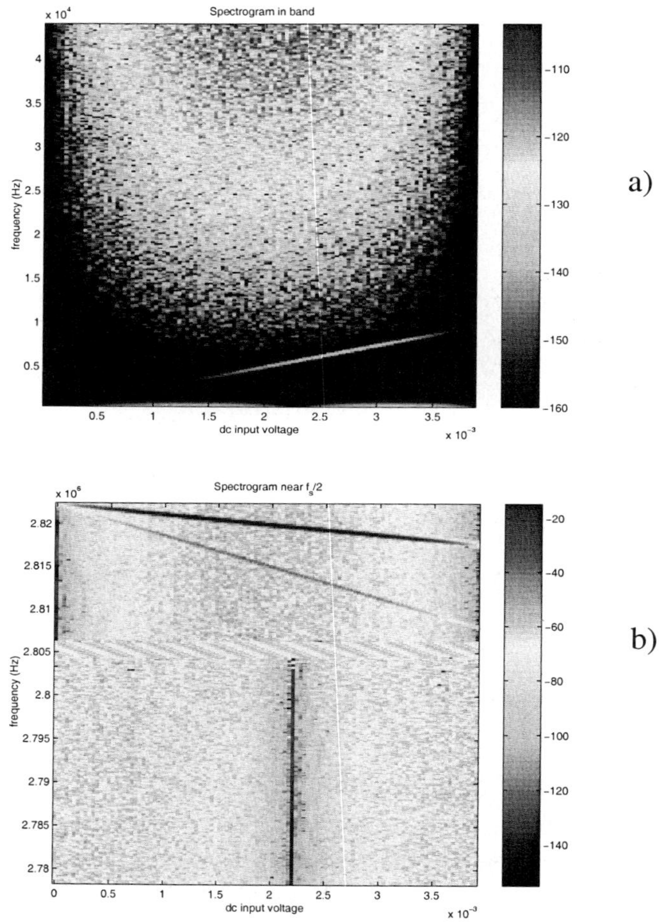

Figure 5 a)In-band and b)out of band spectrograms for a third-order single loop modulator.

This result is in agreement with the theoretical study of Wong and R.M. Gray [5] who have demonstrated that the quantization error that appears to the output of a 1-1 cascaded structure is asymptotically white, uniformly distributed and uncorrelated with the input. It confirms also many empirical observations made on other cascaded topologies, at least for the in-band tonal behavior [8, 9, 11,12 13]. Moreover, the absence of the tones at high frequency make this architecture more suitable to applications in large mixed-mode circuits. The limit of validity of this non tonal behavior of cascaded architectures is the assumption of perfect matching between the noise analog transfer function and the noise canceller digital filter.

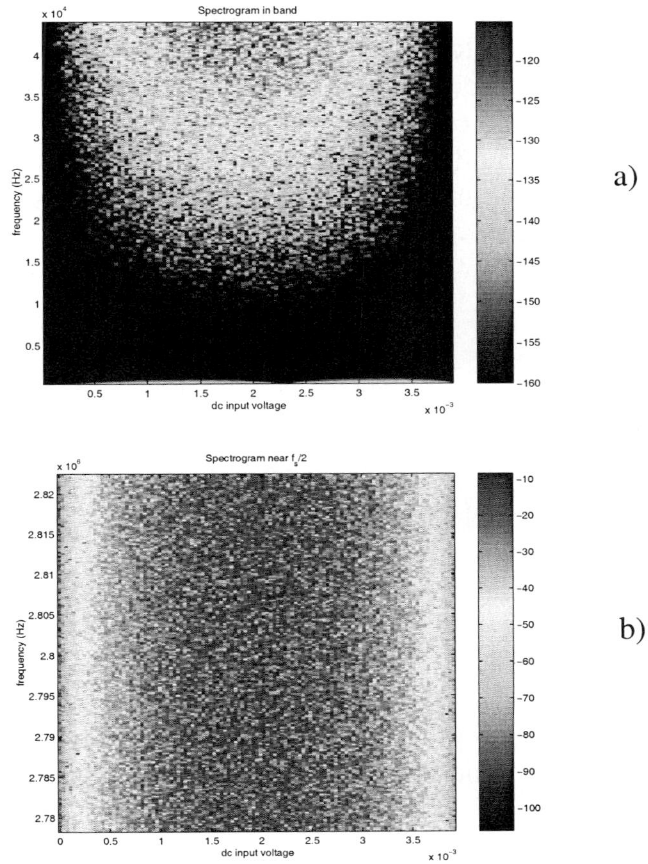

Figure 6 a)In-band and b)out of band spectrograms for a third-order 2-1 cascaded modulator

If this matching is not perfectly achieved then a fraction of the quantization noise of the first stage leaks to the output causing the cascaded architecture not only to have less dynamic range but also to appear tonal. The extent to which this happens depends on circuit parameters such as open loop gain and bandwidth of amplifiers used in the integrators, component matching as well as on the topology of the cascaded architecture itself.

The third-order topology chosen is the one that is less sensitive to analog non idealities and therefore is more robust against tonicity [10, 14].

Extensive behavioral simulations incorporating circuit non idealities have been done on the selected 2-1 architecture and demonstrate that at *OSR*=128, an open loop gain of 80dB , a gain bandwidth product of 35MHz and a coefficient accuracy of more than 0.3% are good enough to keep the leakage of tones under the total output noise floor. Under this assumption the 3D plot of the in-band spectrum as a function of the dc offset at the input of the modulator looks like that of figure 7 where no tones are evident.

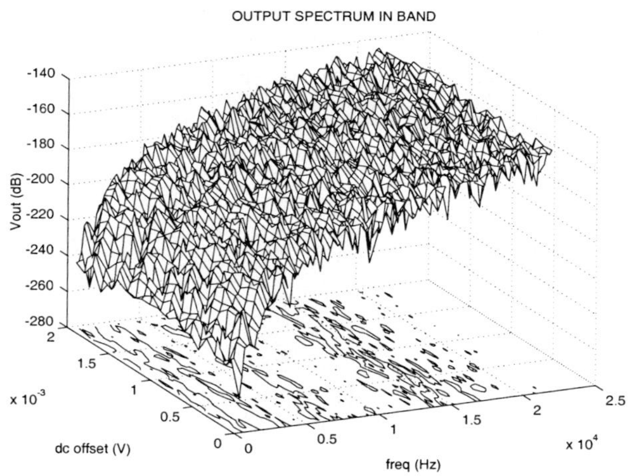

Figure 7 3D plot showing the absence of tones in the signal bandwidth for a 2-1 cascaded modulator.

3. IDLE TONES IN ΣΔ MODULATORS WITH SINUSOIDAL INPUT

Also sinusoidal inputs cause tones to appear in the modulators output [4, 7]. This can be easily inferred since at high oversampling ratio the sinusoidal input can be thought as a slowly varying dc voltage. Figure 8 shows the output spectrum for a second-order single loop and the 2-1 cascaded architecture with a -6 dB 6kHz sine wave as input. Also for sinusoidal signal the cascaded architecture have a smoother output spectrum even at frequencies close to half the sampling rate.

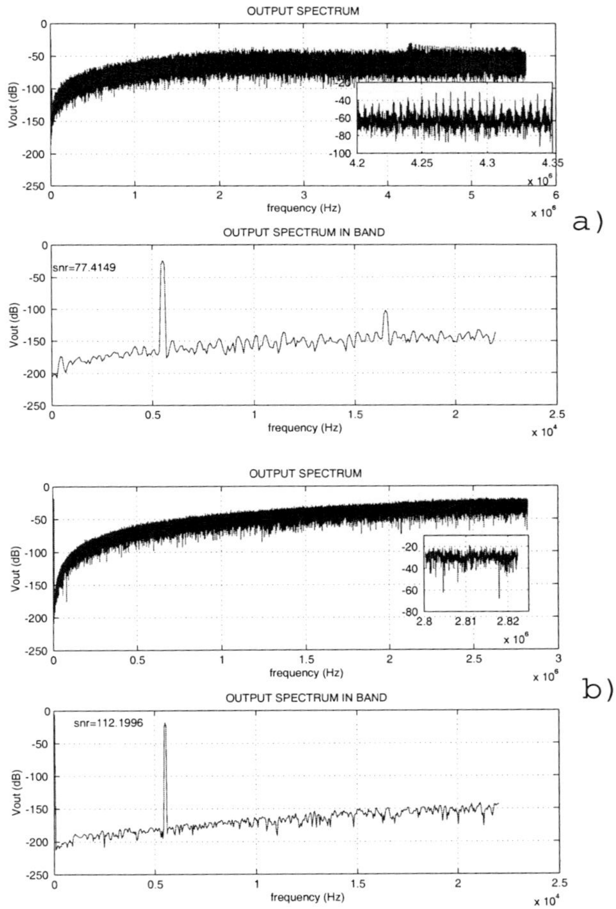

Figure 8 Comparison between second-order a) and 2-1 cascaded b) output spectrum when the input is a -6 dB 6kHz sine wave.

The presence of tones in the signal bandwidth can manifest itself through the so called noise modulation; i.e. the variation of the in-band rms noise level with respect to a dc or ac input level.
Figure 9 compares the noise modulation for the three architectures examined in this paper for an input signal frequency of 5kHz. It is worth to note that human ear is able to perceive a noise modulation of few dB's that therefore should be avoided in high quality audio systems.
Also this simulation confirms the better performance of the cascaded architecture.

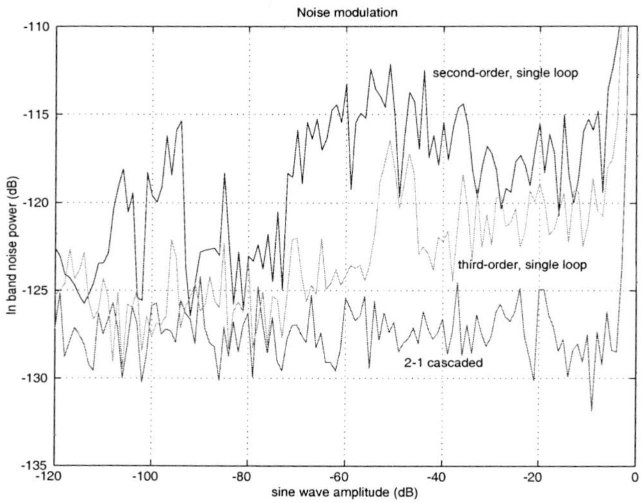

Figure 9 Noise modulation for the three different architectures.

4. 2-1 CASCADED IMPLEMENTATION

The 2-1 cascaded modulator has been implemented in a 0.35μm CMOS double-poly five-metal technology. The simplified circuit diagram of the fully differential switched capacitor modulator is shown in figure 10. By selecting k_{1a} and k_{1b} equal to 1/2 allows a single switched capacitor pair to be used for both input and D/A signals making the difference operation ratioless. As a result, the sensitivity due to component mismatch is further reduced. In addition, the shared input capacitor at the first stage provides the lowest *kT/C* noise.

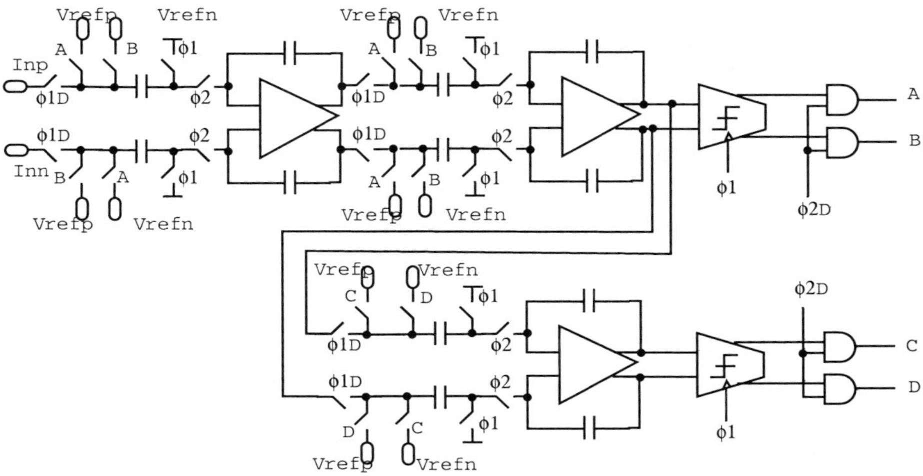

Figure 10 Modulator block diagram.

Since the noise of the first integrator dominates, special care must be taken in the design of this block to improve the performance of the overall modulator. Once the value of the sampling capacitor is chosen to achieve the 16 bit *DR* plus some margin, the bandwidth of the opamp is made large enough to ensure adequate linear settling.

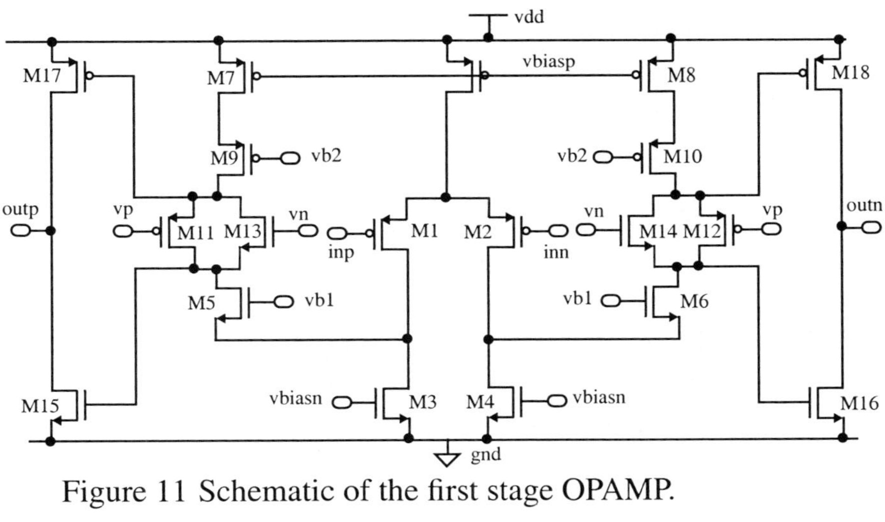

Figure 11 Schematic of the first stage OPAMP.

The schematic of the first stage opamp is shown in figure 11. It consists of a folded cascode PMOS input stage (M1,M2,... M10) followed by a class AB output stage (M11,... M18) that gives the requested slew rate to avoid loss in dynamic range due to non linear settling. The MOS transistor M11,M13 and M12, M14, connected back to back form a floating battery and gives the push-pull capability to the output stage. The *1/f* noise of the opamp is kept low thanks to the use of a large area PMOS input differential pair M1 and M2 whose aspect ratio is 800/0.8. The complete modulator draws a current of 4mA out of a 3.3V power supply.
The modulator is used in a larger mixed mode circuit whose layout is sketched in figure 12. The complete chip includes: three stereo DAC channels, two stereo ADC channels, two DSPs for digital audio effect and interfaces control, three ROMs, two RAMs, a PLL and clocking generation block and sparse logic. The total power consumption is 1.4W.
The chip has dedicated power supply pads for the PLL, the DAC block the ADC block and the digital block. On top of this, the analog substrate is connected to a dedicated clean ground pad.

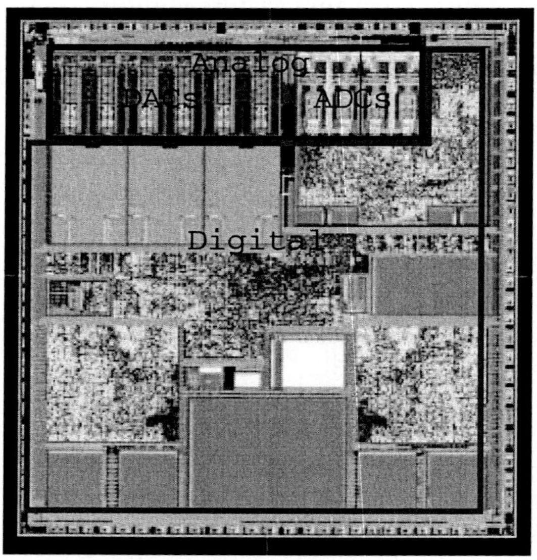

Figure 12 Chip layout

Analog and digital part within each block (ADC & DAC) are separated by a p+ guard ring connected to the clean ground pad. The stereo channel itself is then surrounded by a double isolation ring.

5. EXPERIMENTAL RESULTS

Measurements have been performed on the ADCs embedded into the complete chip while the all system was fully working. Figure 13 compares the output spectrum for the complete third order modulator and the second order first stage both operating at an *OSR*=128 with a -60dB 2kHz sine wave input. It is evident that the second order shows tones in both high and low frequency regions that are absent in the third-order modulator.

To study the tonal behavior of the architecture a sawtooth signal at a repetition of 10Hz has been applied at the input of the modulator and a spectrogram has been plotted against the time. The result of this measurement is plotted in figure 14a for the inband frequency and figure 14b for the frequencies near $f_s/2$.

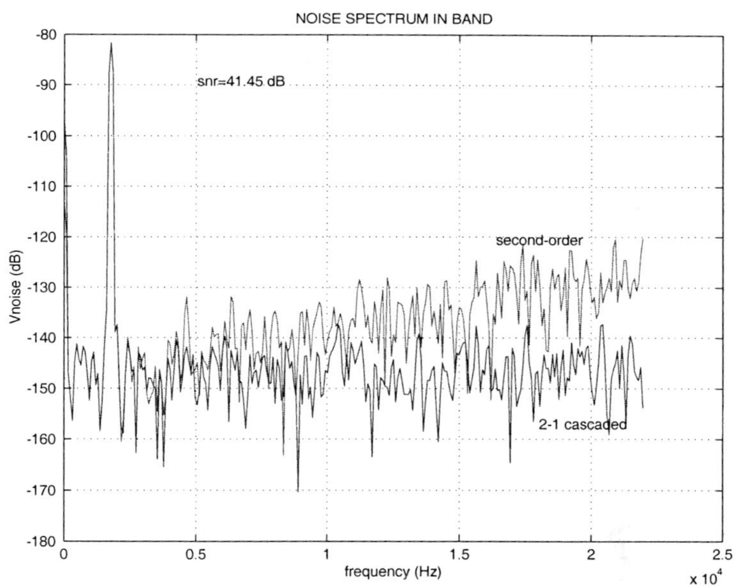

Figure 13 Measured in band output spectrum for the third-order 2-1 cascaded modulator and the second-order first stage at OSR=128 for a -60dB 2kHz input signal.

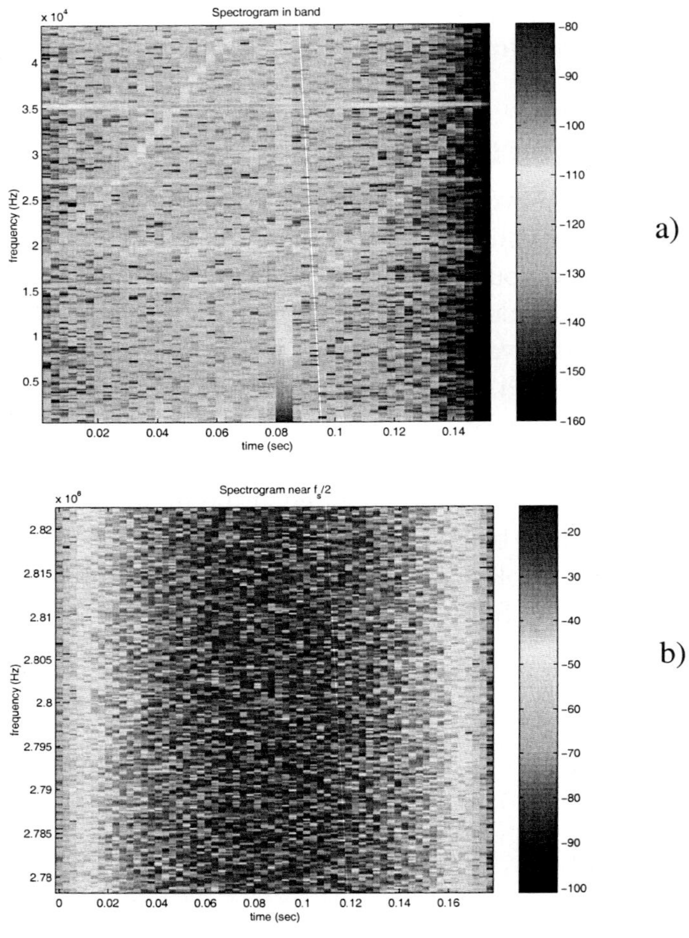

Figure 14 Measured spectrograms for the 2-1 cascaded modulator

From this spectrogram a low energy tone is perceivable in the in band frequency range; this is due to a non complete cancellation of the second order quantization noise. In the in-band spectrogram there are also tones whose frequency is not dependent on the input signal and therefore they appear as a straight line parallel to the x axis. This tones are due to the measurement setup and can be eliminated. The out of band spectrum is more flat and without peaks confirming the behavioral simulations.

6. CONCLUSIONS

The tonal behavior of Sigma Delta ADC has been revised in the view of their application in high quality audio large mixed-mode circuits.
It has been shown that cascaded architectures can provide, for these applications, a better idle channel noise reduction and noise modulation suppression. Moreover high energy tones out of the signal bandwidth, usually present in single loop architectures are absent in a properly designed cascaded modulator. This make this architecture more robust to digital clock coupling and therefore can be efficiently used in mixed-mode circuits.

7. REFERENCES

[1] J. C. Candy and O. J. Benjamin, "The Structure of Quantization Noise from Sigma-Delta Modulation", *IEEE Trans. Comm.*, vol. COM-29, pp. 1316-1323, Sept. 1981.

[2] J. E. Iwersen, "Calculated quantizing noise of single-integration delta modulation coders", *Bell Syst. Tech. J.*, vol. 48, pp. 2359-2389, Sept. 1986.

[3] R.M. Gray, "Spectral Analysis of Quantization Noise in a Single-Loop Sigma-Delta Modulator with dc Input", *IEEE Trans. Comm.*, vol. COM-37, pp. 588-599, Jun. 1989.

[4] R.M. Gray, W. Chou and P.W. Wong, "Quantization Noise in Single-Loop Sigma-Delta Modulator with Sinusoidal Inputs", *IEEE Trans. Comm.*, vol. COM-37, pp. 956-968, Sept. 1989.

[5] P.W. Wong and R.M. Gray, "Two-Stage Sigma-Delta Modulation", *IEEE Trans. Acoust. , Speech, Signal Processing,* vol. AASP-38, pp.1937-1952, Nov.1990.

[6] R.C. Ledzius and J. Irwin,"The Basis and Architecture for the Reduction of Tones in a Sigma-Delta DAC", *IEEE Trans. Circuits Syst. II: Analog and Digital Signal Processing*, vol. CASII-40, pp. 429-439, Jul. 19943.

[7] S. R. Norsworthy, R. Shreier and G.C. Temes, *Delta-Sigma Data Converters, Theory, Design and Simulation*, New York: IEEE Press, 1997.

[8] L. A. Williams III and B. A. Wooley, "Third-Order Cascaded Sigma-Delta Modulator", *IEEE Trans. Circuits Syst.*, vol. CAS-38, pp. 489-498, Mar. 1994.

[9] L. A. Williams III and B. A. Wooley, "A Third-Order Sigma-Delta Modulator with Extended Dynamic Range", *IEEE J. Solid-State Circuits*, vol. SC-29, pp. 193-202, May 1991.

[10] D. B. Ribner et al., "A Third-Order Multistage Sigma-Delta Modulator with Reduced Sensitivity to Nonidealities", *IEEE J. Solid-State Circuits*, vol. SC-26, pp. 1764-1773, Dec. 1991.

[11] K. Uchimura, T. Hayashi, T. Kimura and A. Iwata, "Oversampling A-to-D and D-to-A Converters with Multistage Noise Shaping Modulators", *IEEE Trans. Acoust., Speech, Signal Processing,* vol. AASP-36, pp.1899-1905, Dec.1988.

[12] T. Ritonieni et al., "A Stereo Audio Sigma-Delta A/D-Converter", *IEEE J. Solid-State Circuits*, vol. SC-29, pp. 1514-1523, Dec. 1994.

[13] G. Yin, F. Stubbe and W. Sansen, "A 16-b 320-kHz CMOS A/D Converter Using Two-Stage Third-Order ΣΔ Noise Shaping", *IEEE J. Solid-State Circuits*, vol. SC-28, pp. 640-647, Jun. 1993.

[14] D. B. Ribner, "A Comparison of Modulator Networks for High-Order Oversampled ΣΔ Analog-to-Digital Converters", *IEEE Trans. Circuits Syst.*, vol. CAS-38, pp. 145-159, Feb. 1991.

MIXED-MODE TELECOM DESIGN

Domine M.W. LEENAERTS, Peter W.H. de VREEDE

Philips Research Laboratories,
Eindhoven, the Netherlands

ABSTRACT

This contribution will concentrate on several aspects of mixed-mode telecommunication design. Currently, a strong focus on single chip realisations of relatively simple wireless standards like Bluetooth can be foreseen. However, integrating RF, mixed signal and digital baseband on a single die will confront the designer with design aspects he did normally not take into account.

1. INTRODUCTION

The field of telecommunication design is recently confronted with a (largely unanticipated) explosive growth in wireless mobile phones. The success of standards like GSM and DECT forced industry to put more research in realising more system functionality on a smaller size, while consuming less power and reducing the overall price of the product [1,2]. For instance, in 1994 the second generation of GSM radio came on the market. This monoband system had a total of 270 components. Five years later, this number was reduced to 130 components while the functionality was upgraded to a dual band system (see Fig. 1). A reduction of 50% in components, 50% in PCB area and more than 60% in RF PCB area had been achieved.

R. J. van de Plassche et al. (eds.), Analog Circuit Design, 247–265.

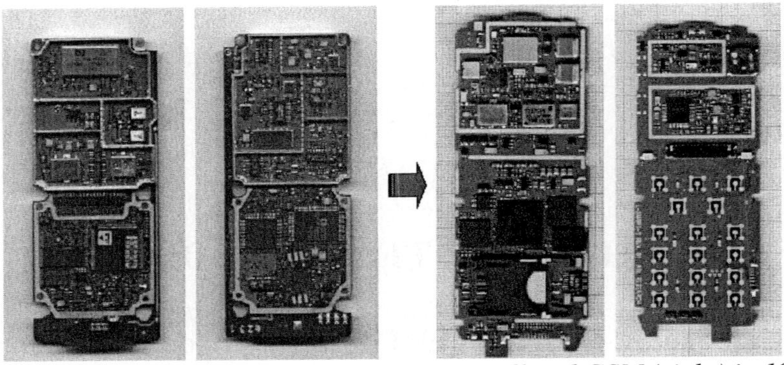

Fig. 1. Monoband GSM in 1994 (left) and Dualband GSM (right) in 1999

This trend of increasing integration density can be observed in other wireless standards too, like for instance the DECT standard (Fig. 2).

The industry has embraced Bluetooth as the most likely candidate to have the highest integration level possible, i.e. a single chip. Obviously, the ultimate goal is to integrate the complete functionality on a single die, leaving only the antenna and battery as external components. However, intending to use Bluetooth chips in (notebook) computers and other devices like cell phones, it has to be cheap (let's say $10) and powerful, and all of that on one chip. Although within this year the first chipsets for Bluetooth will become available, integrating the RF front-end together with large parts of the baseband on a single die, they are far from the final target. Nevertheless, it is for sure that mixed-mode telecommunication design has become the issue in the near future.

If mixed-mode telecommunication systems have to be realised, designers will be confronted with several problems. Having CMOS as

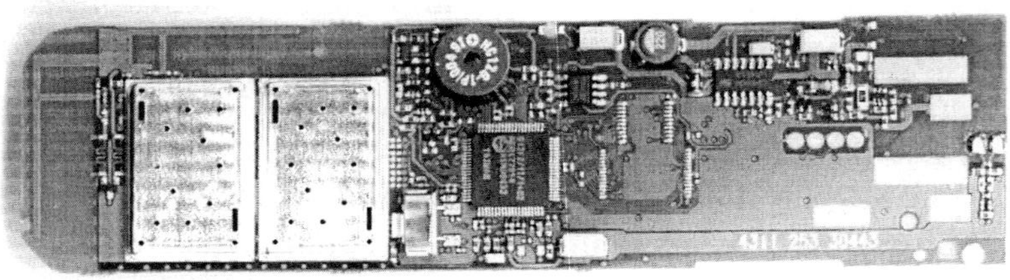

Fig. 2. Philips DECT phone with highly integrated front end.

technology in mind, standard MOS device models predict digital and analogue circuit behaviour accurate enough, but this is certainly not the case for the RF performance. Also designers will be confronted with the fact that the standard library blocks like bondpads and ESD devices are more likely specified for digital purposes rather than for frequencies beyond 1 GHz. Then there is the problem of substrate bounces. Especially for a single die solution this problem is serious because digital, analogue and RF share the same substrate. For instance, the reference sensitivity level for Bluetooth is defined at –70 dBm, leading to a 70μV input signal over 50 Ω. Substrate noise can be much larger than this value, causing serious problems for the LNA at the input of the receiver system to perform well.

This contribution will focus on several design aspects that will show up when trying to integrate RF, analogue and digital circuitry on a single die. This paper is not intended to be complete, but tries merely to confront the reader with design related issues that pop up in mixed-mode telecommunication design. The Bluetooth standard will be used as an example.

2. TECHNOLOGY CHOICE AND TRANSISTOR MODELS

Obviously, the technology choice is an important one. The best choice for the digital baseband, consisting of microprocessors and memory, is clearly CMOS technology. This option is also suitable for the mixed signal parts, like the AD- and DA-converters, PLLs and synthesizers. However, CMOS is not trivial as technology choice for the RF part. Most RF circuits are designed in bipolar-type of processes, CMOS has not yet been proven to be a good candidate for high performance consumer RF. On the other hand, going for single die leaves us no other choice than implementing also the RF part in CMOS. Due to the more relaxed radio specifications in the Bluetooth than in the GSM standard, indeed the Bluetooth standard is the most likely candidate to be integrated on a single die at this moment. Having chosen for CMOS technology, designers sometimes have the possibility to choose between high or low ohmic substrate. The choice is clearly related to substrate bounces and will be discussed later on.

Going for single die raises the question if this is the best way to go from silicon cost point of view; why not having two dies in a single package or simply two separate chipsets? Predictions in cost show that having a high yield (e.g. 95%) for digital but a low yield for RF (e.g. 80%), a single die solution is 15% more expensive than a two die solution at the moment. Clearly an improvement of yield in RF circuitry is mandatory to make a single die solution profitable for consumer applications. At this moment, for high performance RF systems (e.g. GSM, UMTS) a single chip solution is not a realistic target, simply because cost is the main driver and single chip will hardly be the optimum. Again, Bluetooth, as low performance RF system, might be an exception.

Having chosen for CMOS, the next issue is a MOS transistor (MOST) model predicting the behaviour at high frequencies accurately enough. For digital functionality this is most often not a problem, but the analogue RF functionality can not be predicted accurately with standard MOST models. From literature [3,4,5] it becomes clear that MOST models for RF applications must include several properties aside from the intrinsic behaviour. A few of these properties will be discussed briefly.

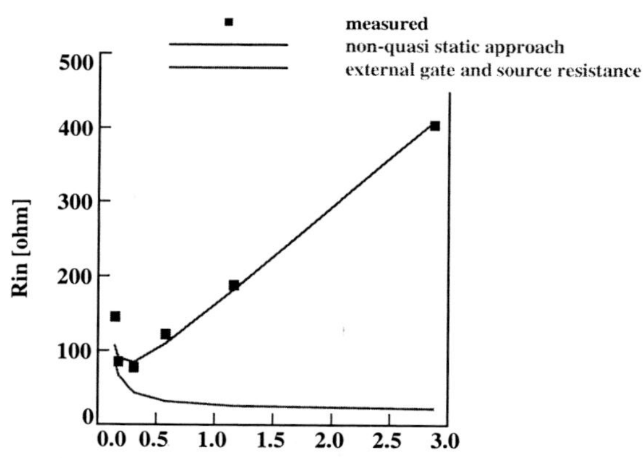

Fig. 3. Differences in measured and simulated input impedance of a MOS device.

- Resistances

The gate resistance is an important parameter and can be approximated by

$$R_g = \frac{1}{3}\frac{W}{L}R_{sq,poly}$$

The value of the gate resistance can be reduced below a few Ohm in case of a proper MOST layout. Resistances in series with drain (R_d) and source (R_s) degrade the drain current and transconductance of the MOST. The source resistance will degrade the noise performance and the maximum oscillation frequency. The drain resistance has impact on the output impedance of the device. Note that these resistances do not cover the complete input impedance behaviour, see Fig. 3.

- Non-quasi static effects

When signal frequencies are in the same order of the maximum oscillation frequencies of the MOS device, non-quasi static effects have to be taken into account. These effects result in an extra real contribution to the input impedance, which can be taken into account by adding an additional resistance to the gate resistance of approximately

$$R_{in} = \frac{1}{ng_m}$$

with n in the range of 5 to 7. The formula is valid when the MOST is operating under small signal conditions and in saturation [5]. The effect of this additional resistance on the overall input impedance can be seen in Fig. 3.

- Bulk resistances

Via the body effect device operations are influenced by signals on the bulk node. These signals come from the substrate as we will see later on. Therefore a proper bulk resistance network is mandatory to include in the intrinsic device model.

An RF MOS device model may therefore look like the one depicted in Fig. 4.

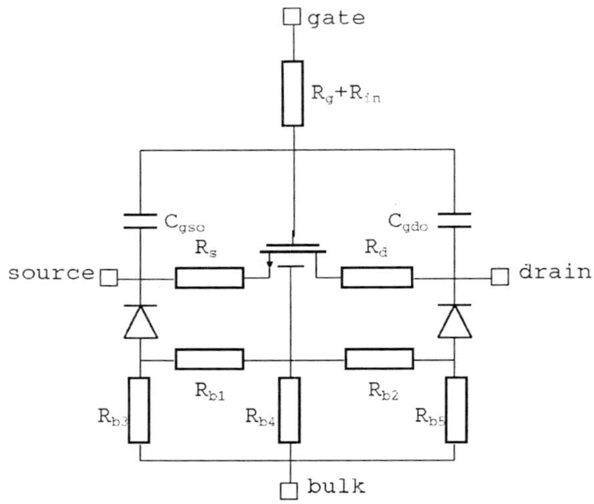

Fig. 4. RF MOST model including bulk model

3. GOING ON/OFF-CHIP: BONDPAD AND ESD

Many publications presented in literature do not mention the problem of bringing signals on and off-chip. However, certainly for RF front-end circuitry this is a major problem and if not solved with care it will deteriorate the performance of the circuitry drastically. Going on-chip means that the signals go from a printed circuit board via pins of the used package, via bondwires to the die. There the signals will first be influenced by the bondpad, connecting the bondwire to the die and then by the ESD protection device, to prevent the circuit being damaged by electrostatic discharge. Here we will concentrate ourselves on the bondpad and the ESD device.

Traditionally a bondpad used in digital design uses several or all metal layers connected to each other and is not shielded from the substrate (see Fig. 5). Although resulting in a large capacitance and high bulk series resistance, the digital performance is not influenced by the bondpad. However, for high frequencies this is not true anymore. The low Q value of the bondpad will degrade the RF performance.

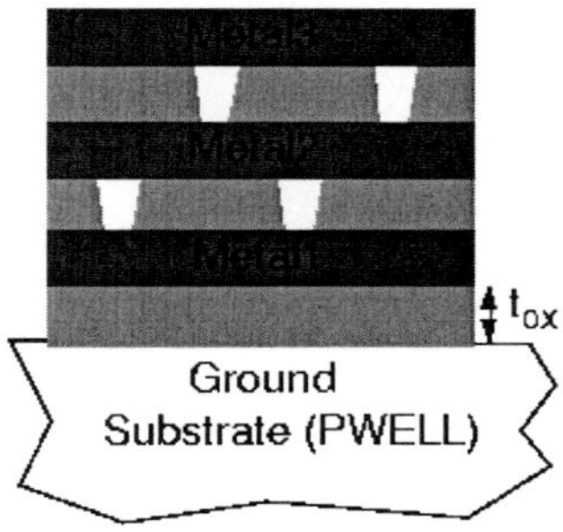

Fig. 5. 'Digital' bondpad using several layers of metal

Shielding the bondpad from substrate and using less metal layers a better performance of the bondpad at radio frequencies can be achieved (see table 1).

Table 1. Conventional (digital bondpad) and special RF bondpad

Parameter	Conventional	RF
$C_{bondpad}$ (fF)	310	85
$R_{bondpad}$ (Ω)	20	4
$Q_{bondpad}$ (@ 900 MHz)	29	520

The traditional ESD devices are not well suited for RF applications, having again a low Q-factor. Normally a grounded gate NMOST (ggNMOST) device is used, as depicted in Fig. 6.

A better performance at high frequencies can be obtained using ESD devices based on diodes as shown also in Fig. 6. When during layout special care is taken on the total resistance of the devices a better quality factor can be obtained as can be seen in Table 2. The replacement circuit of ESD devices is assumed to be a capacitor in series with a resistor. Figure 7 shows some measured performances. A crowbar structure is necessary to prevent that spikes on the power supply rail will damage the circuitry. An ESD device should at least comply with the Human Body Model (HBM), i.e. it should sustain a pulse of 2 kV and 1 Ampere current. Obviously, the voltage across the diode at 1 Ampere current should be as low as possible to prevent the

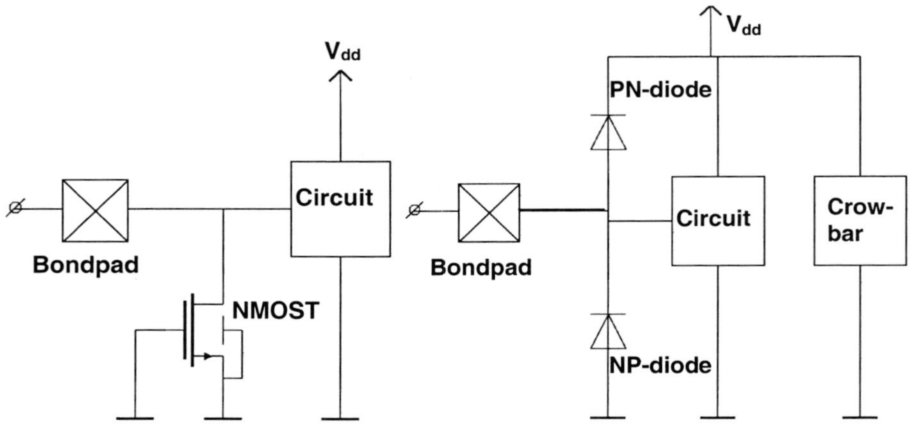

Fig. 6. ggNMOST ESD (left) and diode-based ESD (right)

Table 2. Capacitance, resistance and Q-factor for various ESD devices.

ESD device C050	C (fF)	R (Ω)	Q at 10 GHz	Q at 1.8 GHz
ggNMOST ESD	402	6	8	38
np-diode ESD	314	2	31	~ 165
pn-diode ESD	314	2	40	~ 213

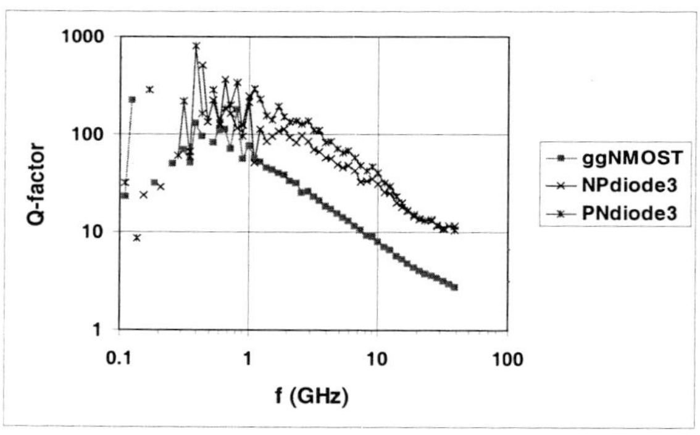

Fig. 7. Measured Q-factor of several ESD devices

gate connected to it to blow up. Figure 8 shows the measurement results during HBM test. The voltage drop is only 2 V. Also the leakage of the device is low, see Fig. 9. Therefore, it is possible to make ESD devices applicable to the RF band.

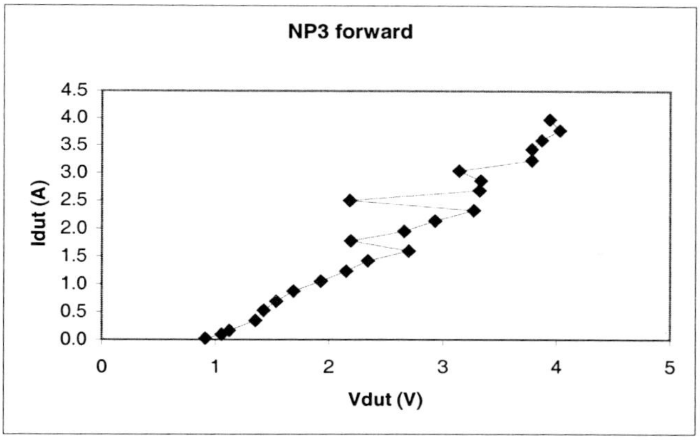

Fig. 8. Current versus forward voltage during HBM test.

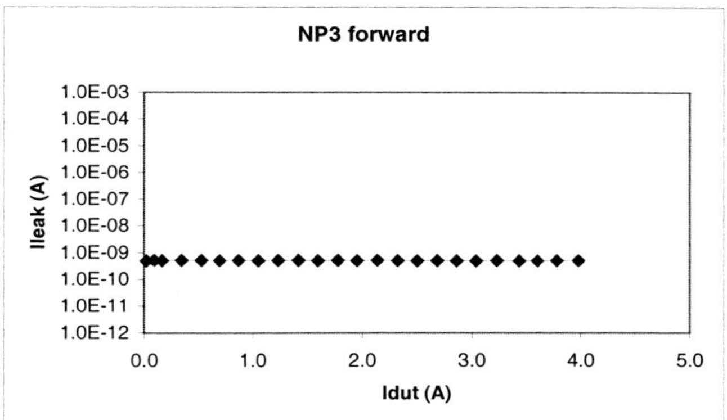

Fig. 9. Leakage current versus DUT current during HBM test.

4. SUBSTRATE BOUNCE

One major problem when bringing analogue circuitry and large digital parts together is the problem of substrate bounce. Recently, this problem has been treated in detail in literature [6,7,8]. Reference articles on substrate noise commonly confuse substrate noise reduction methods for low-ohmic versus high-ohmic bulk processes. For a low-ohmic epitaxial process guard rings do not bring the effective reduction in substrate noise as they do for high-ohmic bulk processes.

One main source of noise is the coupling from the digital circuitry to bulk through the substrate p-n junctions. Each MOSFET forms with its source and drain a p-n junction with the bulk. High speed switching activities in the digital part will be injected as substrate noise due to these junctions. If the analogue ground does not bias the bulk of the analogue part, then the body effect will couple this substrate noise directly into the analogue circuit. This effect is shown in Fig. 10 for a simple circuit [8]. Obviously, this coupling effect is more severe in low-ohmic substrate processes than in high-ohmic processes. If the analogue ground is separated from the substrate, the substrate noise on the backgate will directly modulate the drain current. Therefore, the supply of the analogue circuits should be referred to the substrate. The substrate noise is then present at the gate, source and backgate of the transistor and therefore will not (or to a lesser degree) modulate the drain current.

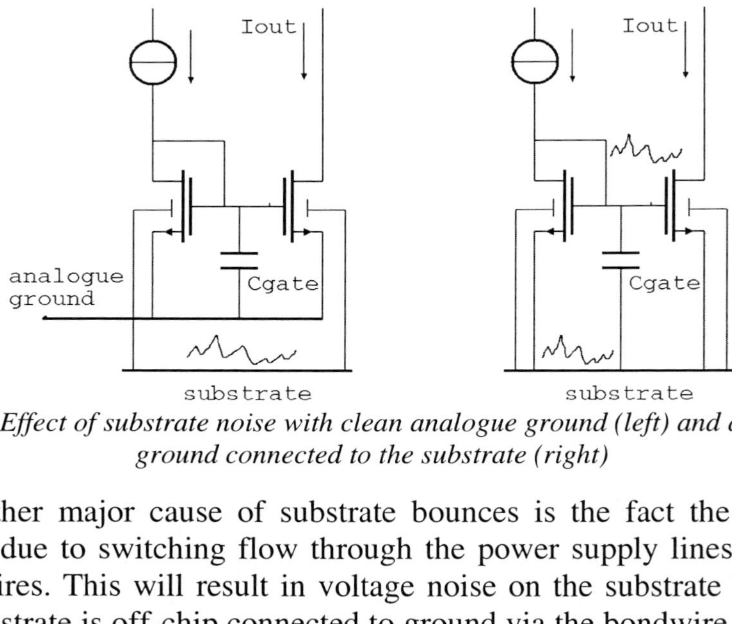

Fig.10. Effect of substrate noise with clean analogue ground (left) and analogue ground connected to the substrate (right)

One other major cause of substrate bounces is the fact the current spikes due to switching flow through the power supply lines via the bondwires. This will result in voltage noise on the substrate because the substrate is off-chip connected to ground via the bondwire as well. This mechanism is shown in Fig. 11.

One other major cause of substrate bounces is the fact the current spikes due to switching flow through the power supply lines via the bondwires. This will result in voltage noise on the substrate because the substrate is off-chip connected to ground via the bondwire as well.

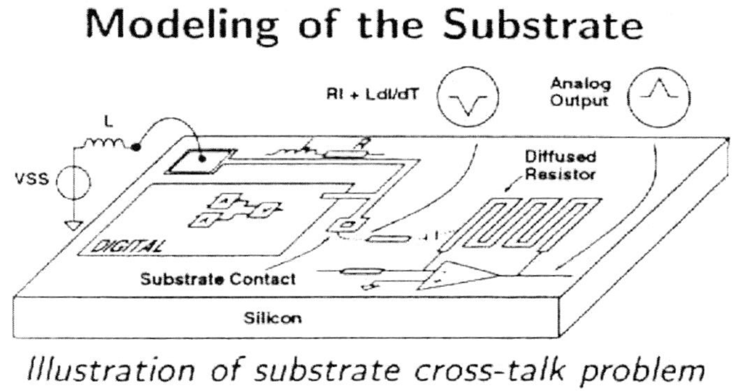

Fig.11. Substrate bounce due to bondwires and current spikes

To show that substrate noise is a serious problem, the following experiment has been carried out. The substrate noise produced by a second order low pass sigma-delta modulator including decimation filter has been measured. The sigma-delta modulator is depicted in Fig. 12.

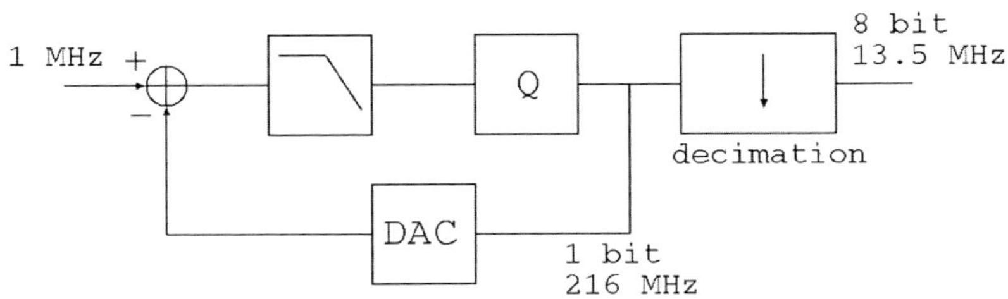

Fig. 12. Second order low pass sigma-delta modulator

The circuit is designed in low-ohmic substrate, 0.5 μm CMOS at 1.8 V power supply with separate analogue and digital supply lines. All design rules to minimise effects of substrate noise have been applied. An external clock at 12 MHz is supplied, and an internal PLL converts this clock up to 216 MHz clock signals. The I/O of the circuit is an 8 bit data stream at 13.5 MHz and a clock line at the same frequency.

Measured is the signal present at the substrate. In Fig. 13 the spectrum of this signal is shown when the power supply of the chip and the clock are turned off. Above 1 GHz, the spectrum is flat (except that we measure some interference from the GSM/DCS band).

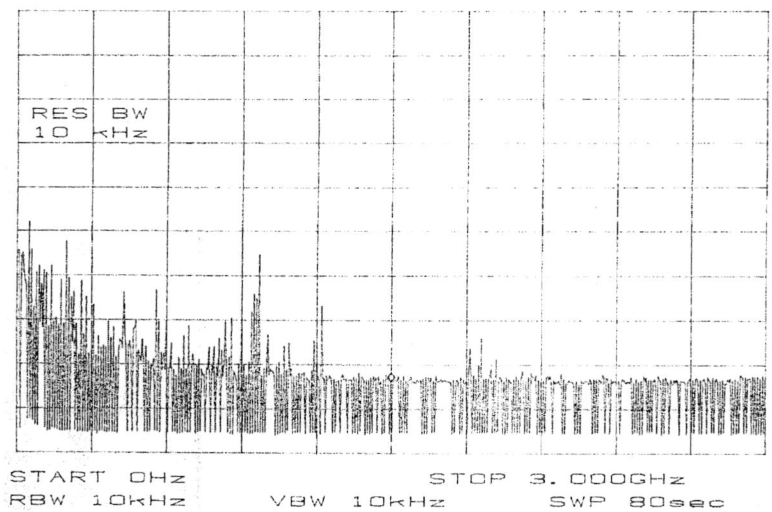

Fig. 13. Spectrum of the signal present at the substrate when system is turned off. (RL: -10 dBm, 10 dB/div)

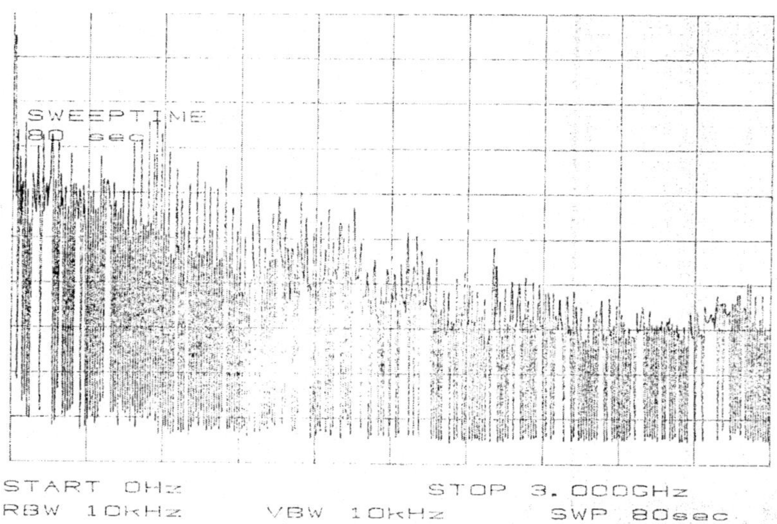

Fig. 14. Spectrum when system is active. (RL: -10 dBm, 10 dB/div)

Figure 14 shows the spectrum of the substrate signal when the system is active. Odd multiples of 13.5 MHz are visible within the whole band up to the RF frequencies. The substrate noise has been increased with 20 dB in the RF band and even more at lower frequencies. According to these measurements, the substrate noise is concentrated at multiples of the digital clock frequency, also observed in other measurements [6]. Needles to say that these clock tones will influence the performance of RF circuitry when located on the same die. Therefore, planning your digital clock frequency and analogue IF frequencies in integrated transceiver front ends is important. The bulk model, as shown in Fig. 4, can be of help in predicting the substrate bounce effects. We have test this assumption using the test circuit in Fig. 15. A clock circuit operating at 775 MHz and an LNA operating at 3 times this clock frequency have been designed in low ohmic 0.25 μm CMOS. Also a guard ring has been provided.

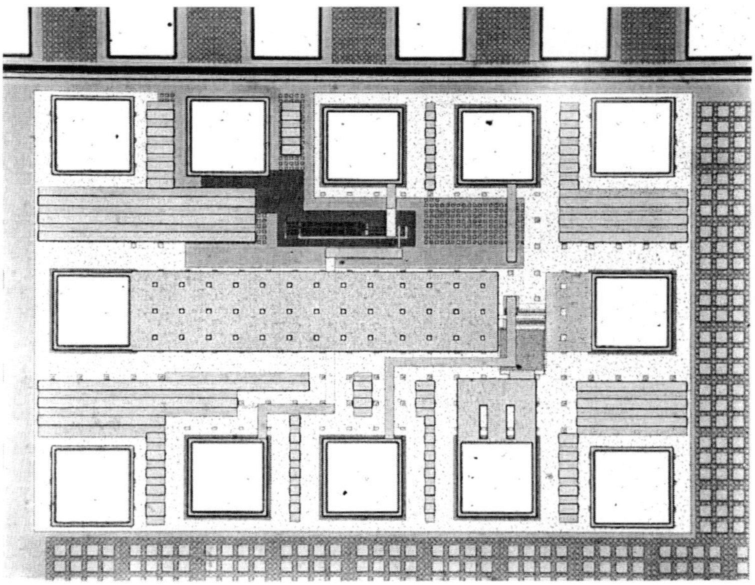

Fig. 15. Test circuit including clock circuitry (top) and LNA (right).

The measured spectrum at the substrate due to the clock operation is shown in Fig. 16 and the measured spectrum at the output of the LNA is shown in Fig.16 as well while at the input of the LNA is available a single tone also at 3 times the clock frequency.

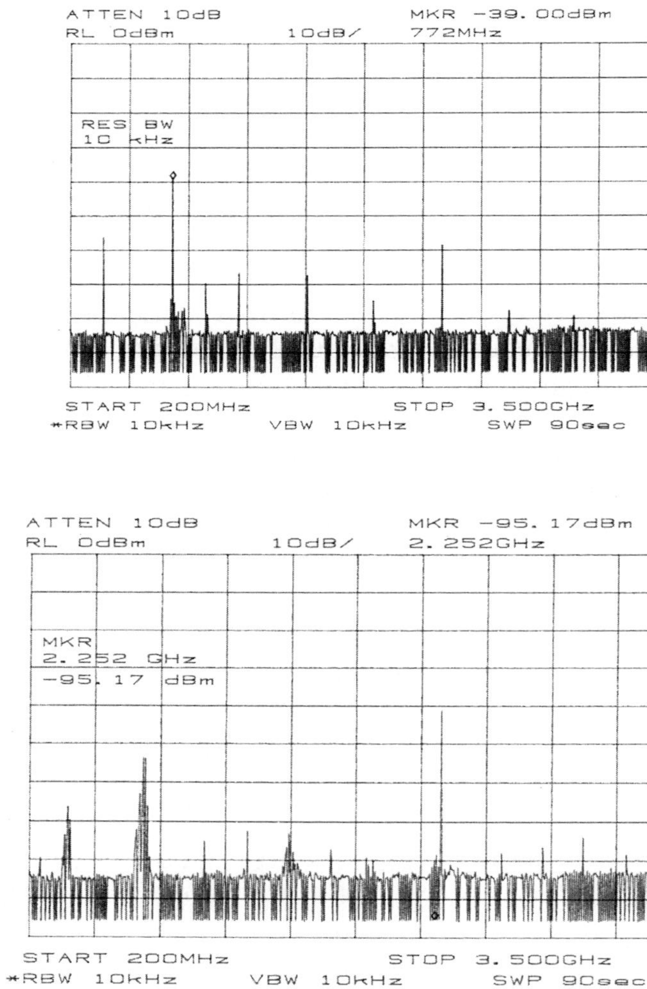

Fig. 16. Spectrum at substrate (top) includes multiples of the clock frequency. Measured spectrum at output LNA (bottom) shows an attenuated copy of the substrate spectrum.

The signal at the output of the LNA is distorted by the attenuated substrate bounce which is also at the output of the LNA available. The third order harmonic of the clock frequency falls into the signal band of the LNA and is indeed available at the output. Clock planning is therefore crucial. It can be shown that the guard ring does not protect the LNA from the clock bounces, in case of low ohmic processes the substrate can be considered as a single node.

Although a lot research has been done on substrate bounce, it is still unclear what the best substrate option is, low or high ohmic substrate. In case of low ohmic substrate, substrate and analogue ground should be connected together to reduce negative effects. Choosing for high ohmic substrate will reduce the noise levels with a factor of 3 compared to heavily-doped epi wafers [9]. Although the measurement results in [9] may tell us that high-ohmic substrate is the best option, we have to be careful. The decision is determined by the structure of the substrate and epi layer together and the injection mechanism of the noise. For certain technologies low ohmic substrate gives better substrate noise results than high ohmic substrate. There are even people who claim, supported by measurements, that it does not matter. Clearly, this subject needs much more attention.

5. SOME SYSTEM ASPECTS

From the above measurements it is clear that clock planning is of major concern. Consider Bluetooth having 78 channels of 1 MHz spacing each, resulting in a bandwidth of 78 MHz. The clock frequency to drive the mixed signal circuitry should therefore be such that odd multiples of the frequency will not fall in this bandwidth. Obviously, the actual clock frequency is also a result of available crystals and the quality of the synthesizer.
The best choice of the architecture is a major problem the designer has to deal with. The architecture must have a.o. low power dissipation, be robust against substrate bounces and cheap in die costs. Having in mind that no external components are allowed, also the antenna filter should be integrated. In the remaining of this sequel we will briefly discuss two possible architectures and focus on the receiver implementation.

The first architecture uses a non-linear receiver concept and is depicted in Fig. 17. Assume that a near zero IF (NIF) concept is used, then the spectrum of interest after the mixer is in the range of 0-1 MHz (NIF frequency is 500 KHz). To overcome the image problem, polyphase (I and Q) signals are used. After the mixer, we still have an analogue signal and the information can be extracted by slicing (min/max operation) this signal.

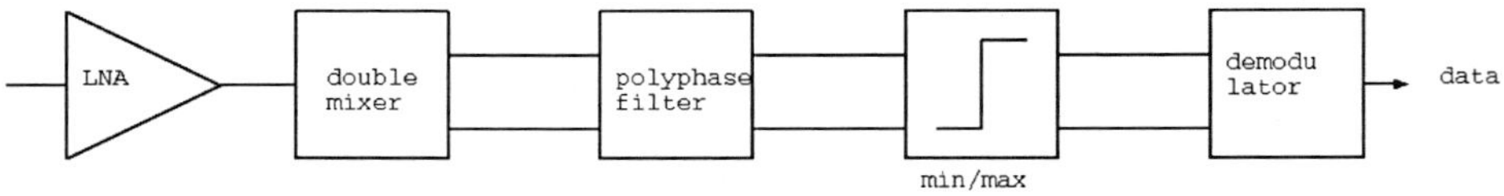

Fig. 17. Non-linear receiver concept

In fact, a clipper is used to clip the analogue signal to a positive (negative) voltage in case the signal is above (below) the zero Volt threshold. The signal after the clipper can be seen as a pulse width modulated signal which then needs to be demodulated into a digital bit stream. Although this is a very crude way to detect zero crossings it is normally a simple way to extract the information. Due to the non-linear nature of the process it is very sensitive to the available signal-to-noise ratio (SNR) at the input of the detector. To improve the bit-error-rate (BER), not only the I and Q signals can be used, but also additional signals can be generated to extract more information regarding the location of the zero crossings of the analogue signal. This will improve the BER considerably, at the cost of increased complexity and power consumption.

To improve the BER for the same SNR, an architecture with linear signal processing can be used. Now, analogue-to-digital converters (ADC) are used to convert the analogue I and Q signals in bit streams, i.e. digital replicas of the analogue signal.

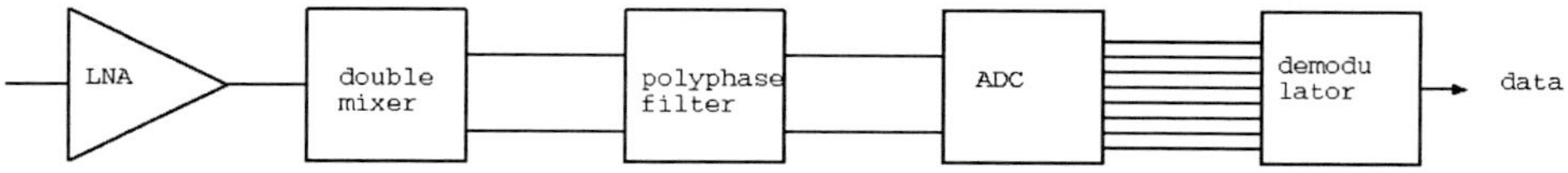

Fig. 18. Linear receiver concept

The conversion is a linear process, the more bits are used, the more accurate we are able to follow the signal and hence the more accurate we can predict the zero crossing. A block diagram of this second architecture is shown in Fig. 18. Obviously, this concept has some advantages over the previous one, but a complication is the implementation of a quadrature ADC. Note also the impact of substrate bounce caused by the decimation filter as has been discussed in the previous section.

One other issue is the voltage supply. For digital circuitry this can be as low as possible, for instance 1.5 V in a 0.18 μm CMOS technology. However, 1.5 V can be cumbersome for some analogue and RF parts. Especially analogue-to-digital converters and the connection between the VCO and mixers become problematic. The latter needs a large swing to drive the switches in the mixers properly and 1.5 V is may be not enough. If RF and digital operate at different supply voltages, this may lead to multiple output DC/DC converters when considering battery operation. DC/DC converters are already a major research topic and having more output makes designs not easier.

As last issue one can think of how to test this mixed mode telecom chip. Standard digital testing will not cover everything and it is not possible to bring out all internal nodes for test purpose. Loop back test is a possibility, however, the RF part may not be loaded by test switches due to their sensitivity to parasitics. Therefore test remains as an important issue as well.

6. CONCLUSIONS

In this contribution several design issues were discussed, all dealing with mixed-mode telecommunication design. We have used Bluetooth as an example, because this standard is believed to be the most likely candidate to have the highest integration level, i.e. a single chip. However, to come to this high level of integration, many design issues have to be solved. We have discussed several of them, not intending to be complete:

- RF CMOS modelling
- Bondpad and ESD
- Substrate bounce
- Architecture of the system

REFERENCES

1. J. Fenk, 'Highly integrated RF-ICs for GSM, DECT, and UMTS Systems, A Status Review and Development Trends', proc. ESSCIRC'99, pp.11-14, 1999
2. G. Weinberger,' The new millennium: Wireless Technologies for a truly Mobile Society', Digest ISSCC 2000, pp.11, 2000
3. J.P. Tsividis, *Operation and Modeling of the MOS transistor,* ,Mc Graw Hill Book Co. New York, 1987
4. A.J. Scholten, et.al. 'A large signalnon-quasi-static MOS model for RF circuit simulation', IEDM-99, pp.163-166, 1999
5. L.F. Tiemeijer, 'RF CMOS modeling', AACD, 1999
6. M. van Heijningen, et.al. 'Modeling of Digital Substrate Noise Generation and Experimental Verification Using a Novel substrate Noise Sensor', proc. ESSCIRC'99, pp.186-189, 1999 Duisburg
7. M. Felder, J. Ganger, 'Analysis of Ground-Bounce Induced Substrate Noise Coupling in a Low Resistive Bulk Epitaxial Process: Design Strategies to Minimize Noise Effects on a Mixed-Signal Chip', *IEEE Trans. Circ. Syst.-II,* vol. 46, pp.14-27-1436, 1999
8. B. Nauta, G. Hoogzaad, 'Substrate Bounce in Mixed-Mode CMOS ICs', AACD'98 Denmark

9. X. Aragones, A. Rubio, 'Experimental comparison of substrate noise coupling using different wafer types,' IEEE Trans. Solid-State Circuits, vol. 34, n0. 10 ,pp 1405-1409, 1999.

PLL's and Synthesizers

Willy Sansen

Most communication systems require single-chip solutions to reduce the cost. Their RF frontends contain frequency synthesizers that are usually made up by Phase-Locked Loops (PLL's) with Voltage-Controlled Oscillators (VCO's) in the GHz range with low phase noise. Six contributions are given which address the design issues and highlight the state-of-the art of these blocks.

The first presentation, given by Jim Parker of Level One Communications, discusses the effects of placing multiple inductor-based VCO's on the same mixed-signal CMOS substrate. The coupling phenomena have been studied, simulated and functionally verified by means of a prototype test set. A careful model of the varactor is shown to be essential for the prediction of the coupling.

It is followed by a presentation by Bram De Muer of KULeuven. He has given a design strategy for fully integrated frequency synthesizers in CMOS technologies. It focuses on the minimization of the loop filter capacitance within a minimum required phase noise and dynamic performance. Measurements are added on a DCS-1800 synthesizer in 0.25 micron CMOS.

The third speaker is Marc Tiebout of Infineon Technologies, Munich. He reports on various CMOS circuits for Synthesizers and PLL's. The first focus is on a low-noise LC tank VCO in 0.35 micron CMOS. The next one is on quadrature generation and finally on mixed-mode high-integration aspects.

The fourth paper is on fractional-N synthesizers. It is given by Sven Mattisson of Ericsson, Lund. He actually focusses on Delta-Sigma-dithered moduli fractional-N synthesizers because they do require no analog spurious-compensation circuits. Design equations are given followed by a discussion on practical implementation aspects.

R. J. van de Plassche et al. (eds.), Analog Circuit Design, 267–268.

Linear models, as explained by Marina de Queiroz Tavares of Philips, Zürich, can predict low phase noise and stability. However the accuracy of such models is limited for wide band loops. The theoretical limits are calculated and verified by the implementation of test chips with multi-loop wide band PLL's up to 2.1 GHz in a 13 GHz/2.4 micron BICMOS technology.

Finally Dirk Pfaff of ETH Zurich, focuses on the low-power limits of RF synthesizers. The main culprits are the VCO and the prescaler. Different resonator circuits are analyzed and compared. Measurement results are presented on a 1 GHz VCO, quadrature signal generator and dual modulus prescaler.

It can be concluded from these presentations that the design of low-noise VCO's and synthesizers can be handled in various ways, all leading to excellent results with respect to low power consumption and high-frequency performance, as now commonly required in frontends of present -day communication systems.

On Placing Multiple Inductor-Based VCOs on the Same Mixed-Signal Substrate

James Parker and Michael Altmann

Level One Communications, an Intel Company

Abstract

The effects of placing multiple inductor-based voltage-controlled oscillators (VCOs) on the same mixed-signal CMOS substrate have been studied, simulated and functionally verified using a prototype test chip. It is concluded that by carefully modeling the varactor as a nonlinear element capable of phase detection, interaction between two RF PLLs may be accurately predicted.

I. Introduction

The demand for single-chip communication ICs (integrated circuits) has increased over the last several years. This is due primarily to cost-reduction requirements demanded by a competitive marketplace. Single-chip communication ICs are typically mixed-signal as they require the integration of both analog and digital circuits on the same silicon die. For instance many system solutions for communication standards that implement ATM/Sonet, 10/100/1000-BaseT Ethernet, DSL, etc. require a mixed-signal solution to one degree or another.

As baud periods continue to decrease, system jitter requirements will become more stringent. Communication standards such as the IEEE 802.3ab for 1000-BaseT Ethernet require that total transmit jitter (in certain cases) be less than 300ps pk-pk [1]. This jitter specification is largely driven by the need to operate in a full-duplex environment. Full-duplex, high-speed communication requires cancellation of high-frequency echo. Accurate high-frequency echo cancellation dictates a very low-jitter clock for waveform sampling.

In many RF systems, PLLs based on voltage-controlled LC-tank oscillators are commonly used as frequency synthesizers. This is due to

R. J. van de Plassche et al. (eds.), Analog Circuit Design, 269–285.

superior phase-noise characteristics, and therefore excellent jitter performance relative to ring oscillators that have similar power and swing requirements. Because of the natural noise filtering afforded by an LC-tank circuit, LC-based VCOs make an attractive alternative for use in non-RF applications, such as the mixed-signal systems described above.

Often, a need exists for multiple PLLs on the same silicon die. This gives rise to the possibility of undesired interaction between the PLLs. In this paper, the interaction between two inductor-based VCOs on a common substrate is explored. The paper is divided into four remaining sections. Section II gives background information. Section III examines the major elements of an LC-based PLL, and where necessary derives appropriate mathematical models that describe the VCO operation. Section IV presents a model of two PLLs that interact with one another. The model is used as a basis for HSPICE simulations and the derivation of both small- and large-signal equations. Finally, section V gives both measured results from a test chip and some conclusions.

II. Background

Figure 1 shows a generalized block diagram of a single-chip,

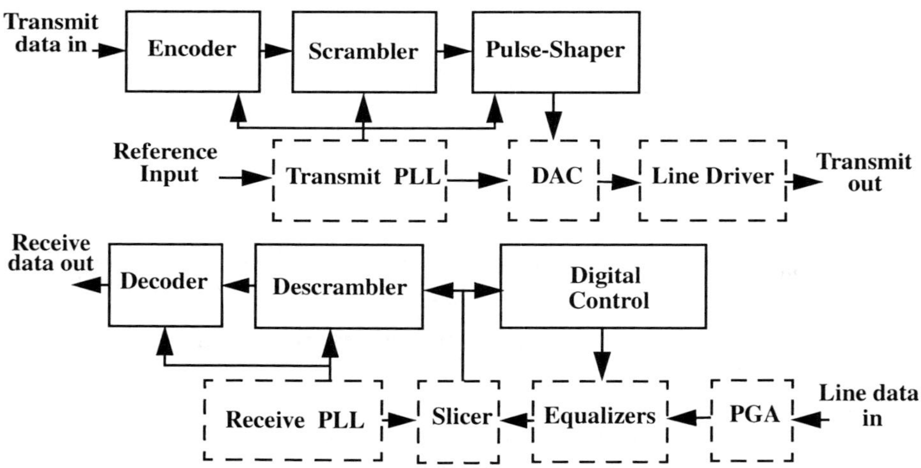

Figure 1: Mixed-signal communication system

mixed-signal communication system. It consists of both a transmit path and a receive path. Blocks typically implemented in the digital domain are shown with solid-lines, while blocks that involve analog functions are shown with dashed lines. The transmit path usually consists of an encoder, scrambler, pulse-shaper, DAC and line driver. A low-jitter (high-bandwidth) transmit PLL is used primarily for synchronization and frequency synthesis. In today's high-speed, wireline communication systems, symbol intervals often exceed 100 MBaud. Thus, it is advantageous to have a frequency synthesizer that generates a high-speed clock from a low-cost, low-frequency external crystal. The receive path shown in Figure 1 consists of a programmable gain amplifier (PGA), analog equalization with digital control block [2], slicer, descrambler and decoder. The timing recovery for such a channel can be realized in a variety of ways. For purposes of this paper, the receive PLL is assumed to consist of a separate PLL from the one used in the transmit path and is primarily analog in nature.

Typically, the equalizers used in the receive path consist of multiple feedback loops. To maintain stability and allow an optimal system solution with maximum SNR at the input to the slicer, these feedback loops require time constants that range from hundreds to thousands of baud intervals. To avoid undue equalization and timing-recovery loop interaction, the time constants of the receive PLL are made larger than the time constants used for adapting the equalizers.

In summary, the timing requirements for a mixed-signal communication IC will generally consist of two PLLs. The first is a high-bandwidth loop used to provide low-jitter frequency synthesis for symbol transmission. The second is a low-bandwidth PLL used for symbol recovery in the receive path. However, if the transmit PLL doubles as the receive PLL using a digital phase interpolation scheme [3], a second, separate PLL may still be needed if multiple transmit/receive paths are integrated on the same die. In summary, with integration levels ever increasing, multiple PLLs operating on the same silicon substrate is a necessity. This gives rise to a possibility of interaction between the PLLs.

III. The Inductor-Based PLL

Figure 2 shows a simplified block diagram of the two RF PLLs

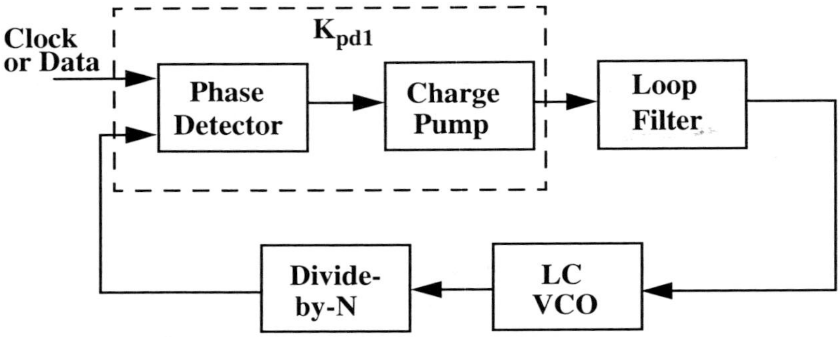

Figure 2: Block diagram of transmit and receive PLLs

described in this work. As drawn, the PLL may be used either for frequency synthesis or for data-directed timing recovery. The loop consists of a phase detector, charge pump, loop filter, LC-based VCO and arbitrary divide-by-N circuit. Note that the divider is necessary because the incoming data is sampled at the baud rate which is assumed to be less than the oscillator output frequency.

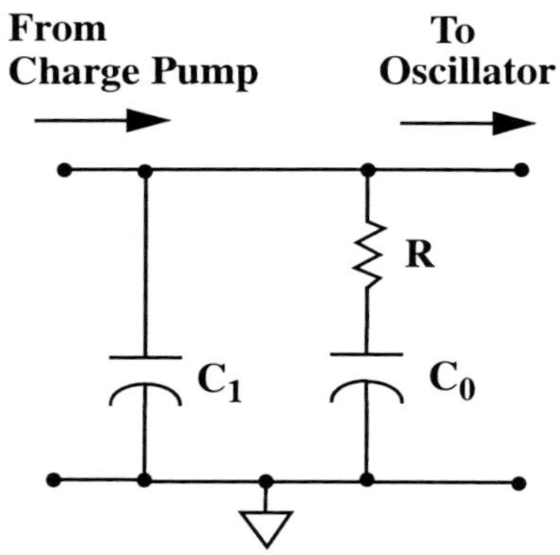

Figure 3: Loop Filter

When the PLL is used as a frequency synthesizer, the input to the phase detector is a crystal-based clock at a sub harmonic of the baud period. The phase detector is both a phase and frequency detector that uses several RS-latches [4]. When combined with a charge pump with pump current I_p, the small-signal gain of the phase detector and charge pump is $K_{pd1}=I_p/2\pi$. The transmit loop filter in Figure 3 consists of a resistor R, and two capacitors, C_0 and C_1 that creates a 1-zero, 2-pole network. The zero frequency is $1/(RC_0)$ rad/s. The two pole frequencies are at DC and $1/(C_0$ series $C_1)R$ rad/s. The oscillator, which has a small-signal gain of K_o/s, is described below. The divider, with small-signal gain 1/N, uses differential current-mode logic based on a replica-bias circuit to control the output swing over process, voltage and temperature [5]. Because the crystal clock is assumed to have very low jitter over a wide frequency range, the small-signal bandwidth (in rad/s) of the PLL, equal to $K_oK_{pd1}R/N$, is made as large as possible to minimize effects due to both the oscillator's noise and its finite power-supply and substrate noise rejection. In this application, the transmit PLL bandwidth is approximately 750 kHz.

When used for timing recovery, the input to the phase detector is represented by sampled data. In this application, the phase detector consists of a set of digital gates that realize a linear phase transfer characteristic with a small-signal gain of $K_{pd1}=0.625*I_p/2\pi$. The loop filter is the same as shown in Figure 3. For simplicity, both the divider and oscillator are the same as those used in the frequency synthesizer. As before, the bandwidth of the receive PLL is $K_oK_{pd1}R/N$. To keep the bandwidth small, the product of I_pR is made small. In this application, the bandwidth is approximately 160 kHz.

A. LC Oscillator

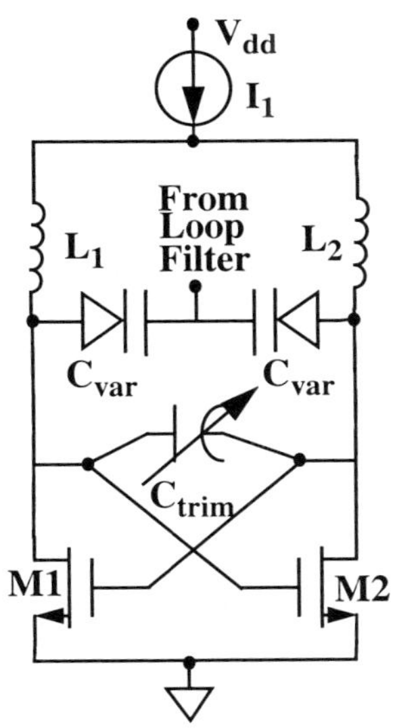

Figure 4: Simplified schematic of LC oscillator

Figure 4 shows a simplified schematic of the LC oscillator [6]. It consists of cross-coupled transistors M1-M2, current source I_1 and a LC-tank circuit. M1-M2 create a small-signal negative resistance to insure initial oscillation start-up and also provide large-signal amplitude regulation. The tank circuit capacitance consists of varactors, C_{var}, and a segmented capacitor array, C_{trim}. The inductors (L_1-L_2) are approximately 3 nH.

The pull range of the oscillator must cover variations due to processing and temperature. The varactors, C_{var}, which consist of multiple p+ diffusion strips in a single n-well, are used to compensate for temperature variations in the VCO frequency. In Figure 4, the oscillation frequency, ω_o, is defined as:

$$\omega_o = \left(\frac{1}{L_{eq}C_{eq}}\left(1 - \frac{R_{eq}^2}{\omega_o^2 L_{eq}^2}\right)\right)^{0.5} = \left(\frac{1}{L_{eq}C_{eq}}\left(1 - \frac{1}{Q^2}\right)\right)^{0.5} \quad \text{Eq. 1}$$

where L_{eq} and R_{eq} are the equivalent inductance and series resistance of L_1-L_2. C_{eq} is the equivalent capacitance in parallel with the inductors. R_{eq} is dominated by the metal-line resistance of the inductors, and varies by roughly 30% over temperature. In Eq. 1, the small-signal sensitivity of ω_o to small changes in R_{eq} is $-1/Q^2$. Thus, if the nominal Q-factor of the inductor is 3.3, the oscillation frequency will vary by roughly -3% for a positive 100 deg. C shift in temperature.

In Figure 4, C_{trim} is used to overcome process variations in the integrated LC tank circuit. Assuming that the total tank capacitance varies by as much as ±15% over process, C_{trim} is used to compensate for this factor and keep the control voltage on the varactor centered at the nominal operating temperature. In this application, C_{trim} consists of a segmented array of capacitors controlled by a 4-bit digital word. This bank of capacitors may be realized in a variety of ways [7,8].

i. Phase Detector Model of the Varactor

With respect to oscillator interaction, the most critical element to be modeled is the varactor. This is because the varactor is a nonlinear element that can act as a multiplying-type phase detector.

The varactor diode consists of two capacitors, the area component, C_a, and C_p, which represents the varactor periphery capacitance. To first order, both capacitances are modeled by the generic equation:

$$C_j = \frac{C_{j0}}{\left(1 + \frac{V_{DC}}{\phi}\right)^n} \quad \text{Eq. 2}$$

In the equation, V_{DC} is the DC voltage across the nwell/p+ junction, ϕ

is the built-in potential between the doped regions, n is dependent on the doping profile (typically $0.3 < n < 0.6$) and C_{j0} is the capacitance when V_{DC} is identically 0.

In the LC-oscillator of Figure 4, the voltage across the varactor is not constant, but also consists of a time-varying component. The DC component represents the final steady-state voltage driven by the negative feedback of the PLL. The time-varying component consists of the RF sinusoidal output signal. This signal is well- approximated as sinusoidal because of the filtering of the LC-tank circuit. Thus, using Eq. (2), the average varactor capacitance over one cycle of oscillation is modeled by the following:

$$\overline{C_{var}} = \frac{1}{T}\int_0^T \left(\frac{C_{ao}}{\left(1 + \frac{V_{DC} + A_1 \sin\left(\frac{2\pi t}{T}\right)}{\phi}\right)^n} + \frac{C_{po}}{\left(1 + \frac{V_{DC} + A_1 \sin\left(\frac{2\pi t}{T}\right)}{\phi}\right)^m} \right) dt \qquad \text{Eq. 3}$$

In Eq. 3, A_1 is the amplitude of the sinewave oscillation across the varactor. Assuming the following values for the process and operating-point dependent variables: ϕ=0.92, T=800ps, V_{DC}=0.5V, C_{a0}=3.575pF, C_{p0}=590fF, n=0.55, and m=0.39, the average varactor capacitance may be calculated relative to the oscillation amplitude. For A_1 ranging from 0.15Vp to 0.35Vp, $\overline{C_{var}}$ ranges from 3.326pF to 3.36pF.

The phase detector characteristics of the varactor are now derived. Suppose that the simple sinewave in Eq. 3 is replaced with the sum of two sinusoids: $A_1\sin(2\pi/Tt)+A_2\sin(2\pi/Tt+\theta)$. As before, $A_1\sin(2\pi/Tt)$ represents the desired oscillator signal across the varactor. The phase of $A_1\sin(2\pi/Tt)$ is arbitrarily set to zero radians because it is used as the reference phase in the following description. $A_2\sin(2\pi/Tt+\theta)$ represents the crosstalk component due to the transmit PLL that appears on the varactor of the receive PLL. A_1 is fixed at 0.25V and C_{var} is plotted as a function of θ while fixing A_2 at 7.5mVp, 15mVp and 30mVp. These three results are shown in Figure 5a. To obtain a result in terms of V/rad, the plot in Figure 5a is scaled by using the following relation:

$$\frac{\partial V_{DC}}{\partial \theta} = \frac{\partial \overline{C_{var}}}{\partial \theta} \cdot \frac{\partial V_{DC}}{\partial \overline{C_{var}}} \qquad \text{Eq. 4}$$

In other terms, Figure 5a is scaled by a factor that, for a given change in θ, relates changes in the varactor capacitance to equivalent changes in DC voltage across the varactor. These results are shown in Figure 5b.

There are two interesting results obtained from Figures 5a-b. First, the slope of the curves is either positive or negative depending on the phase θ. Second, when θ is ±nπ/2 rad. (n=1,3,5...) the net change in voltage across the varactor is zero. This may be generalized by stating that only signals that are correlated to $A_1\sin(2\pi/Tt)$ may result in a DC voltage shift across the varactor.

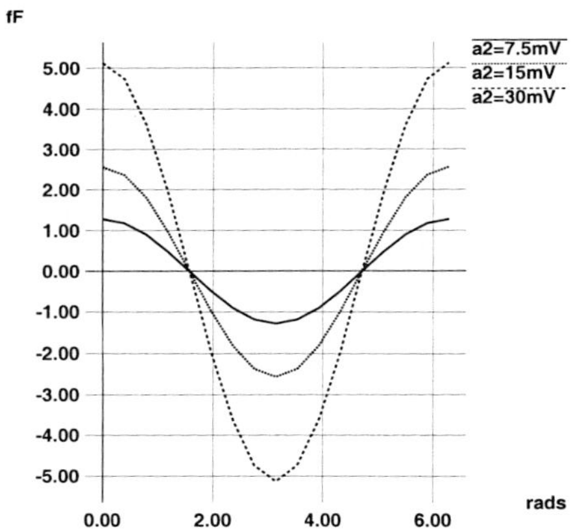

Figure 5a: ΔC_{var} vs. Δθ

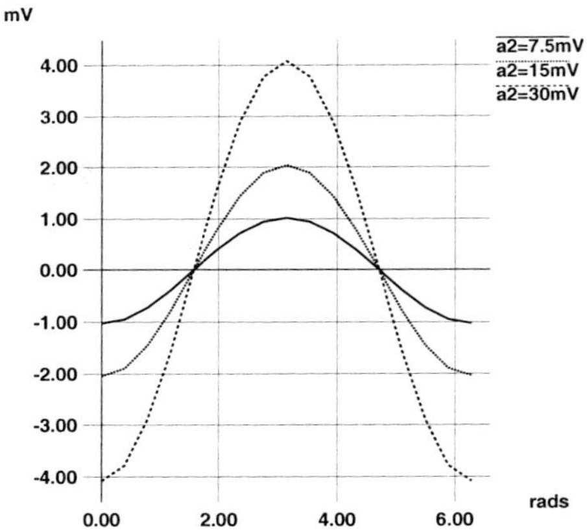

Figure 5b: Equivalent ΔV vs. Δθ

IV. Simulation of PLL Interaction

A. The Parasitic PLL

The above analysis shows that if significant energy from one

LC-based oscillator is coupled onto the varactor of another identical oscillator, a parasitic PLL is created. In practice, this can happen in a variety of ways. The most direct paths are through the power supply and substrate connections. In this analysis, it is assumed that RF energy of the transmit PLL is coupled onto the control node of the varactor for the receive PLL. However, it should be noted that the reverse can also occur, but is not examined here. In Figure 6, a block diagram of this parasitic PLL is shown. The input to the loop is θ_{xmt}, the phase of the crosstalk energy from the transmit PLL. θ_{xmt} is then phase compared with the output of the oscillator, θ_{osc}. The resulting phase difference is multiplied by the varactor's phase transfer characteristic derived earlier and shown in Figure 5b. If one of the phase transfer characteristics in Figure 5b is approximated with a slope of K_{pd2} (measured where the sinewave crosses the x-axis), the transfer characteristic of the parasitic PLL is:

$$\frac{\theta_{osc}}{\theta_{xmt}} = \frac{1}{1 + \frac{s}{K_o K_{pd2}}} \qquad \text{Eq. 5}$$

The bandwidth of this loop is K_oK_{pd2}. Furthermore, the single pole lies either in the right-half plane or the left-half plane depending on the phase relationship of θ_{xmt} relative to θ_{osc}. From Figure 5b, if θ_{xmt}-θ_{osc} is between 0 and -π, the slope of K_{pd2} is negative and the pole lies in the right-half plane. If θ_{xmt}-θ_{osc} is between 0 and π, the slope is positive and the pole lies in the left-half plane. As described previously, the magnitude of K_{pd2} is a function of the amplitude of the crosstalk signal from the transmit PLL that is coupled onto the varactor of the receive PLL. Assuming the crosstalk due to the transmit PLL is 30mVp, Figure 5b shows the slope of the curve as it crosses the x-axis is K_{pd2} = 4 mV/rad.

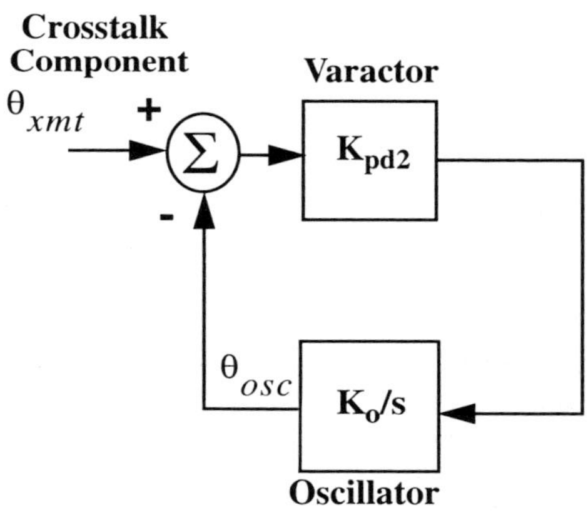

Figure 6: Block diagram of parasitic PLL

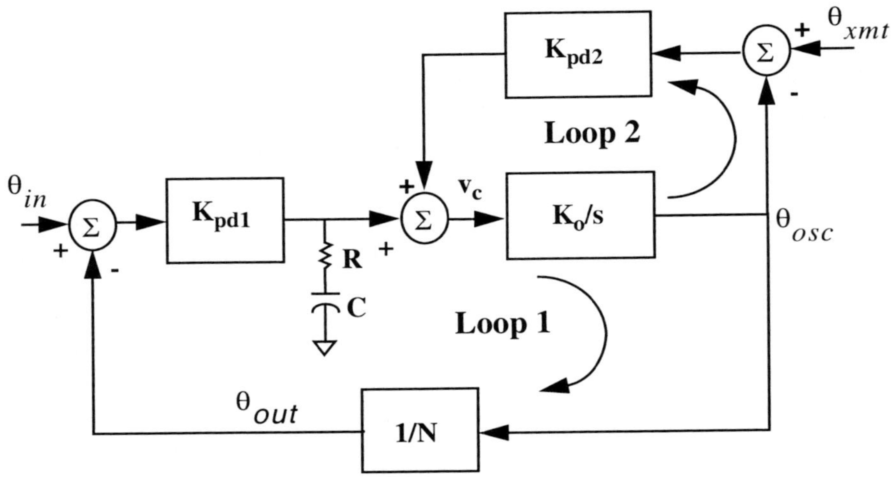

Figure 7: Model used for small- and large-signal analysis

B. The Small-Signal Model of Two Interacting Oscillators

Figure 7 shows a block diagram of the model used to derive the small-signal behavior of the two interacting oscillators. The model consists of two separate PLLs. The largest loop, labeled as Loop 1, consists of the low-bandwidth, timing-recovery loop for the receiver. It is hereafter referred to as Loop 1 for brevity. Loop 1 consists of all the linear elements described previously in Section II. However, to simplify the analysis, the loop-filter capacitor that is used for jitter suppression (C_1 in Figure 3) is omitted, and the capacitor C_0 is hereafter referred to as C. The input to Loop 1 is θ_{in} and the output is labeled θ_{out}, the output of the divider. The second loop in Figure 7, hereafter referred to as Loop 2, consists of the elements shown in Figure 6 that were described above. Small-signal analysis of this system reveals how the two loops interact with one another. If θ_{xmt} is set to zero, the transfer function of Loop 1, θ_{out}/θ_{in}, is calculated:

$$\frac{\theta_{out}}{\theta_{in}} = \frac{\frac{K_o K_{pd1}}{NC}(1 + sRC)}{s^2 + s\left(K_o K_{pd2} + \frac{K_o K_{pd1} R}{N}\right) + \frac{K_o K_{pd1}}{NC}} \qquad \text{Eq. 6}$$

If θ_{in} is set to zero, the transfer function of Loop 2, $\theta_{osc}/\theta_{xmt}$, is:

$$\frac{\theta_{osc}}{\theta_{xmt}} = \frac{sK_oK_{pd2}}{s^2 + s\left(K_oK_{pd2} + \frac{K_oK_{pd1}R}{N}\right) + \frac{K_oK_{pd1}}{NC}} \qquad \text{Eq. 7}$$

The denominator, D(s), of both loop equations is equivalent. The term in s in D(s) contains the sum of the loop bandwidths for Loop 1 ($K_oK_{pd1}R/N$) and Loop 2 ($K_{pd2}K_o$) in Figure 7 assuming both loops are operating independently of one another. If K_{pd2} is non-zero, the pole positions of Loop 1 are shifted from their desired locations, affecting its desired step response. In the extreme, if $K_{pd2}K_o$ becomes much larger than $K_oK_{pd1}R/N$, the bandwidth of Loop 1 begins to approach zero, effectively looking like an open circuit.

From D(s), it is also seen that if the magnitude of K_{pd2} is greater than $K_{pd1}R/N$, and K_{pd2} is negative, the poles of D(s) enter the right-half plane and a phase oscillation occurs. The startup frequency of this oscillation is predicted by Eq 6. This is done by determining where the phase of the equation is equal to zero. This may be shown to occur at a frequency of K_oK_{pd1}/NC rad/s, which is the natural frequency (ω_n) of D(s). However, if the loop gain is greater than one, the oscillation frequency will shift down from ω_n.

C. Large-Signal Modeling through Simulation

To better understand the large-signal time-domain behavior of Figure 7, several simulations were performed in HSPICE. To simulate the nonlinear phase-detector transfer function of Loop 2 (corresponding to the curve for A_2=30mVp in Figure 5b) a voltage-controlled piece-wise linear source was used as an approximation. In the simulation, K_{pd1}=0.625*12/2π μ A/rad, R=7.75kΩ, C=39nF, Ko=2π*136 Mrad/s/V, and N=8.

Figure 8 shows the superposition of the transfer characteristics for the phase detectors in Loop 1 and Loop 2. For Loop 1, the x-axis represents the relative phase between θ_{osc} and θ_{in}. For Loop 2, the x-axis represents the relative phase between θ_{xmt} and θ_{osc}. Note that the slope of the phase transfer characteristic of Loop 1 is equal to (R/N)K_{pd1} and passes through the origin. Thus, without interference from Loop 2 (i.e. K_{pd2}=0), Loop 1 will drive θ_{osc} to a steady-state solution that is in-phase with θ_{in}. Scaling K_{pd1} by the factor R/N is done for two reasons. First, scaling by R converts the units of K_{pd1} to mV/rad for compatibility with K_{pd2}. Second, scaling by the factor 1/N allows changes in θ_{osc} in Figure 7 to be translated to an equivalent correction voltage across R in Loop 1. In other terms, as Loop 2 adjusts θ_{out}, the Loop 1 correction to this phase change will be scaled by a factor of 1/N.

In Figure 8, a key characteristic of the Loop 2 phase transfer characteristic is its steady-state operating point. In the absence of interference from Loop 1 (i.e. K_{pd1}=0), Loop 2 will force θ_{xmt}-θ_{osc}=π/2 ± 2nπ (n=1,2,3...) radians in the steady-state. Thus, if the phase of the interfering signal from the transmit PLL happens to lie π/2 radians from θ_{osc}, both Loop 1 and Loop 2 will converge to the same operating point. However, for all other phase relationships between θ_{xmt} and θ_{osc} the steady-state operating points of the loops are different, leading to interesting time-domain behavior. For convenience in the remaining description, a shorthand notation is adopted. The expression: $\theta_{xmt} = Phase|_{Kpd2 = 0}$ is hereafter used to quantify the phase offset of

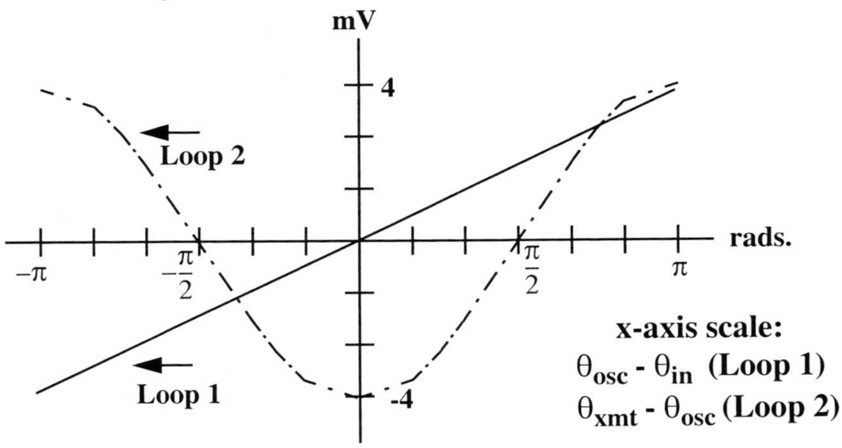

Figure 8: Loop 1 and Loop 2 phase detector characteristics

θ_{xmt} relative to θ_{osc} under the condition that K_{pd2} is equal to zero. For instance, $\theta_{xmt} = \left.\frac{\pi}{2}\right|_{Kpd2 = 0}$ means that if Loop 1 and Loop 2 were not interacting with one another, the phase difference θ_{xmt} - θ_{osc} would be equal to a constant $\pi/2$ radians.

Figure 9 shows θ_{osc} as a function of time if $\theta_{xmt} = -\left.\frac{\pi}{2}\right|_{Kpd2 = 0}$. At this phase alignment, the single pole of Loop 2 is in the right-half plane. θ_{osc} has been scaled by $T/2\pi$ (T=800ps) so that the y-axis has units of ps. The phase oscillation in Figure 9 shows two distinct regions of operation. The steep portion of the curve between ±200ps represents the time spent on the negative slope of the Loop 2 phase transfer characteristic in Figure 8. During this time, positive feedback in Loop 2 controls the response of the system. Once the magnitude of the oscillation amplitude exceeds 200ps, the amplitude is driven by negative feedback in Loop 2 (since K_{pd2} is now positive) towards ±400ps, which satisfies the steady-state criteria for Loop 2, i.e. θ_{xmt}-θ_{osc}=$\pi/2$, $-3\pi/2$. During the slow recovery time, Loop 1 drives θ_{out} towards zero (assuming θ_{in}=0), the desired steady-state solution for Loop 1. This slow recovery occurs due to the second integrator in Loop 1. Without the additional integrator, Loop 1 would not have the loop gain at any frequency to overcome the affects of Loop 2. The time-constant, τ, of the Loop 1 recovery time is set by the location of the lowest frequency pole in Eq 6. Using the quadratic equation, and the assumption that $K_oK_{pd1}R/N + K_{pd2}K_o >> 2\omega_n$, τ is approximated by:

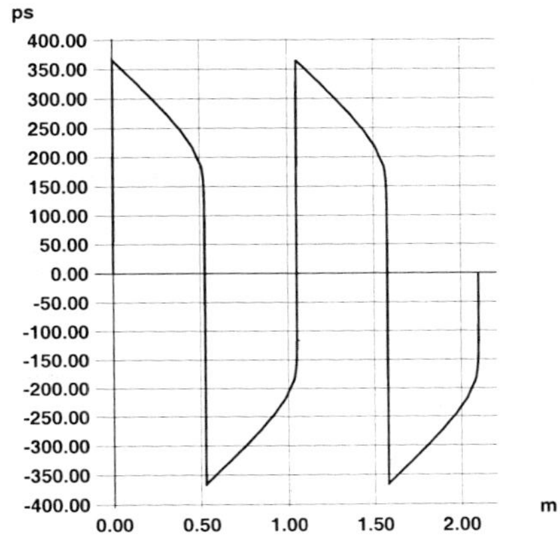

Figure 9: Oscillation inθ_{osc} for $\theta_{xmt} = -\left.\frac{\pi}{2}\right|_{Kpd2 = 0}$

$$\tau \approx RC\left(1 + \frac{N\overline{K_{pd2}}}{RK_{pd1}}\right) \qquad \text{Eq. 8}$$

In Eq. 8, $\overline{K_{pd2}}$ is used because the instantaneous value of K_{pd2} is not constant over the oscillation cycle. Since the phase transfer characteristic of Loop 2 is sinusoidal, $\overline{K_{pd2}}$ is approximated as 0.707 times the peak amplitude of the phase transfer characteristic in Figure 8.

The frequency of the phase oscillation in Figure 9 is related to the time constant derived in Eq. 8. However, an exact closed-form solution for this oscillation frequency is difficult to derive. This is because the frequency of oscillation is also a function of the steady-state oscillation amplitude. Further, it is a function of θ_{xmt}-θ_{osc} when K_{pd2}=0. Generally, the steady-state amplitude is found by finding the instantaneous phase of θ_{osc} that generates equal but opposite correction voltages to the oscillator by Loop 1 and Loop 2. In other terms, the control voltage v_c in Figure 7 must be zero under these conditions. For the case where $\theta_{xmt} = -\frac{\pi}{2}\Big|_{Kpd2 = 0}$, the steady-state amplitude is approximated by the following:

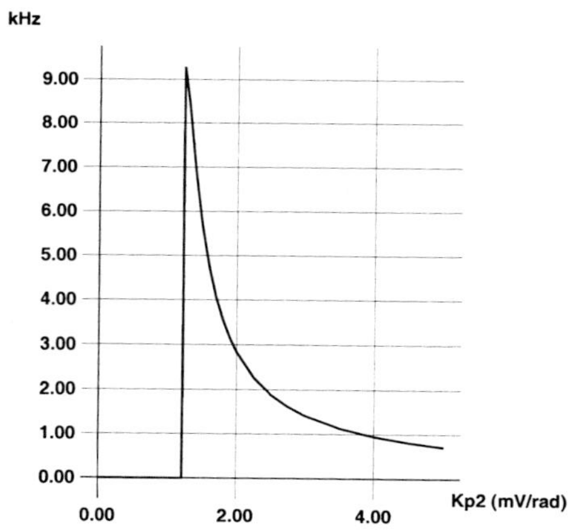

Figure 10: Frequency vs. K_{p2}

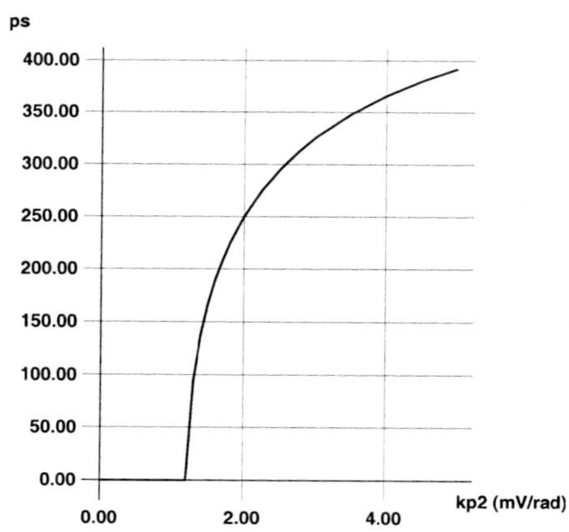

Figure 11: Amplitude vs. K_{p2}

$$Amplitude = \frac{T}{2\pi} \frac{\pi + \dfrac{\alpha(\tau)}{K_{pd2}}}{1 + \dfrac{RK_{pd1}}{NK_{pd2}}} \text{ ps.} \qquad \text{Eq. 9}$$

As with Eq. 8, the average value of K_{pd2} over the oscillation cycle is used. The function $\alpha(\tau)$ represents the steady-state peak voltage integrated on C during the slow recovery time predicted by Eq. 8. Thus, both Eqs. 8 and 9 must be iteratively solved to find an exact expression for the phase-oscillation frequency. If it is assumed that $K_oK_{pd1}R/N \le |K_oK_{pd2}| \le 2K_oK_{pd1}R/N$ (i.e. the bandwidths of Loop 1 and Loop 2 are within a factor of two of each other), $\alpha(\tau) \approx 0$ and the frequency of oscillation may be found by evaluating Eq. 9 directly and then solving for the time required for the system to relax to a magnitude of 200ps based on the time constant of Eq. 8. Figures 10 and 11 summarize the simulated values for the frequency of phase oscillation and its steady-state amplitude (in ps) as a function of K_{p2}.

So far, the interaction of the system has only been evaluated for $\theta_{xmt} = -\frac{\pi}{2}\Big|_{Kpd2 = 0}$. Figure 12 shows θ_{osc} plotted when $\theta_{xmt} = -\frac{3\pi}{4}\Big|_{Kpd2 = 0}$. Note that both the frequency of oscillation and duty cycle have shifted. This is due to the large-signal phenomenon predicted by Eqs. 8 and 9. Again, due to the second integrator in Loop 1, the average phase error over time between θ_{osc} and θ_{in} is 0.

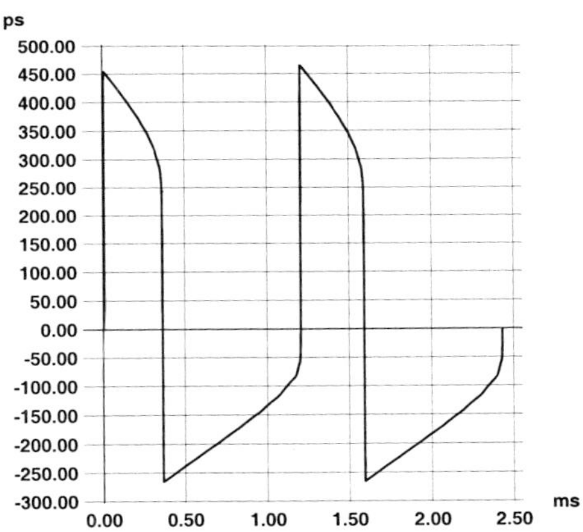

Figure 12: Oscillation in θ_{osc} for $\theta_{xmt} = -\frac{3\pi}{4}\Big|_{Kpd2 = 0}$

V. Measured Results and Conclusions

A test chip was fabricated to evaluate the effects of the interaction between two PLL's using LC-based VCOs. The Loop 1 closed-loop bandwidth was approximately 160 kHz, which was verified with jitter transfer function measurements. When triggering on the rising edge from the output clock of the transmit PLL, the phase of the reference frequency to the transmit PLL was adjusted until the jitter on the rising edge of the receive clock broke into two distinct jitter bands that were equally populated. This behavior is indicative of a phase oscillation in the receive VCO output clock. In order to directly observe the VCO output phase versus time, the receive clock (after the divide-by-8) was oversampled at a sufficiently high rate (4GHz) by a sampling oscilloscope. This data was analyzed to find the cycle-to-cycle jitter of all the clock transitions. This jitter was then integrated over time to produce Figure 14. As can be seen, the duty-cycle is roughly 50% and the waveform correlates well in both amplitude and frequency with respect to Figure 9. Thus it is estimated that approximately 30mVp of transmit PLL signal is coupling onto the varactor of the receive PLL.

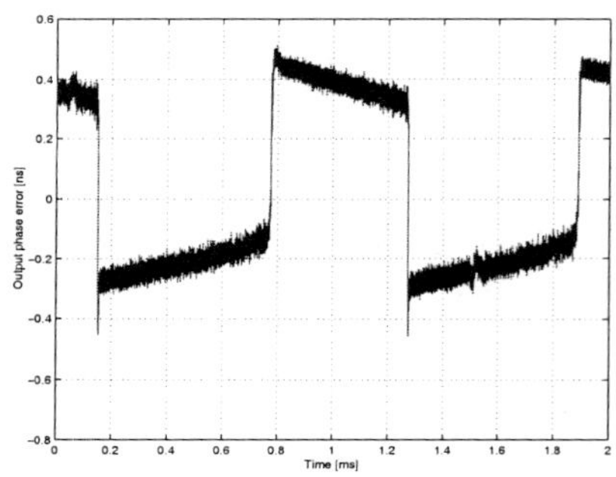

Figure 14: Measured phase oscillation in lab

In conclusion, the potential interaction between two LC-tank based PLLs has been modeled, simulated and verified with a prototype test chip. The model is useful for predicting the amount of crosstalk between unrelated PLLs. Further, it provides the designer insights into how much isolation is necessary to achieve proper system performance. It shows that as the bandwidth of the receive PLL is decreased, the amount of isolation required between the transmit and receive PLL must increase. In the event of undesired coupling between LC-based PLLs, the equations give the designer a feel for how much the coupling must be reduced to guarantee stability over all possible phases of θ_{xmt}.

VI. References

[1] Draft Supplement to IEEE Standard 802.3 1000Base-T, P802.3ab, Draft 6.0, p. 40.78, March 1999.

[2] J. Everitt, et. al. "A CMOS Transceiver for 10-Mb/s and 100-Mb-s Ethernet," I*EEE Journal of Solid-State Circuits*, vol. 33, pp 2169-2177, Dec. 1998.

[3] Larsson, Patrik. "A 2-1600MHz 1.2-2.5V CMOS Clock Recovery PLL with Feedback Phase-Selection and Averaging Phase-Interpolation for Jitter Reduction," I*SSCC 1999 Digest of Technical Papers*, pg. 356, Feb. 1999.

[4] F.M Gardner. *Phaselock Techniques*, 2nd ed. New York: Wiley, 1979.

[5] B. Kim, et. al. "A 30 MHz Hybrid Analog/Digital Clock Recovery Circuit in 2-μm CMOS," *IEEE Journal of Solid-State Circuits*, vol. 25, pp. 1385-1394, Dec. 1990.

[6] Craninckx, J and Steyaert, M, "A 1.8-GHz Low-Phase-Noise CMOS VCO Using Optimized Hollow Spiral Inductors," *IEEE Journal of Solid-State Circuits,* vol. 32, pp 736-744, May 1997.

[7] Parker, J and Ray, D, "A 1.6-GHz CMOS PLL with On-Chip Loop Filter," *IEEE Journal of Solid-State Circuits*, vol. 33, pp 337-343, March 1998.

[8] Castello, R. et al. "A 30% Tuning Range Varactor Compatible with Future Scaled Technologies," *Symposium on VLSI Circuits, 1998, Digest of Technical Papers*, pp 34-35, June 1998.

Fully Integrated CMOS Frequency Synthesizers for Wireless Communications

Bram De Muer and Michiel Steyaert

Katholieke Universiteit Leuven, ESAT-MICAS
Kardinaal Mercierlaan 94, 3001 Heverlee, Belgium
bram.demuer@esat.kuleuven.ac.be

Abstract

This work presents a design strategy for fully integrated CMOS frequency synthesizers, using the DCS-1800 class I/II standard as driving application. The design of the loop filter is emphasized. The goal is to minimize the integrated capacitance while maintaining the required phase noise and dynamic performance by proper filter topology selection and optimization of the loop parameters. Based on these results, a type-II, fourth-order PLL frequency synthesizer is implemented in a standard 0.25μm CMOS technology, including a monolithic voltage controlled oscillator and a fully integrated loop filter. The measurements show full compliance with the DCS-1800 class I/II specifications, while the occupied area is only 2x2 mm^2. To conclude, the influence of ΔΣ fractional-N division control on the performance of the PLL frequency synthesizer is discussed.

1 Introduction

The end of the 20th century will be remembered for the unrivalled growth of the telecommunication industry. Not only the wide-spread use of Internet and wireline communication has emerged, but also wireless communication means have become available for the man in the street. Driven by economical forces, the consumer telecom market has pushed the telecommunication system designers to smaller and cheaper solutions to allow mass volume production.

R. J. van de Plassche et al. (eds.), Analog Circuit Design, 287–323.

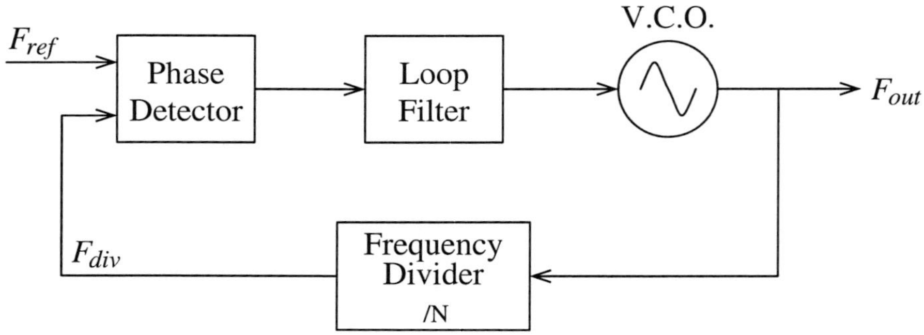

Figure 1: A functional block diagram of a PLL frequency synthesizer

To comply with those demands, integration of the analog RF part of wireless transceivers in cheap digital CMOS technology seems a viable solution. However, some obstacles remain.

A major challenge in the design of future CMOS single-chip transceivers for mobile communication systems is the building block that generates the Local Oscillator (LO) carrier signal: the frequency synthesizer. Generally, three common frequency synthesizer types exist: the table-look-up synthesizer (or digital synthesizer), the direct synthesizer and the indirect or phase-locked loop (PLL) synthesizer [1]. Other synthesizer types such as the Direct Digital Frequency Synthesis (DDFS) [2] or multi-loop synthesizers [3] are combinations of the previous types. Digital synthesizers cannot be used for high frequency applications, whereas direct synthesis is too bulky and power-hungry for full integration. On the other hand, DDFS can only synthesize limited frequency ranges and multi-loop synthesizers suffer from unwanted sidebands due to the limited accuracy of the quadrature mixing process [3]. The indirect or phase-locked loop (PLL) synthesizer is the most interesting solution for integration, which explains its popularity in mobile communication systems. The following sections discuss the basic operation principles of a PLL frequency synthesizer. The presented PLL design flow is mainly focused on the loop parameter choice and the filter topology selection. The different performance trade-offs are clarified using the DCS-1800 class I/II standard as an example. The consequences of extending the classical PLL concept to ΔΣ fractional-N synthesis are briefly discussed.

2 Basic Operation Concept of a PLL

A phase-locked loop synthesizer, as shown in Figure 1, basically consists of four building blocks: a phase detector, a loop filter, a voltage controlled oscillator (VCO) and a frequency divider. The output frequency F_{out}, generated by the VCO is divided by a number N in the frequency divider. The phase of the divided frequency is compared in the phase detector to the phase of the clean reference frequency F_{ref} , generated by a Xtal oscillator. The voltage output signal of the phase detector is proportional to the phase difference between both inputs and consists of a AC and a DC component. The DC component is accumulated and the AC component is filtered by the low-pass loop filter. The resulting signal is a DC signal with a small superimposed AC signal, which serves as the control voltage for the VCO. The change in control voltage is such that it pushes the VCO in the direction to reduce the phase error at the input of the phase detector. When the phase relationship between the inputs is constant, meaning that their frequencies are equal, the PLL is said to be in lock and the output frequency equals :

$$F_{out} = N \cdot F_{ref} \tag{2.1}$$

An important feature of the PLL is the suppression of the AC signal by the loop filter, meaning that noise and undesired signals will be attenuated by the loop. The narrower the loop bandwidth, the more suppression is achieved, but the PLL's dynamic behavior is deteriorated and the acquisition of lock, total capacitance, loop stability,... is affected. The mathematical description will be given in the following sections. It can be concluded that the loop filter is critical in the design of PLLs.

The output spectrum of the PLL at frequencies outside the loop bandwidth is determined by the VCO spectrum. Therefore, for fully integrated frequency synthesizers, a high-quality, low-noise VCO must be integrated. The design of integrated VCOs in CMOS technology is already elaborately discussed [4-6], and is therefore beyond the scope of this work, as is the design of high frequency CMOS frequency dividers [4,7] and accurate phase detectors [1,8,9].

3 PLL Data Sheet

3.1 Phase Noise

Phase noise is a measure for the spectral purity of the PLL output signal and is the most stringent specification for PLL design for wireless transceivers. The output of an ideal oscillator is a perfect sinusoidal wave of frequency ω_o. In the frequency domain this corresponds to a Dirac impulse at ω_o. In practice, the output of an oscillator suffers from amplitude and phase fluctuations due to noise. In the frequency domain, the fluctuations cause sidebands close to ω_o, called phase noise. To quantify phase noise, a unit bandwidth is considered at a certain offset frequency $\Delta\omega$ from ω_o, its noise power is calculated and divided by the carrier power (Figure 2(a)). This results is the single sided power spectral noise density in dBc/Hz:

$$L\{\Delta\omega\} = 10 \cdot \log\left(\frac{noise\ power\ in\ 1Hz\ \ BW\ at\ \ \omega_o + \Delta\omega}{carrier\ power}\right) \tag{3.1}$$

The equivalent of phase noise in the time domain is jitter on the zero-crossings of the output signal. A first-order formula to relate jitter to phase noise is, with ΔT_c the cycle jitter:

$$L\{\Delta\omega\} = 10 \cdot \log\left(\frac{\omega_o^3}{\Delta\omega^2} \cdot \Delta T_c\right) \tag{3.2}$$

Figure 2(b) shows the importance of phase noise in a wireless receiver. The noisy LO signal mixes the small wanted signal as well as the large unwanted signal to the IF frequency. The tail of the downconverted unwanted signal significantly corrupts the downconverted wanted signal [6]. To get an idea about the phase noise in wireless systems, the phase noise specification for a DCS-1800 system is calculated (Table 1) at the most important offset frequencies, determined by the blocking specifications [10].

The in-band phase noise specification of the PLL is determined by the needed signal-to-noise ratio (SNR) after downconversion. The PLL phase noise is superimposed on the downconverted wanted signal by mixing with the wanted signal itself (Table 1), causing a deterioration of the SNR.

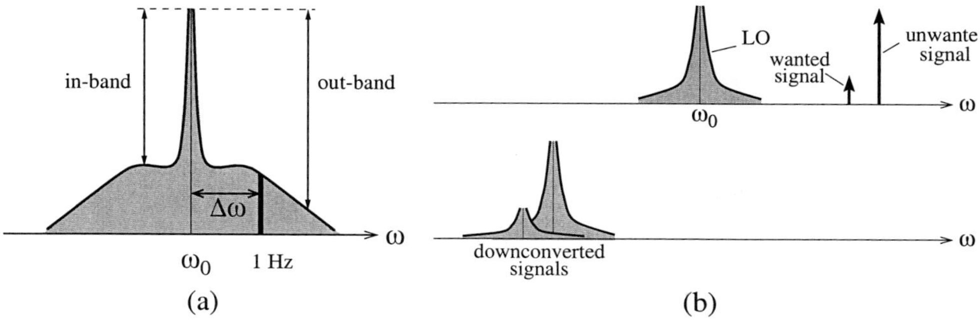

Figure 2: (a) Definition of a PLL phase noise spectrum and (b) the influence of phase noise in a receiver

To study phase noise in a PLL synthesizer, the PLL can be approximated by a linear-time-invariant (LTI) feedback system with the phase of the signals as the system variable (Figure 3). K_{pd} is the phase detector gain in V/rad, K_{vco} the VCO gain in MHz/V, N the division factor and $G_{lf}(s)$ the filter transfer function. The open loop tranfer function is, with $G(s)$ the forward transfer function and $H(s)$ the feedback factor:

$$GH(s)=\frac{K_{pd}\cdot G_{lf}(s)\cdot K_{vco}}{N\cdot s} \tag{3.3}$$

Based on the open loop transfer function the phase output response due to reference and VCO phase noise can be calculated:

$$\frac{\theta_{out}}{\theta_{ref}}=\frac{G(s)}{1+GH(s)}=\frac{N\cdot K_{pd}\cdot G_{lf}(s)\cdot K_{vco}}{N\cdot s+K_{pd}\cdot G_{lf}(s)\cdot K_{vco}} \tag{3.4}$$

$$\frac{\theta_{out}}{\theta_{vco}}=\frac{1}{1+GH(s)}=\frac{N\cdot s}{N\cdot s+K_{pd}\cdot G_{lf}(s)\cdot K_{vco}} \tag{3.5}$$

Knowing that $G_{lf}(s)$ has a low-pass characteristic, the reference phase noise is low-passed to the output and multiplied by a factor N. On the other hand, phase

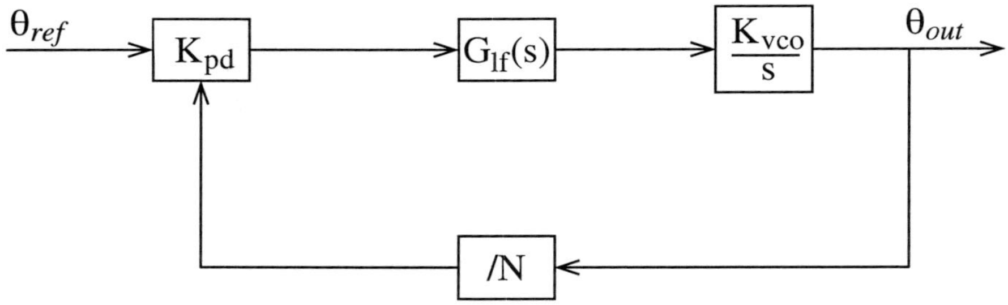

Figure 3: The PLL as LTI feedback system

noise generated by the VCO is high-passed to the output. Therefore, the out-band phase noise is mainly determined by the VCO, while in-band noise originates from noise sources in the phase detector, filter and divider. The loop filter design is thus critical to achieve the wanted phase noise performance.

3.2 Transient Behavior

To study the response of a PLL to a specified input $\theta_{ref}(t)$, the phase error $\theta_{err}(t)= \theta_{ref}(t) - \theta_{div}(t)$ is defined.
First, the steady-state error $\theta_{err,ss}$, i.e. the value of $\theta_{err}(t)$ when transients have died away, is calculated. It is assumed that the loop is in lock and that the phase error is small enough to use the LTI approximation.

$$\lim_{t\to\infty} \theta_{ref}(t) = \lim_{s\to 0} s\theta_{ref}(s) \tag{3.6}$$

If the reference input is a phase step of magnitude $\Delta\theta$, the resulting steady state error is :

$$\theta_{err,ss} = \lim_{s\to 0}\left(s \cdot \frac{\Delta\theta}{s}\frac{1}{1+GH(s)}\right) \tag{3.7}$$

If on the other hand the division factor N is changed, which corresponds to a frequency step $\Delta\omega$, the steady-state error equals:

$$\theta_{err,ss} = \lim_{s\to 0}\left(s \cdot \frac{\Delta\omega}{s^2}\frac{1}{1+GH(s)}\right) \tag{3.8}$$

If a first-order loop is assumed, i.e. $G_{lf}(s)$ is a constant K_{lf}, the loop will behave as a type-I system (one pole at zero frequency) due to the integration in the VCO. The steady-state error due to a phase step is zero, but the error due to a frequency step will be $\Delta\omega/\omega_c$ with $\omega_c=N/(K_{pd}.K_{vco}.K_{lf})$, the crossover frequency, i.e. the frequency where the open loop gain has a magnitude equal to one. If a second integration is present in the loop filter, the system behaves as a type-II system and the steady-state error is zero for both a phase and a frequency step.

Another important issue is the transient response of the loop or the settling of the loop, which comes down to determining the transient phase error due to a frequency step. For a first order loop the following calculation can be made :

$$\theta_{err}(t)=\frac{\Delta\omega}{\omega_c}\cdot\left(1-e^{-\omega_c t}\right) \tag{3.9}$$

So the final frequency is obtained after an exponential behavior with time constant inversely proportional to the loop gain. The time T_ε it takes to settle to a certain accuracy ε (e.g. 100 Hz) can be calculated to be:

$$T_\varepsilon=-\frac{\ln\left(\frac{2\pi\cdot\varepsilon}{\Delta\omega}\right)}{\omega_c} \tag{3.10}$$

From eq. (3.10), it can be seen that the design of the loop filter plays an important role in the transient behavior of a PLL. If a small loop bandwidth is chosen, the time constants in the filter become large, increasing the settling time. The simple equation cannot be used to study the settling time of more complex, higher-order loops, so the actual settling behavior must be examined by simulation of a PLL model.

The process of lock acquisition is beyond the scope of this work. More information can be found in [1,8,9].

3.3 Other PLL Specifications

Spurious suppression: The spurious suppression is in first order determined by the suppression of the loop filter. Second order effects are mismatches in the phase detector and injection of signals into the VCO LC-tank due to substrate and power supply line coupling.

Tuning range: A wireless transceiver uses predefined frequency bands to transmit and receive, which must be accommodated by the frequency synthesizer. The tuning range, i.e. the frequency range that can be synthesized, is mainly determined by the tuning range of the VCO.

Integratability: Integration of the VCO's LC-tank is beyond the scope of this work, making the loop filter the most critical part to integrate in silicon. In order for the PLL to occupy a moderate area, the integrated capacitance of the loop filter must be minimized, maintaining the required dynamic behavior and phase noise performance.

Power consumption: The power consumption must of course be as low as possible for mobile telecommunication system. The power consumption is mainly determined by the integrated low-phase-noise VCO.

Frequency resolution: The frequency resolution specification is set by the required channel spacing of the intended application, the DCS-1800 standard in this case. From eq. (2.1), it can be seen that the frequency resolution is determined by the reference frequency. However, when $\Delta\Sigma$ fractional-N synthesis is implemented, fractional multiples of the reference frequency can be synthesized, enabling higher reference frequency values [11,12]. By using $\Delta\Sigma$ fractional-N synthesis, frequency resolution can be exchanged with digital complexity, the implementation of which is the natural biotope of CMOS.

Table 1: DCS-1800 class I/II specifications [10]

Phase Noise (out-band)	-119 dBc/Hz at 600 kHz offset
	-136 dBc/Hz at 3 MHz offset
Phase Noise (in-band) for 30dB SNR	-80 dBc/Hz over a 200 kHz BW
Spurious suppression	-80 dBc
Settling time	865 μs
	288 μs high-speed data comm.
Frequency resolution	200 kHz
Tuning range	9.5 % of 1.8 GHz

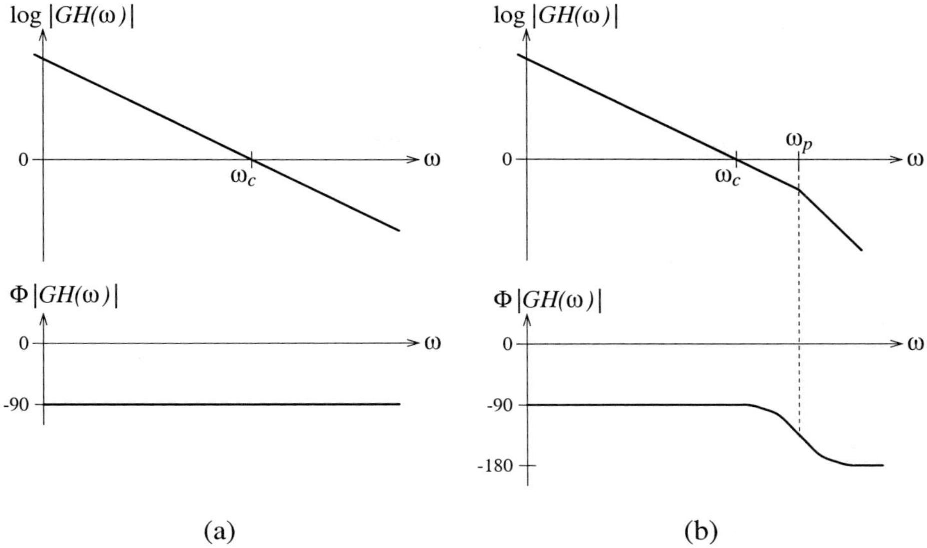

Figure 4: Open loop gain and phase for (a) a first-order and (b) a second-order type-I PLL

4 Loop Filter Topology Selection

As mentioned in the previous section, the loop filter choice is critical for the determination of the specifications of the PLL. The first-order PLL was used to illustrate the different specifications, but additional poles and zeros can be introduced in the open loop transfer function to set the noise and transient performance of the loop.

4.1 First-order PLL

In fact, every PLL is at least a first order, type-I PLL because of the integration, present in the VCO. In the case of a genuine first-order PLL, the loop filter transfer function $G_{lf}(s)$ is a constant, K_{lf}. The open loop transfer function is given by eq. (3.3) and shown in Figure 4(a). Only one pole at zero is present, giving a magnitude drop of 20 dB/dec and a constant phase of –90°. The dynamic behavior was investigated in the previous section.

4.2 Second-order PLL

To ensure sufficient spurious suppression, an extra pole can be added at a frequency larger than the crossover frequency. Because the loop bandwidth is typically chosen 10 times smaller than the reference frequency, the second pole must be low enough to be able to suppress reference spurs. The PLL is now a second-order, type-I PLL. The open loop gain and phase are plotted in Figure 4(b).

The loop transfer function is now

$$G_{lf}(s) = \frac{K_{lf}}{1 + s/\omega_p} \tag{4.1}$$

The resulting closed loop transfer function for θ_{ref} equals

$$\frac{\theta_{out}}{\theta_{ref}} = \frac{G(s)}{1+GH(s)} = \frac{\omega_p K_{pd} K_{lf} K_{vco}}{s^2 + 2\zeta\omega_n s + \omega_n^2} \tag{4.2}$$

with ω_n the natural frequency

$$\omega_n = \sqrt{\frac{\omega_p K_{pd} K_{lf} K_{vco}}{N}} = \sqrt{\omega_p \omega_c} \tag{4.3}$$

and ζ the damping factor

$$\zeta = \frac{1}{2}\sqrt{\frac{\omega_p N}{K_{pd} K_{lf} K_{vco}}} = \frac{1}{2}\sqrt{\frac{\omega_p}{\omega_c}} \tag{4.4}$$

To ensure loop stability, the phase margin of the system must be high enough, meaning that the second pole must be a factor higher than the loop bandwidth. If ω_p is too low, the transient response of the loop becomes less damped (eq. (4.4)), which can cause ringing. So the placement of the second pole, determines the transient behavior and the phase noise and spurious suppression.

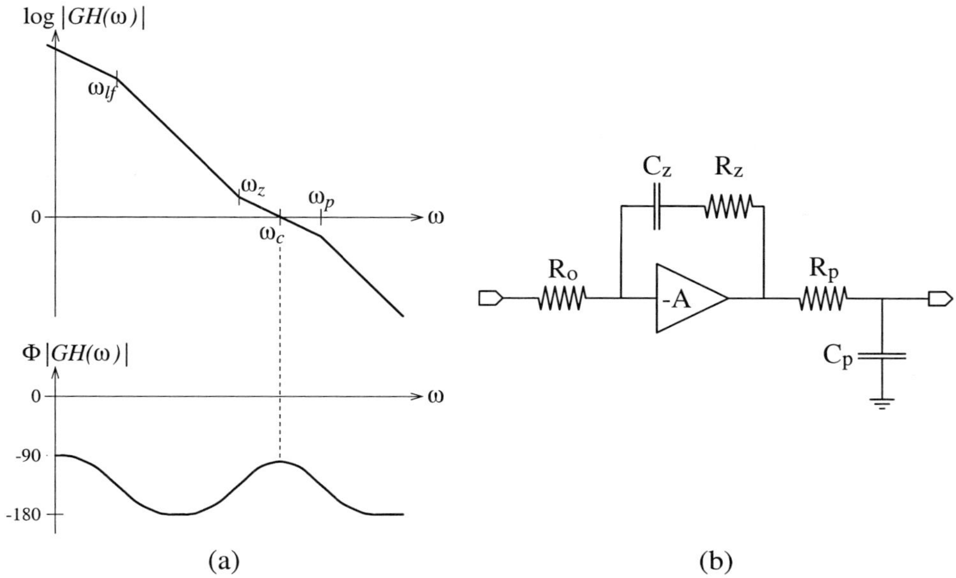

Figure 5: (a) Open loop gain and phase for a type-I, third-order PLL and (b) a type-II, third-order loop filter implementation

4.3 Third-order PLL

To improve the transient behavior of a second-order loop, a low-frequency pole ω_{lf} can be added to the loop filter [1]. To ensure a phase margin of at least 60°, the additional -90° phase shift must be compensated by introducing an extra zero, lower than the crossover frequency. The bode plot of a third-order, type-I PLL is shown in Figure 5(b). The slope of the transfer function at higher frequencies is -40 dB/dec, which is the same as for the second-order loop, such that the out- band phase noise and spurious suppression is similar. The higher low-frequency gain enables the PLL to better track input frequency steps (eq.(3.8)). In fact, ω_{lf} can be set to zero, introducing an extra integration. The loop becomes a type-II loop with a zero steady-state error for a frequency step. A typical type-II, third-order filter is shown in Figure 5(b). The zero is defined by C_z and R_z, while the pole is given by R_p and C_p.

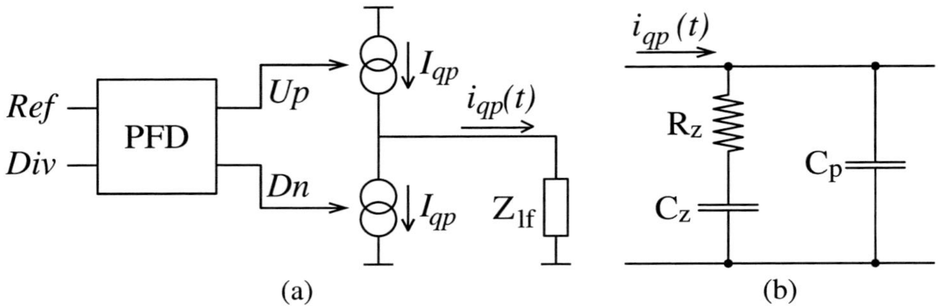

Figure 6: (a) Phase-frequency detector and charge pump and (b) a loop filter for a type-II, third-order charge-pump PLL

4.4 Charge-pump PLL

Three categories of phase detectors (Figure 1) can be distinguished [1,8]: analog phase detectors or multipliers, sequential phase detectors such as EXOR gates, flipflops and phase-frequency detectors (PFD). The combination of a phase-frequency detector and a charge-pump (Figure 6(a)) is widely used in PLLs [1], because of its frequency-sensitive error signal, that can aid acquisition when the loop is out of lock. If the pump currents of the charge-pump are both I_{qp}, the phase detector gain is

$$K_{pd} = \frac{I_{qp}}{2\pi} \tag{4.5}$$

A loop filter implementation for a charge-pump PLL is shown in Figure 6(b). The open loop transfer function becomes

$$GH(s) = \frac{I_{qp} \cdot K_{vco}}{2\pi \cdot N} \cdot \frac{G_{lf}(s)}{s} \tag{4.6}$$

$$G_{lf}(s) = \frac{1 + s\tau_z}{s(C_z + C_p)(1 + s\tau_z)} \tag{4.7}$$

with $\tau_z = R_z C_z$ and $\tau_p = R_z \cdot \left(C_z^{-1} + C_p^{-1}\right)^{-1}$.

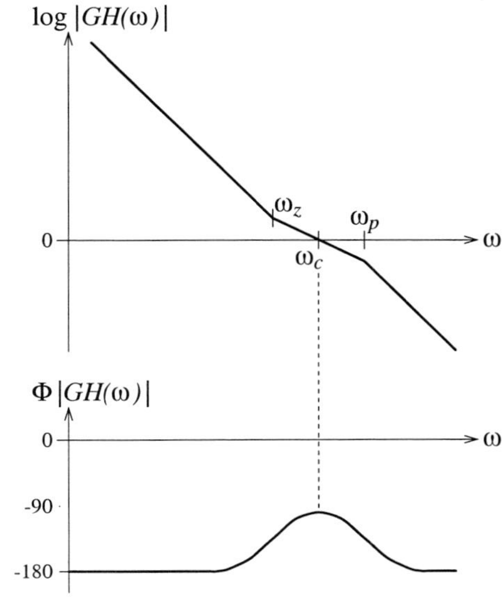

Figure 7: Open loop gain and phase for a type-II, third-order charge-pump PLL

This is the transfer function of a third-order, type-II PLL (Figure 7). There are two poles at zero frequency; One caused by the VCO and one caused by the charge pump, under lock condition. A charge-pump PLL is the only way to realize a type-II behavior with only a passive filter. The crossover frequency is approximately

$$\omega_c \approx \frac{I_{qp} \cdot K_{vco} \cdot R_z}{2\pi \cdot N} \cdot \frac{C_z}{(C_z + C_p)} \tag{4.7}$$

The placement of the pole ω_p is a trade-off between the necessary spurious suppression, which requires a low value and the dynamic behavior of the loop. As a rule of the thumb, the zero is placed at one fourth of the crossover frequency and the pole at four times the crossover frequency. This placement guarantees sufficient phase margin (around 60°) for loop stability and sufficient damping to suppress ringing. Based on this rule, the size of the passive filter elements can be calculated for a chosen ω_c.

To ensure that the loop can track even very small phase errors, it is advised to use a zero-dead-zone PFD [8]. In the zero-dead-zone PFD circuit, the charge pump current sources always conduct current, even when the loop is in lock. This means that the charge pump never is in its high-impedance state and the loop is always closed under lock condition. When a dead zone is present, the phase noise is not suppressed at very low offset frequencies proportional to the size of the dead zone. Another important issue in PFD-charge pump design, is the minimization of spur generation. Therefore, the charge pump current sources must be carefully matched and the charge injection due to transients when the pumps are switched must be minimized. The analysis of the PFD and charge pump transistor circuits is beyond the scope of this work [8].
To check whether the type-II, third-order charge-pump PLL can be used for integration in CMOS wireless transceiver, the phase noise contributions of the charge pumps and the loop filter are calculated. The noise of the two currents sources of the charge-pump can be written as

$$di_{qp}^2 = 2\alpha_{qp} \cdot di_n^2 \tag{4.8}$$

with α_{qp} the time fraction that the current sources are active.
The transfer function for this noise source will be :

$$\frac{\theta_{out}}{i_{qp}} = \frac{G_{lf}(s) \cdot K_{vco} \cdot N}{s \cdot N + K_{pd} \cdot G_{lf}(s) \cdot K_{vco}} = H_{qp}(s) \tag{4.9}$$

Knowing that the noise current of current source is $di_n^2 = 4kT \cdot g_{m,n} \cdot df$ with k the Boltzmann constant, T the temperature and $g_{m,n}$ the transconductance of the current source transistor, the phase noise can be calculated at a $\Delta\omega$ offset:

$$\begin{aligned} L_{qp}\{\Delta\omega\} &= \frac{\theta_{out}^2(\Delta\omega)}{2} \\ &= \frac{1}{2} \cdot \left|H_{qp}(\Delta\omega)\right|^2 \cdot 2\alpha_{qp} \cdot 4kT \cdot g_{m,n} \\ &= \left|H_{qp}(\Delta\omega)\right|^2 \cdot \alpha_{qp} \cdot 4kT \cdot \frac{2I_{qp}}{(V_{GS} - V_T)_{qp}} \end{aligned} \tag{4.10}$$

The noise is now calculated for a charge pump current of 10 μA, an on-time fraction α_{qp} of 0.1and an overdrive voltage $(V_{GS}\text{-}V_T)_{qp}$ of 0.3 V. With $\omega_c = 2\pi.100$ kHz, N=64 and a $K_{vco} = 2\pi.350$ MHz/V, the passive elements can be calculated to be R_z= 11.5 kΩ, C_z = 555 pF and C_p = 35 pF. The resulting phase noise at 600 kHz can be calculated to be $L_{qp}\{600\text{kHz}\} = -121$ dBc/Hz. This is too high to achieve the specification in Table 1 with enough margin, taking into account the noise contributions of the other building blocks, especially of the VCO. The charge pump current noise contribution can be lowered by increasing the overdrive voltage or decreasing the on-time, but both action are prawn to deteriorate the spurious suppression. The same calculation can be performed for the thermal noise from R_z : $di_{Rz}^2 = 4kT/R_z \cdot df$. The transfer function is:

$$\frac{\theta_{out}}{i_{Rz}} = \frac{K_{vco} \cdot N}{s \cdot N + Z_{lf}(s) \cdot K_{vco} \cdot K_{pd}} \cdot \frac{R_z C_z}{sR_z C_z C_p + (C_z + C_p)} \tag{4.11}$$

The first part is the transfer function from the output of the filter to the PLL output and the second factor is the transfer function of the resistor current noise to voltage filter output noise. The phase noise at 600 kHz can be calculated in the same way as above. $L_{Rz}\{600 \text{ kHz}\}$ = -109.5 dBc/Hz, making it impossible to achieve the DCS-1800 specification. To reduce the phase noise, the loop bandwidth can be lowered, causing an increase of integrated capacitance. For example, to lower the noise to –125 dBc/Hz, the loop bandwidth must be lowered by a factor 3.3. The resulting increase in integrated capacitance is a factor 3.3^2 or from 555 pF to almost 6 nF, which is too large to integrate. Using the first order approximation of eq. (3.10), the settling time is increased by 3.3.

4.5 Fourth-order PLL

To reduce the phase noise contributions of the charge pump and the loop filter resistor R_z, an extra pole can be added at the same frequency or a little further than ω_p. This further suppresses the phase noise at high offset frequencies and might allow to relax the sizing of the loop parameters.

To implement the fourth-order, type-II charge-pump PLL, an active loop filter implementation is chosen. As mentioned in the previous section, the charge-

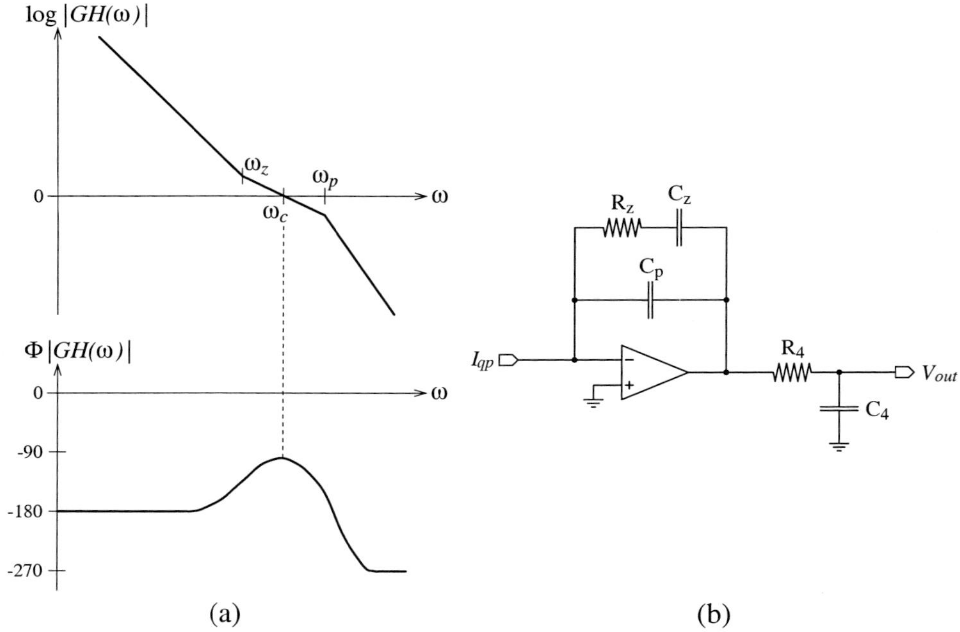

Figure 8: (a) Open loop gain and phase of a type-II, fourth-order charge-pump PLL (b) an active loop filter implementation

pump PLL offers the advantage that a type-II loop can be realized without an active filter. However, when using only passive components, some problems arise in the charge pump. In the configuration of Figure 6, the output voltage of the pump varies with the VCO control voltage. As a result, the overdrive voltages of the switches and the current sources will not be the same for the top source and the bottom part of the charge pump. The resulting mismatches in up- and down currents cause a net charge injection in the loop filter, even when the loop is locked and thus reference frequency spurs. Another problem is that a large output voltage range is required to allow VCO tuning without loosing the saturation condition in the transistors. But as deducted in eq. (4.10), the $(V_{GS}\text{-}V_T)_{qp}$ must be high for low noise. So despite the charge-pump PLL design, an active filter will have to be used to set the charge pump output to a fixed level. A straightforward implementation is shown in Figure 8, together with the bode plot of its open loop gain and phase.

The extra pole is placed on top of ω_p, by making $R_4C_4 = \tau_p$. The resistor R_4 introduces an extra noise source, but this noise contribution can be made arbitrarily small by increasing C_4. To ensure sufficient phase margin the placing of both ω_4 and ω_p, must be altered with respect to the third-order loop. Both poles must be placed six times higher than the crossover frequency instead of four times. The high frequency phase noise and spurious suppression is then increased with a factor $(4/6)^2(\Delta\omega/6\omega_c)^2$. To make the phase noise contribution of R_4 smaller than that of R_z, C_4 must be chosen large, so that the total integrated capacitance once again becomes too large.

5 Dual-path Fourth-order PLL

It is shown in the previous chapter that it is difficult to realize the phase noise specifications imposed by the DCS-1800 standard (Table 1), without the need for several nF of integrated capacitance. A fourth pole can help to reduce the phase noise, but the best remedy remains the reduction of the loop bandwidth. When the loop bandwidth is reduced, the influence on the loop's dynamic behavior and the explosive increase in capacitance must be taken into account.

A solution to integrate a relatively low-bandwidth, higher-order PLL with limited capacitance is the dual-path filter topology developed in [13]. The next sections discuss the topology implementation and optimization of a dual-loop filter that achieves the DCS-1800 specifications.

5.1 Dual-path Filter Topology

The principle of the dual-path filter topology is explained in Figure 9. In a conventional passive filter of a third-order charge-pump PLL (Figure 6(b)), the capacitor C_z that realizes the compensating zero is the most area-expensive component. The dual-path filter topology circumvents the need to physically integrate the zero with an RC combination at the desired frequency. By combining two signal paths, a virtual zero is realized. The placement of the zero is controlled by the relative amplification between both paths.

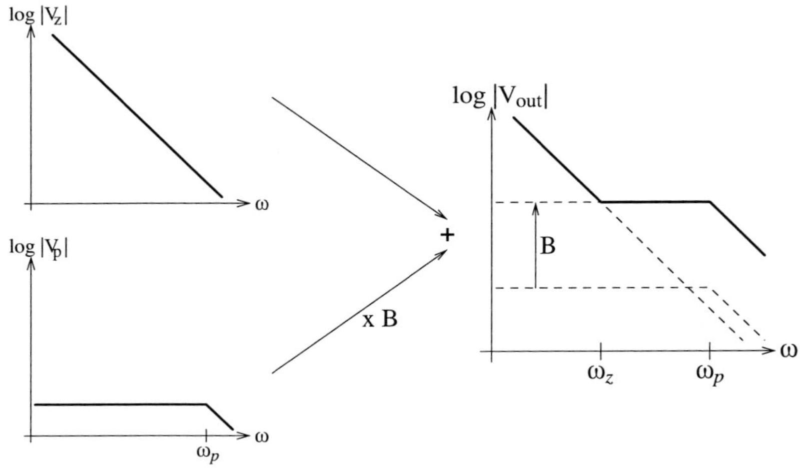

Figure 9: The dual-path loop filter principle

The first signal path V_z is an integration of the input current and the second is a low pass filter V_p, with a pole at ω_p.

$$V_z = \frac{1}{sC_z} \cdot I_{in} \tag{5.1}$$

$$V_p = \frac{R_p}{1 + sR_pC_p} \cdot I_{in} \tag{5.2}$$

These two signal paths are now combined to form the virtual zero, with a relative amplification by a factor B.

$$\begin{aligned} V_{out=} &= V_z + B \cdot V_p \\ &= \left[\frac{1}{sC_z} + B \cdot \frac{R_p}{1 + sR_pC_p} \right] \cdot I_{in} \\ &= \frac{1}{sC_z} \cdot \frac{1 + s\tau_z}{1 + s\tau_p} \cdot I_{in} \end{aligned} \tag{5.3}$$

with $\tau_z = R_z \cdot (C_p + B \cdot C_z) \approx BR_pC_z$ and $\tau_p = R_pC_p$

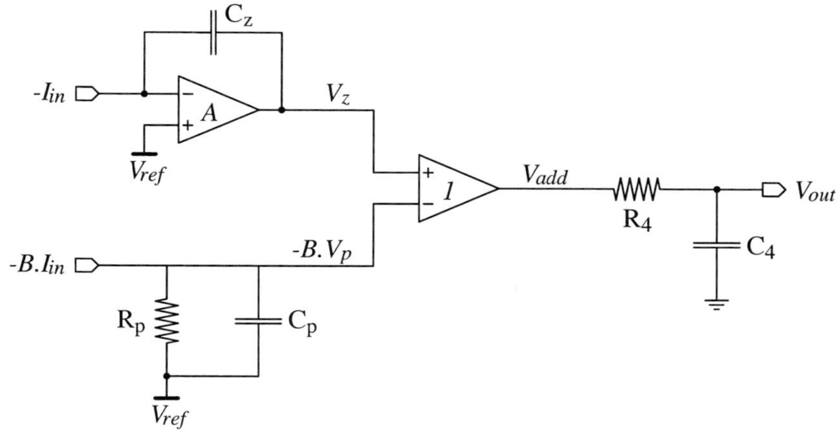

Figure 10: The dual-path loop filter implementation

So, a low-frequency zero is realized, without the need for a large integrated capacitor, because τ_z is multiplied by a factor *B*. The dual-path filter can be implemented as shown in Figure 10. The circuit has two current inputs; one is integrated to form V_z, the other is low-passed to V_p. The input currents are delivered by two charge-pumps with identical switch control signals. The implementation of the multiplication factor *B* is done in the current domain by using a pump current that is a factor *B* larger for the charge pump driving the signal V_p. The integration in the top path is done actively, to make sure that the charge pump output is biased at V_{ref} for spurious reduction. The bottom path can be implemented passively, because no current will run through R_p, when the loop is in lock, holding the charge pump output at V_{ref}. The adding of the two signal paths is done by converting both voltages to currents, which are summed in a diode-transistor, such that the gain in both paths is one. The output transistors of the filter adder have a small overdrive voltage to be able to accommodate the VCO with an as large as possible voltage range without loosing the saturation condition.

5.2 Transfer Functions

The different noise transfer functions for the PLL with a dual-path loop filter are calculated in this section to enable the optimization of the different loop filter parameters.

5.2.1 Open Loop Gain

The loop filter impedance of Figure 10 can be calculated as follows:

$$\begin{aligned} V_{out} &= V_{add} \cdot \frac{R_4}{1+sR_4C_4} \\ &= \left[V_z + B \cdot V_p\right] \cdot \frac{R_4}{1+sR_4C_4} \\ &= \left[\left(\frac{-1}{s \cdot C_z} \cdot -I_{in}\right) - \left(B \cdot \frac{R_p}{1+sR_pC_p} \cdot I_{in}\right)\right] \cdot \frac{R_4}{1+sR_4C_4} \\ &= \frac{1}{s \cdot C_z} \cdot \frac{1+s\tau_z}{(1+s\tau_p)(1+s\tau_4)} \cdot I_{in} \end{aligned} \tag{5.4}$$

with $\tau_z = R_z \cdot (C_p + B \cdot C_z) \approx BR_pC_z$

$\tau_p = R_pC_p$

$\tau_4 = R_4C_4$

The open loop gain is than:

$$GH(s) = \frac{I_{qp} \cdot K_{vco}}{2\pi \cdot N} \cdot \frac{1+s\tau_z}{s^2 \cdot C_z \cdot (1+s\tau_p)(1+s\tau_4)} \tag{5.5}$$

with a crossover frequency :

$$\omega_c = \frac{I_{qp} \cdot K_{vco}}{2\pi \cdot N} \cdot \frac{(C_p + B \cdot C_z) \cdot R_p}{C_z} \tag{5.6}$$

To ensure sufficient phase margin, the zero $\omega_z = 1/\tau_z$ is placed a factor α below the crossover frequency. The two high-frequency poles $1/\tau_p$ and $1/\tau_4$ are chosen to coincide in order to have the best trade-off between out-of-band noise suppression and phase margin. They are placed a factor β above the crossover frequency. An extra degree of freedom can be introduced by defining a factor γ. If R_4 is made γ times smaller than R_p, C_4 must be made γ times larger than C_p. This enables the trade-off between phase noise and integrate capacitance. The passive filter elements can now be calculated using the following equations:

$$R_p = \frac{1}{B} \cdot \frac{2\pi \cdot N}{I_{qp} \cdot K_{vco}} \cdot \omega_c \tag{5.7}$$

$$C_z = \frac{\alpha}{B \cdot R_{p\cdot} \cdot \omega_c} = \frac{I_{qp} \cdot K_{vco}}{2\pi \cdot N} \cdot \frac{\alpha}{\omega_c^2} \tag{5.8}$$

$$C_p = \frac{1}{B \cdot R_{p\cdot} \cdot \omega_c} = \frac{I_{qp} \cdot K_{vco}}{2\pi \cdot N} \cdot \frac{B}{\beta \cdot \omega_c^2} \tag{5.9}$$

$$R_4 = \frac{R_p}{\gamma} \tag{5.10}$$

$$C_4 = C_p \cdot \gamma \tag{5.11}$$

From the open loop gain, the transfer function for the VCO noise and the Xtal reference noise can be calculated straightforwardly using the LTI PLL model.

$$\frac{\theta_{out}}{\theta_{vco}} = \frac{s^2 C_z \cdot N \cdot (1 + s\tau_p)(1 + s\tau_4)}{s^2 C_z \cdot N \cdot (1 + s\tau_p)(1 + s\tau_4) + K_{pd} \cdot K_{vco} \cdot (1 + s\tau_z)} \tag{5.12}$$

$$\frac{\theta_{out}}{\theta_{ref}} = \frac{N \cdot K_{pd} \cdot K_{vco} \cdot (1 + s\tau_z)}{s^2 C_z \cdot N \cdot (1 + s\tau_p)(1 + s\tau_4) + K_{pd} \cdot K_{vco} \cdot (1 + s\tau_z)} \tag{5.13}$$

5.2.2 Charge Pump Noise

The noise contribution of both charge-pumps to the output phase noise can again be calculated. The transfer function for both pumps is given by:

$$\frac{\theta_{out}}{i_{qp,z}} = \frac{sC_z \cdot N \cdot K_{vco} \cdot (1+s\tau_p)}{s^2 C_z \cdot N \cdot (1+s\tau_p)(1+s\tau_4) + K_{pd} \cdot K_{vco} \cdot (1+s\tau_z)} \cdot \frac{1}{s \cdot C_z} \tag{5.14}$$

$$\frac{\theta_{out}}{i_{qp,p}} = \frac{sC_z \cdot N \cdot K_{vco} \cdot (1+s\tau_p)}{s^2 C_z \cdot N \cdot (1+s\tau_p)(1+s\tau_4) + K_{pd} \cdot K_{vco} \cdot (1+s\tau_z)} \cdot \frac{R_p}{(1+s\tau_p)} \tag{5.15}$$

with the noise magnitudes given by the same expression as eq. (4.8) and $di_{qp,p}^2 = B \cdot di_{qp,z}^2$. When both charge pump noises are added quadratically, it can be shown that the phase noise contribution here is a factor B smaller than in the conventional fourth-order filter implementation.

5.2.3 Loop Filter Noise

The noise transfer functions for the different loop filter current noise sources are calculated in the same way as above.

$$\frac{\theta_{out}}{i_{Rp}} = \frac{sC_z \cdot N \cdot K_{vco} \cdot (1+s\tau_p)}{s^2 C_z \cdot N \cdot (1+s\tau_p)(1+s\tau_4) + K_{pd} \cdot K_{vco} \cdot (1+s\tau_z)} \cdot \frac{R_p}{(1+s\tau_p)} \tag{5.16}$$

In the conventional loop filters, the implementation of R_z was critical for noise. Since R_z is here replaced with $B.R_p$, the phase noise contribution of R_p is also a factor B smaller with respect to the contribution of R_z.
The same applies to the noise contribution of R_4. Z_a is the output impedance of the loop filter adder.

$$\frac{\theta_{out}}{i_{R4}} = \frac{sC_z \cdot N \cdot K_{vco} \cdot (1+s\tau_p)(1+s\tau_4)}{s^2 C_z N \cdot (1+s\tau_p)(1+s\tau_4) + K_{pd} \cdot K_{vco} \cdot (1+s\tau_z)} \cdot \frac{R_4}{sC_4(Z_a + R_4) + 1} \tag{5.17}$$

Two active elements are present in the loop filter, both generating noise. Their noise sources are approximated by the equivalent input noise voltage sources $4kT/g_m$.
The shapes of their noise transfer functions are identical.

$$\frac{\theta_{out}}{v_{gm}} = \frac{sC_z \cdot N \cdot K_{vco} \cdot (1+s\tau_p)}{s^2 C_z \cdot N \cdot (1+s\tau_p)(1+s\tau_4) + K_{pd} \cdot K_{vco} \cdot (1+s\tau_z)} \tag{5.18}$$

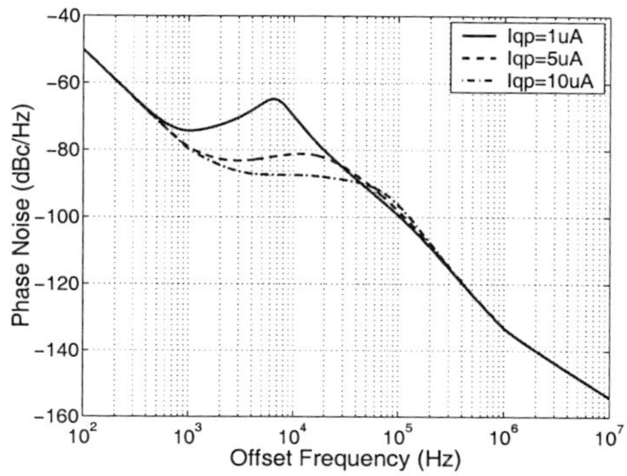

Figure 11: Influence of I_{qp} on the PLL phase noise performance

The g_m of both the adder and the integrator amplifier allow a power-noise trade-off.

5.3 Filter Optimization

To start the optimization of the loop filter, the fixed parameters of the PLL are determined. In this application, the reference frequency is set to 26 MHz. This means that the divider modulus range must be between 65 and 73, to synthesize the DCS-1800 frequency band. Therefore, a 64/79 16-modulus prescaler is implemented. The parameters of the VCO, implemented in the PLL, are considered to be known [14]. The VCO gain K_{vco} is between 2π.300 MHz/V and 2π.120 MHz/V in the desired frequency range. The phase noise of the VCO (Figure 14 (a), gray) is –127.5 dBc/Hz at 600 kHz and –142.5 dBc/Hz at 3 MHz offset from a 1.82 GHz carrier.

The optimization is performed in Matlab [15], using eq. (5.7)-(5.11) to determine the loop components and the total integrated capacitance. The noise transfer functions of eq. (5.12)-(5.18) are used to evaluate the phase noise contributions, using the relationship $L\{\Delta\omega\} = \theta_{out}^2(\Delta\omega)/2$. The model allows to rapidly explore the effect of the loop parameters on the overall PLL performance.

The loop parameters that are incorporated in the optimization process are the loop bandwidth ω_c, the charge pump current I_{qp}, the charge pump current factor B and the fourth pole factor γ. The optimization goal is achieving the

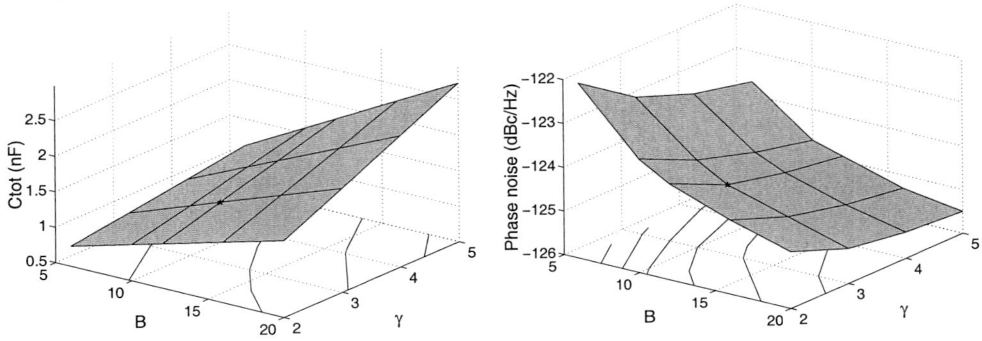

(a) (b)

Figure 12: Influence of the charge pump current factor B and the fourth pole factor γ on (a) the capacitance and (b) the phase noise

phase noise specification within two constraints. The first constraint is low power consumption, mainly reflected in the choice of the g_m of the adder and the integrator amplifier .The second and most stringent constraint is the occupied chip area. The total integrated capacitance, which is the sum of C_z, C_p and C_4, must have a value that allows integration in silicon.
The factors α and β in eq. (5.8)-(5.9) are no optimization parameters, because they set the phase margin and thus the loop stability. The chosen values are $\alpha = 4$ and $\beta = 6$.

The optimization can be started by setting the charge pump current I_{qp} as low as possible, since this results in the smallest integrated capacitance (eq. (5.8)-(5.9)). However, by decreasing the charge pump current, the settling time is decreased and the phase noise is increased. Not only the out-of-band phase noise is affected but also the in-band phase noise, which must remain below –80 dBc/Hz (Table 1) will be increased. The influence of I_{qp} on the phase noise is plotted Figure 11, with fixed passive loop parameter values. The minimum charge pump current is also limited by the parasitic charge injection of the switching transistors in the charge pump. They are chosen minimum size and will therefore inject a known, fixed charge in the charge pump. When the charge pump current is decreased, this parasitic charge injection becomes more and more dominant, giving rise to increased spurs at the reference frequency.

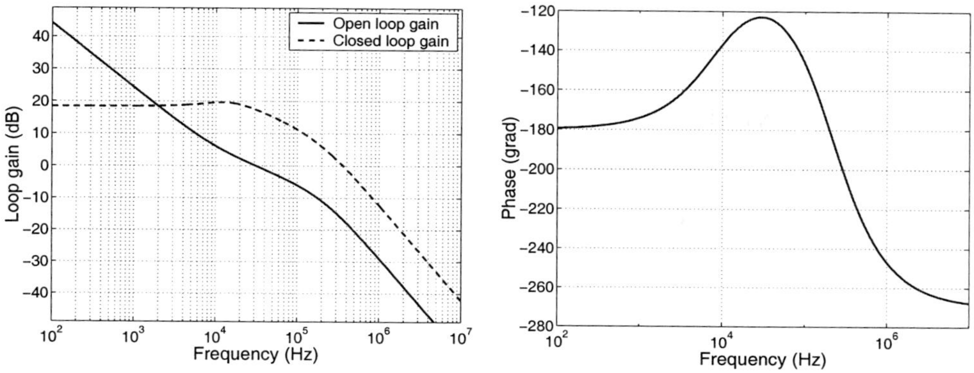

(a) (b)

Figure 13: (a) Open and closed loop gain and (b) phase of the optimized PLL

The second optimization step involves the choice of the charge pump current factor B and the fourth pole factor γ. Increasing the values for B and γ, will lead to better phase noise, but also to more integrated capacitance. The influence of both parameters is plotted in Figure 12. As a trade-off, the values $B = 12$ and $\gamma = 3$ are chosen (the marker in Figure 12).

Last but not least, the loop bandwidth is set. The influence of the loop bandwidth on the phase noise performance, loop dynamics and integrated capacitance is already elaborately discussed. The results of the optimization are shown Figure 14. A summary of the final PLL parameters and the results is given in Table 2 . The settling time is determined by simulation of an ideal behavioral model in the mixed-signal simulator Saber [16] for a frequency step of 104 MHz and a settling accuracy of 100 Hz . A loop bandwidth of 35 kHz and a charge pump current of 1 μA is chosen for the optimization of the loop components. However, as can be seen in Figure 11, the in-band phase noise is too high in that case. Therefore, in further simulation the passive loop parameter values are fixed but the PLL charge pump current will be set to 4 μA. For the charge pump, an on time fraction α_{qp} of 0.1 and a $(V_{GS}\text{-}V_T)_{qp}$ of 0.3 V is chosen. Figure 13 shows the simulated open loop gain and phase. The crossover frequency is indeed 35 kHz. The phase margin is 57° and the damping factor ζ is 0.77, which is sufficient to avoid excessive ringing and slow settling. Figure 14 show the respective phase noise contributions of all loop components. Figure 14(a) shows clearly that the VCO phase noise is high-passed to the output and determines the PLL for higher offset frequencies. However, the overall phase noise performance is also greatly affected by the VCO. Firstly, if the VCO exhibits much $1/f^3$ phase noise, the in-band PLL phase noise is deteriorated, especially at offset frequencies around to the loop bandwidth. Flicker noise upconversion minimization techniques can be used to overcome this problem [17]. Secondly, a variation in the VCO gain K_{vco}, changes the open loop gain and thus the loop bandwidth, the phase margin and the phase noise performance of the loop. To ensure stability and good noise performance over the total frequency range, the large possible VCO gain variation (2π.300 MHZ/V to 2π.120 MHZ/V) must be kept under control. A linearization circuit is implemented that tries to keep the open loop gain constant by controlling the product K_{vco} x I_{qp} (eq.(5.5)). The circuit analysis of the linearization block is beyond the scope of this work.

Table 2: Final PLL loop parameters and performance

Design parameters		**Passive elements**		**Performance**	
ω_c	35 kHz	R_p	3.2 kΩ	$L_{qp,z}$ {600 kHz}	-163.8 dBc/Hz
I_{qp}	4 μA	R_4	1.07 kΩ	$L_{qp,p}$ {600 kHz}	-148.1 dBc/Hz
α	4	C_p	240 pF	L_{Rp} {600 kHz}	-136.6 dBc/Hz
β	6	C_4	710 pF	L_{add} {600 kHz}	-138.9 dBc/Hz
γ	3	C_z	450 pF	L_{int}{600 kHz}	-138.9 dBc/Hz
B	12	C_{total}	1.4 nF	L_{R4} {600 kHz}	-135.3 dBc/Hz
				L_{vco} {600 kHz}	-127.5 dBc/Hz
PM	57°			L_{tot} {600 kHz}	-125.9 dBc/Hz
ζ	0.77			Settling time	293 μs

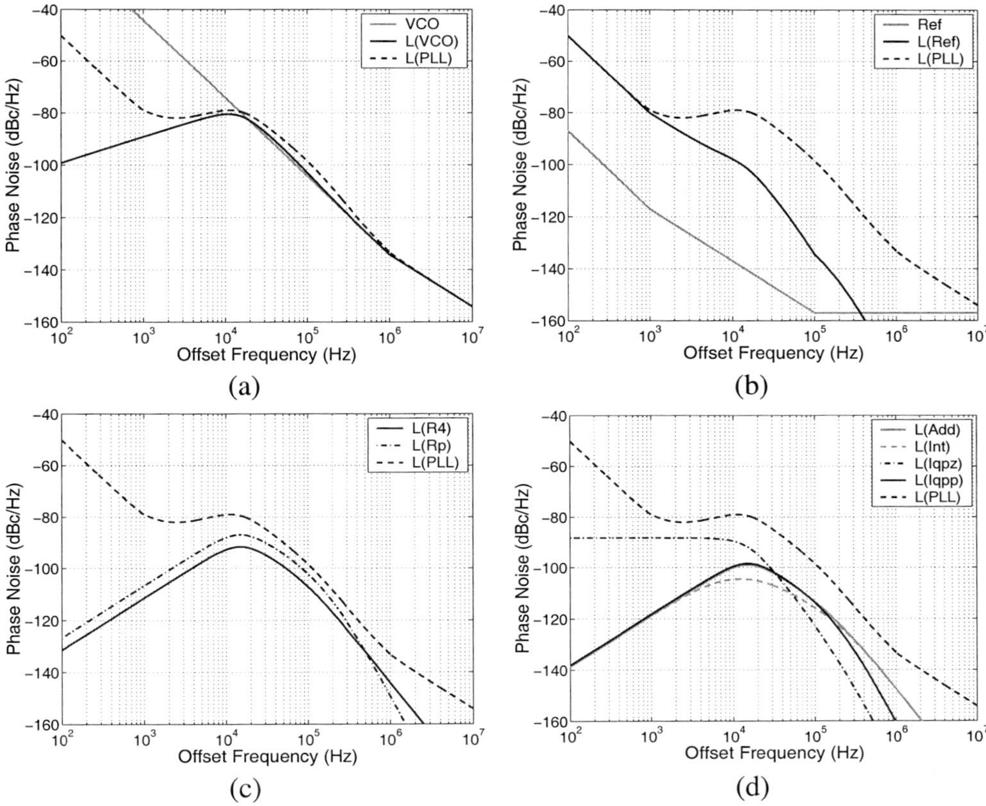

Figure 14: Phase noise contributions of (a) the VCO, (b) the reference noise, (c) the passive loop elements and (d) the active loop elements and charge pumps

6 Measurement Results

Based on the results of the optimization process from in the previous chapter, a phase-locked loop frequency synthesizer is integrated in a standard, two metal layer 0.25μm CMOS process. The IC includes a fully integrated loop filter, a fully integrated LC-tank VCO, a high-speed 16-modulus prescaler, a zero-dead-zone phase-frequency detector, a linearization block and two charge pumps [18]. An IC microphotograph is shown in Figure 16. The measurements are performed with a power supply voltage of only 2 V. An HP

Table 3: Measurement results

Phase Noise (out-of-band)	-125.8 dBc/Hz at 600 kHz offset
	-143.7 dBc/Hz at 3 MHz offset
Phase Noise (in-band)	< -80 dBc/Hz over a 200 kHz BW
Spurious suppression	-84 dBc
Settling time	226 μs
Tuning range	10.5 %
Area	2x2 mm^2

signal generator is used as the 26 MHz reference frequency source. The measurement results are listed in Table 3. The PLL complies with the DCS-1800 class I/II system specifications in Table 1. The phase noise is measured with a dedicated phase noise measurement system employing the delay line method. The result is plotted in Figure 15(a) and compared with the simulated phase noise at 1.82 GHz. The simulations are in excellent agreement with the measurements, except for the in-band phase noise. This can be explained by mismatches between the implemented and actual absolute values of the passive loop components. The phase noise at 3 MHz offset is better than the VCO phase noise mentioned in the previous chapter. The bias current of the VCO implemented in the PLL is higher than in [14], resulting in higher power consumption, but better phase noise at high offset frequencies, due to the increased output amplitude. The phase noise measurement at offsets higher than 5MHz is not accurate due to the delay line.

The settling behavior of the PLL is shown in Figure 15(b). When synthesizing a 104 MHz step (4 prescaler moduli), the PLL settles to a 100 Hz accuracy in 226 μs. This is faster than simulated, because the simulations where performed with a charge pump current of 1 μA, while the measurements are performed with 4 μA. The tuning range of the PLL is 10.5 %, which corresponds to 8 prescaler moduli. The spurious suppression is measured to be –84 dBc, which is the noise floor of the spectrum analyzer. The power consumption of the PLL is 70 mW from a 2 V power supply. The fully integrated, low-phase-noise VCO is responsible for almost 70% of the total power consumption. The IC area is 2x2 mm^2, including bonding pads and bypass capacitors.

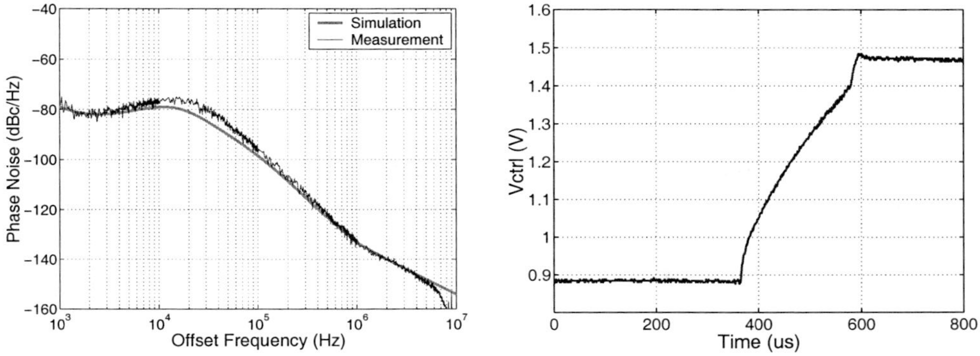

(a) (b)

Figure 15:(a) Measured phase noise and (b) measured settling time

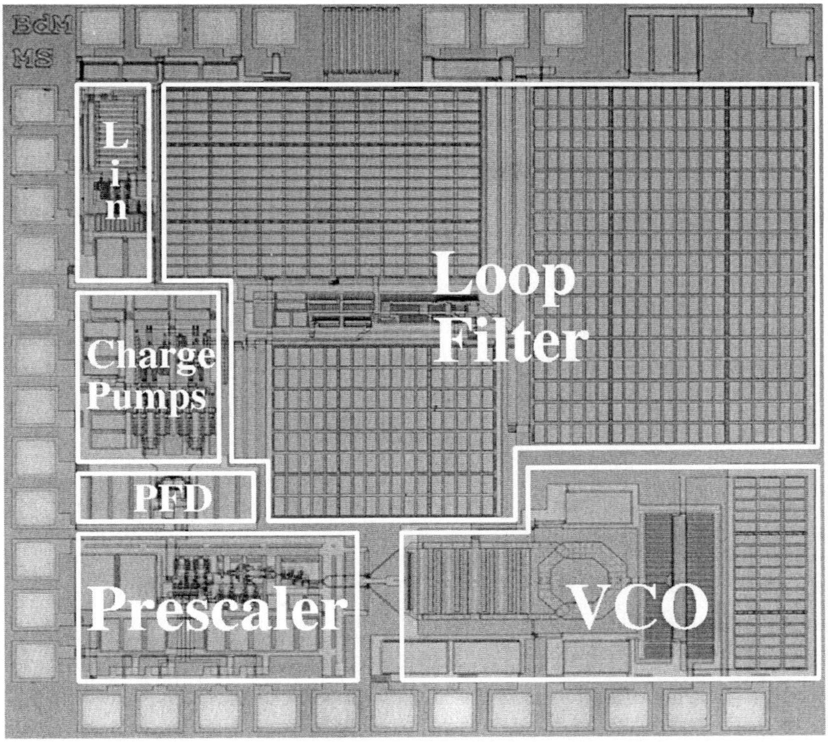

Figure 16: IC Microphotograph

7 ΔΣ Fractional-N Synthesis

The presented PLL achieves a frequency resolution of 26 MHz, since the PLL can only synthesize integer multiples of the 26 MHz reference frequency (eq.). To satisfy the frequency resolution specification imposed by the DCS-1800 standard, the PLL must be able to divide by fractional numbers. This can be done with ΔΣ fractional-N division control [11,12]. A block diagram of a ΔΣ fractional-N synthesizer is shown in Figure 17. The basic operation can be explained using the single digital accumulator in the right corner of Figure 17; When a digital word K is applied to the k-bit digital accumulator, a carry bit is produced on average every $K/2^k$ cycles of the reference frequency F_{ref} . This carry bit controls the divider modulus. A dual-modulus prescaler will divide K cycles by N+1 and 2^k-K cycles by N, resulting in an average division number:

$$\begin{aligned} N_{fract} &= \frac{\left((N+1)\cdot K + N\cdot\left(2^k - K\right)\right)}{2^k} \\ &= N + \frac{K}{2^k} \end{aligned} \tag{7.1}$$

As can be seen, the generation of the carry bit and thus the modulus switching is a periodic event, creating large spurs in the PLL output spectrum. To alleviate these spurs, the accumulator is replaced by an all-digital ΔΣ modulator, which performs a similar function, but with higher frequency switching and more randomization. The resulting ΔΣ bitstream, that controls the divider number, consists of a DC-value that corresponds to the required fractional division number and of shaped quantization noise. A first-order modulator with a DC-input cannot randomize the quantization error sufficiently, giving rise to pattern noise [19] and thus spurs in the PLL output. Therefore, higher order modulator topologies are needed to remove all possible periodicity. Higher order ΔΣ modulators remove all the quantization noise around DC and multiples of F_{ref} . The resulting quantization noise is mixed by the phase detector at a rate F_{ref}, such that the quantization noise is not folded into the signal band and is maximal at multiples of $F_{ref}/2$. This noise is then removed by the averaging action of the loop filter. It is advised to choose the ratio between the loop bandwidth and the sampling frequency greater than 100

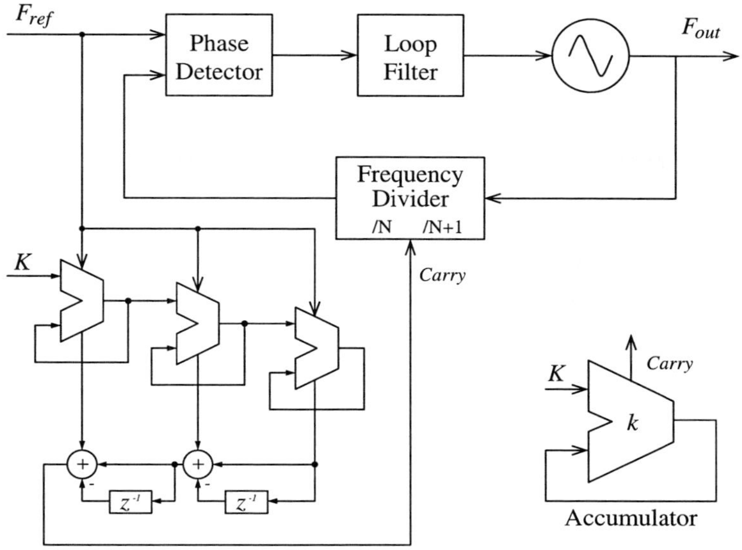

Figure 17:ΔΣ fractional-N synthesizer controlled by a 3-th order MASH converter (right under: control by a single accumulator)

to ensure enough filtering of the low-frequency quantization noise [20].A possible modulator implementation is a MASH (Multi-stage noise-shaping) converter (Figure 17), because of its easy integration in CMOS and its unconditional stability, as opposed to other higher order modulators [21]. A more detailed overview of the impact on the overall spectral purity of the PLL is given in the following sections.

7.1 Impact on Phase Noise

The use of fractional-N synthesis makes it possible to choose high reference frequencies while maintaining the wanted frequency resolution. As a result, the division modulus can be decreased and the in-band noise is less amplified by the loop (eq.(3.4)). However, the modulator also contributes to the overall phase noise. To model the impact of ΔΣ control on the PLL, the same PLL LTI model as before is used, but the model of the prescaler must be adapted. The prescaler-modulator combination can be seen as a digital-to-phase (D/P) converter. A change in digital input causes a change of the prescaler modulus and thus the output frequency. A dual-modulus prescaler can be seen as a 1-bit phase quantizer, which subtracts 0 or 2π rad every reference cycle.

Because the modulator and the prescaler are inherently digital circuits, the analysis is performed in the z-domain. The prescaler modulus is:

$$N(z) = N + \frac{K}{2^k} + H_{qnoise}(z) \cdot Q(z) \tag{7.2}$$

with $H_{qnoise}(z)$ the noise shaping function of the ΔΣ modulator and $Q(z)$ the quantization noise. The resulting output frequency of the PLL equals:

$$F_{out}(z) = \left(N + \frac{K}{2^k} \right) \cdot F_{ref} + H_{qnoise}(z) \cdot Q(z) \cdot F_{ref} \tag{7.3}$$

The frequency fluctuation caused by the quantization noise is given by the last term of eq. (7.3). When a n-th order MASH converter is used, the noise transfer function is given by $H_{qnoise}(z) = \left(1 - z^{-1}\right)^n$ [21]. The quantization noise is that of a one bit quantizer, i.e. $\Delta^2/12$ and $\Delta = 1$, under the assumption that the quantization noise can be approximated by uniformly distributed white noise, which is a safe approximation for higher order modulators, even with non-busy inputs [19]. The power spectral density (PSD) of the quantization noise is than $(12.F_{ref})^{-1}$, with F_{ref} the sampling frequency. The PSD of the frequency fluctuation is then:

$$\begin{aligned} S_f(z) &= \left| H_{qnoise}(z) \cdot F_{ref} \right|^2 \cdot \frac{1}{12 \cdot F_{ref}} \\ &= \frac{\left| 1 - z^{-1} \right|^{2n} \cdot F_{ref}}{12} \end{aligned} \tag{7.4}$$

To obtain the PSD of the phase noise, an simple rectangular integration is performed in the z-domain.

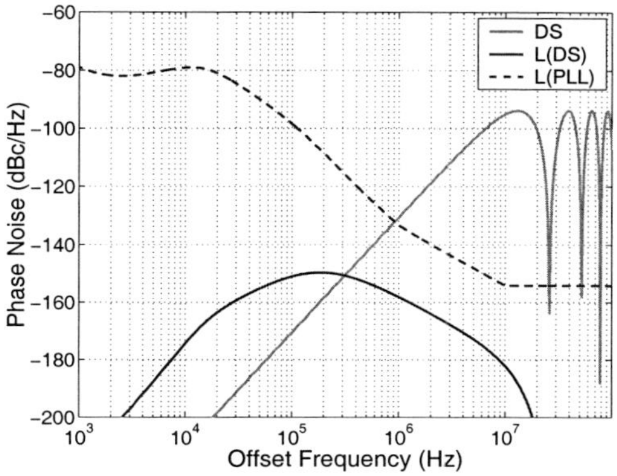

Figure 18: Phase noise at the prescaler output (DS) and the resulting output phase noise at the PLL output (L(DS)).

$$S_\theta(z) = \frac{\left|1 - z^{-1}\right|^{2n} \cdot F_{ref}}{12} \cdot \frac{(2\pi)^2}{\left|1 - z^{-1}\right|^2 \cdot F_{ref}^{\ 2}} = \frac{\pi^2}{3 \cdot F_{ref}} \cdot \left|1 - z^{-1}\right|^{2(n-1)} \tag{7.5}$$

Knowing that $z = e^{\frac{j2\pi f}{F_{ref}}}$ and that the prescaler divides by N the phase noise PSD at the output of the prescaler becomes

$$S_\theta(f) = \frac{\pi^2}{3 \cdot F_{ref}} \left[2 \cdot \sin\left(\frac{\pi f}{F_{ref}} \right) \right]^{2(n-1)} \cdot \frac{1}{N^2} \tag{7.6}$$

This noise is transferred to the output in the same way as the reference noise (eq. (5.13)). To ensure enough noise suppression at higher offset frequencies, care has to be taken to make the resulting noise contribution rolls off in the same way as the normal PLL phase noise, i.e. 20 dB/dec. When the loop filter has an order m, the resulting order of the output should be one: m-(n-1)=1 or m=n. In this case, the filter order is 3, thus the order of the $\Delta\Sigma$ modulator is chosen to be 3. A higher order modulator can provide even more noise shaping at the expense of a higher power consumption. The formulas above are derived for a dual-modulus prescaler, but the same applies to multi-modulus prescalers. The increased number of bits enable more quantization levels of the division modulus but the modulus range itself is extended with the same amount. This means that the distance between two quantization levels Δ remains equal. Consequently the above derivation still holds. The resulting noise contribution is plotted in Figure 18. It can be seen that the quantization noise has little effect on the overall PLL phase noise, but that it could deteriorate the phase noise at high offsets, when not properly designed.

7.2 Impact on Spurs

The main issue in design of fractional-N synthesizers is not the phase noise, but the spurs that can be present in the output spectrum. This section gives a short overview of the possible spur sources in a fractional-N synthesizer.
Firstly, when the input of the modulator is not a busy signal, such as DC in this case, spurs can be generated by the ΔΣ modulator itself. However, for second and higher order modulators, the noise shaping is active for DC-inputs as well [19]. If the randomization by the modulator is not effective enough, the number of bits in the accumulators and thus the number of internal quantization levels can be increased to decrease the periodicity. Also, additional dithering by a pseudo-random bit sequence can be introduced in the first accumulator of the modulator [19], which is the main spur source in case of a MASH converter.

A second source of spurs are non-linearities in the PLL components. The most critical block in terms of non-linearities, is the PFD. In this block the noise is mixed at a rate F_{ref} such that non-linearities, e.g. a dead zone, cause intermodulation. For instance, spurs that should be randomized by the subsequent ΔΣ stages, re-appear in the PLL output spectrum. Another effect is the shaping effect is lost at small offset frequencies. Instead of a roll off of (n-1).20 dB/dec, a flat noise floor is present. This is due to folding of high-frequency noise in the signal bandwidth by the non-linear PFD mixing. [12]

The third source of spurs and the most difficult to remedy is digital coupling through the substrate and power supply lines of the IC. Apart form layout tricks and shielding, the spurs can be reduced by careful design. The decision making in the phase detector, that determines the length of the output pulses and hence the charge injected in the loop filter, must be as accurate as possible. By assigning a different time interval to divider switching and computational events and to the phase detector switching, the influence of digital coupling on the accuracy of the PFD and hence spur generation can be reduced [11].

8 Conclusions

A design strategy for fully integrated CMOS frequency synthesizers, using the DCS-1800 system as driving application is presented. First, the most important specifications of a frequency synthesizer are discussed. It was concluded that apart from the VCO design, the design of the loop filter is critical to achieve a wanted performance. The loop filter design emphases the minimization of the integrated capacitance while maintaining the wanted phase noise performance by filter order and topology selection and optimization of the loop parameters. Based on these results, a type-II, fourth-order PLL frequency synthesizer is implemented in a standard, 2-metal layer 0.25μm CMOS technology, including a monolithic voltage controlled oscillator and a fully integrated loop filter. The measurements show compliance with the stringent DCS-1800 class I/II specifications. The chip area is only 2x2 mm^2 with 1.4 nF of integrated capacitance. To conclude, the influence of ΔΣ fractional-N division control on the spectral purity of the PLL frequency synthesizer is discussed.

References

[1] W.F. Egan, "Frequency synthesis by Phase Lock", J. Wiley & Sons, New York, USA, 1981

[2] A.A. Abidi, "Radio-frequency integrated circuits for portable communications", *Proc. Of the IEEE Custom Integrated Circuits Conference (CICC 94)*, pp. 151-158, 1994.

[3] B. Razavi, "Challenges in the design of frequency synthesizers for wireless applications", *Proc. Of the IEEE Custom Integrated Circuits Conference (CICC 97)*, pp. 151-158, 1997.

[4] J. Craninckx, "Low-phase-noise fully integrated CMOS frequency synthesizers", Ph.D. Dissertation, Katholieke Universiteit Leuven, 1997.

[5] P. Kinget, "Integrated GHz Voltage Controlled Oscillators", Analog Circuit Design, (W.Sansen, J.H.Huijsing and R.J. van de Plassche, eds), pp.353-381, New York : Kluwer Academic Publishers, 1999.

[6] B. Razavi, "A study of phase noise in CMOS oscillators", *IEEE Journal of Solid-State Circuits*, vol. 31, no. 3, pp. 331-343, March 1996.

[7] B. De Muer, M. Steyaert, "A single-ended 1.5GHz 8/9 dual-modulus prescaler in 0.7m CMOS with low phase noise and high input sensitivity", *Proc. of the European Solid-State Circuits Conference (ESSCIRC 98),* pp. 256-259, The Hague, September 1998.

[8] F.M. Gardner, "Phaselock Techniques", J. Wiley & Sons, New York, USA, 1979

[9] R.E. Best, "Phase-Locked Loops: design, simulation &application", Third Edition, McGraw-Hill, 1997.

[10] -, "European Digital Cellular Telecommunications System (Phase 2+): Radio transmission and reception", ETS(I) 300 910 (GSM 05.05 version 5.4.1).

[11] B. Miller and R.J. Conley,"A multiple modulator fractional divider", *IEEE Trans. On Instrumentation and Measurement*, vol. 40, no. 3, pp. 578-583, June 1991.

[12] T.A.D. Riley, M.A. Copeland and T.A. Kwasniewski, "Delta-sigma modulation in fractional-N frequency synthesis", *IEEE Journal of Solid-State Circuits*, vol. 28, no. 5, pp. 553-559, May 1993.

[13] D. Mijuskovic, M.J. Bayer, et.al, "Cell based fully integrated CMOS frequency synthesizers", *IEEE Journal of Solid-State Circuits*, vol. 29, no. 3, pp.271-279, March 1994.

[14] B. De Muer, N. Itoh, M. Borremans and M. Steyaert "A 1.8 GHz highly-tunable low-phase-noise CMOS VCO", *Proc. Of the IEEE Custom Integrated Circuits Conference (CICC 00)*, to be published, 2000.

[15] The Mathworks Inc., "Matlab user's guide, version 5", 1997.

[16] -, "Saber User Mnaual", Analogy, 1996.

[17] B. De Muer, C. De Ranter and M. Steyaert, "A fully integrated 2GHz LC-VCO with phase noise of -125dBc/Hz at 600kHz", *Proc. of the 1999 European Solid-State Circuits Conference (ESSCIRC '99)*, pp. 206-209, 1999.

[18] M. Steyaert, J. Janssens, B. De Muer, M. Borremans, N. Itoh, "A 2V CMOS cellular transceiver front-end", *ISSCC Digest of technical papers*, pp. 142-143, Feb. 2000

[19] S.R. Norsworthy, R. Schreier, G.C. Temes, "Delta-sigma data converters: theory, design and simulation", IEEE Press, 1996.

[20] Bar-Giora Goldberg, "Digital techniques in frequency synthesis", McGraw-Hill, 1995.

[21] A.I. Yasuyki, K. Uchimura, A. Iwata, et al., "A 16-bit oversampling A-to-D conversion technology using triple-integration noise shaping", *IEEE Journal of Solid-State Circuits*, vol. 22, no. 6, pp. 921-928, December 1987.

Design and Optimization of RFCMOS-Circuits for Integrated PLL's and Synthesizers

Marc Tiebout

Infineon Technologies AG,
Department Wireless Products,
Technology and Innovation, Systems
P.O. Box 800949, D-81609 Munich, Germany

Abstract

Voltage controlled oscillators and prescalers are critical high speed building blocks in RF synthesizers and PLL's. For wireless applications, major VCO-design criteria are low voltage, low power, low phasenoise, wide and/or linear tunability and small area. Primary key to LC-VCO optimization is optimal LC-tank design. 0.35μm CMOS VCO designs, fulfilling GSM and DECT phasenoise requirments, are presented. Solutions to generate quadrature outputs are discussed. Main prescaler design goal is low power and low voltage design. Multi modulus prescalers using dynamic CMOS flipflops can be used for minimal power consumption. Finally some mixed-mode aspects of RFCMOS circuit design for single-chip RF synthesizers and PLL's are discussed.

1. Introduction

In fig. 1, some block diagrams for typical frequency synthesizers or PLL's used in mobile telecommunications are shown. Circuits running at RF-frequencies are the voltage controlled oscillator (VCO), programmable frequency divider (prescaler) and the 0/90° phase shifting for I-Q paths. PLL's for a large, and still growing, number of telecommunication systems have to be designed (DECT, GSM900,

R. J. van de Plassche et al. (eds.), Analog Circuit Design, 325–337.

DCS1800, Bluetooth, HomeRF, remote keyless entry, ...). As these mobile systems are battery-operated, lowest power consumption is always an issue.

In section 2, prescaler design is discussed. Main target for this digital circuit is lowest power consumption. As the phasenoise requirements for VCO's in mobile communication standards cannot be met by ring oscillators at a reasonable power consumption level, section 3 concentrates on LC-VCO design. In Section 4, a low power 1.8GHz LC-VCO in 0.35μm is presented. Section 5 describes a similar design optimized for low phasenoise aiming at GSM. In section 6 follows a discussion about the generation of 0°/90° signals required in I-Q transceivers. Finally some mixed-mode aspects of integrated VCO's are discussed in section 7.

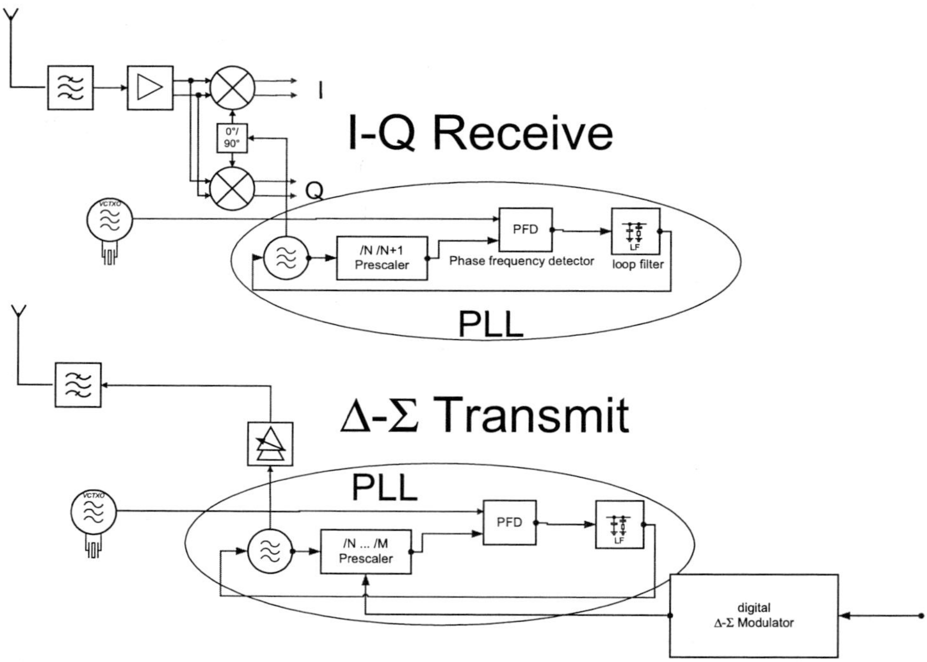

Fig. 1: selected examples of PLL's in mobile transceivers.

2. Prescaler Design

A dual modulus or multi-modulus prescaler, as used in fig.1, is a frequency divider with a programmable divider ratio, e.g. 128/129. Although a prescaler is a digital circuit, most realizations in production are bipolar because of their lower power consumption. Prescaler circuits can be realized using CML-logic fliplops (fig. 3) in bipolar processes [1] or in CMOS [2, 4-7], using architectures as can be found in fig.2 ([11]). The ongoing downscaling of CMOS-processes, makes it possible to design prescalers for the typical mobile communications frequency ranges from 900MHz up to 2.4GHz using dynamic CMOS flipflops like e.g. the structure of Yuan and Svenson [3], (fig. 4) and to optimize them for low power operation.

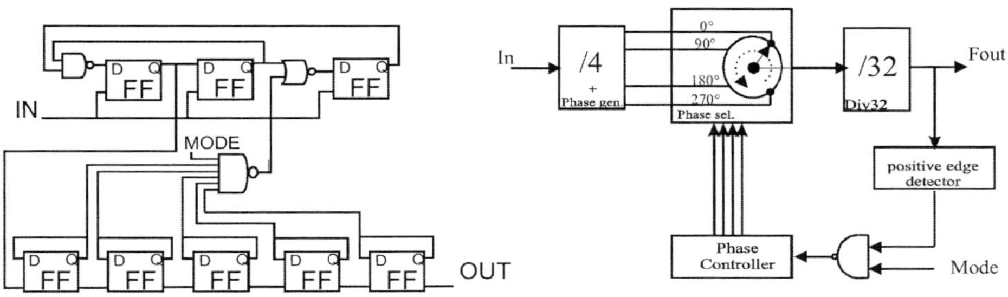

Fig.2: dual modulus 128/129 prescaler architectures,
left: classical architecture, right: phase switching architecture.

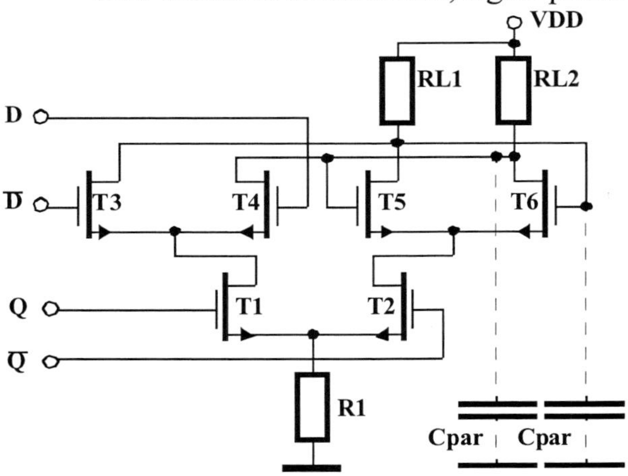

Fig. 3: CML logic flipflop.

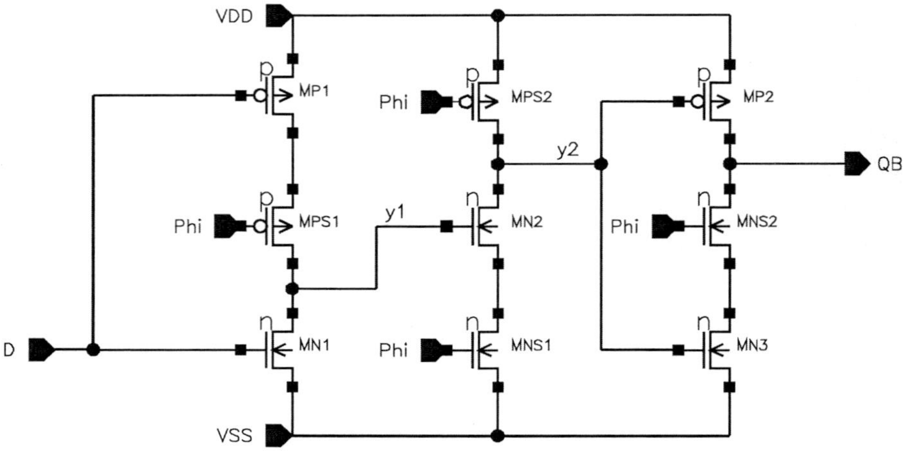

Fig. 4: dynamic CMOS flipflop.

A low power CMOS 128/129 dual modulus prescaler using this dynamic flipflops in the phase switching architecture was designed in INFINEON 0.25μm standard CMOS process [7]. Measured performance for an input signal power of 12dBm is presented in Fig.5, chifoto is shown in Fig. 6.

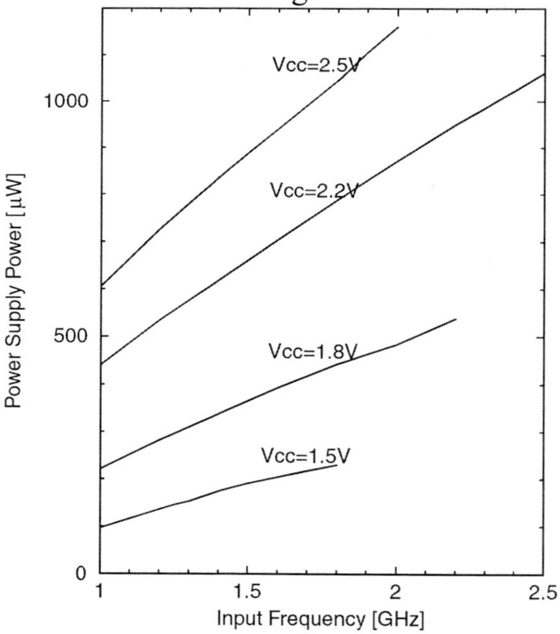

Fig. 5: 128/129 prescaler [7] power consumption.

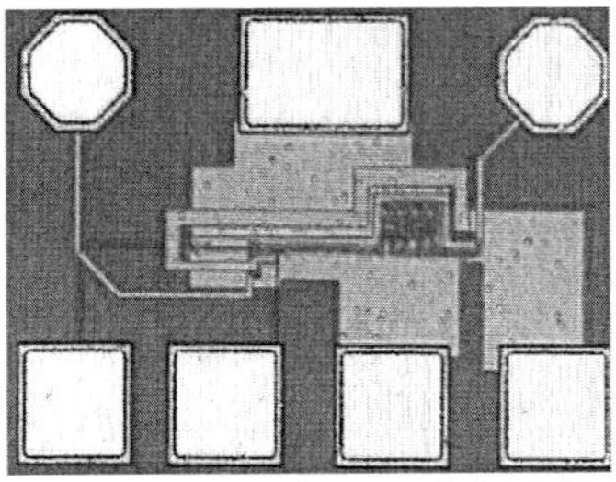

Fig. 6: chipfoto of 128/129 dual modulus prescaler.

A figure of merit to compare different prescalers can be defined as

$$FOM = \frac{P_{tot}}{F_{in}} \quad (1)$$

where P_{tot} is the prescaler power consumption and F_{in} is the input frequency. In Table 1, some recent prescaler publications are listed sorted by the FOM, the bipolar prescaler design [1] outperforms nearly all CMOS-design's.

Prescaler	Technology [μm]	Power Supply [V]	Speed [GHz]	FOM (1) [μW/MHz]
[4]	1.0 cmos	5	1.6	31.2
[6]	0.8 cmos	5	1.22	20.9
[2]	0.7 cmos	3	1.75	13.7
[5]	0.8 cmos	5	1.59	8
[1]	0.8 bip	2.7	2.5	1.52
[7] (fig. 6)	0.25 cmos	1.8	2.0	0.24

Table 1: prescaler performance comparison.

Main disadvantage of dynamic CMOS prescalers, using full power supply swings, is the large voltage needed at the input (up to 12dBm input power in a 50Ω system) and the possible crosstalk of the digital full swing levels into other sensitive RF circuits. To demonstrate the practical use of these prescalers, the design of fig. 6 was co-integrated with a fully integrated LC-VCO (chipfoto in Fig. 8) in the same 0.25μm technology. The LC-VCO topology and design procedure is discussed in section 5 of this paper. The prescaler was directly driven from the VCO. Fig. 7 shows the phasenoise performance of the VCO with prescaler switched off and on. No difference can be found.

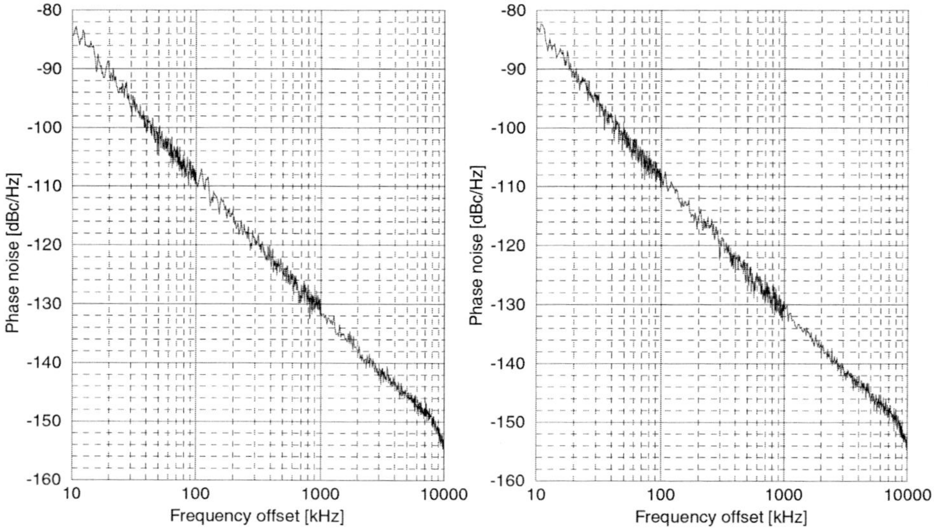

Fig. 7: measured VCO phasenoise, left: prescaler disabled, right: prescaler enabled (VCO VCC=1.5V, 6mA, frequency=1.22GHz).

Fig. 8: Chipfoto

3. LC-VCO Design

A LC-VCO consists of a resonance LC-tank and an active element cancelling the losses in the tank to maintain oscillation. Integrated VCO's use bipolar or MOS transistors as active element. NMOS, PMOS or a combination of NMOS and PMOS topologies are possible as proposed respectively in fig.9, 10 and 11.

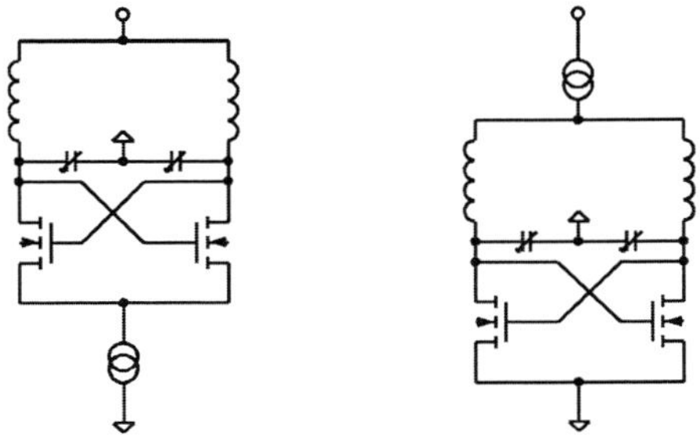

Fig. 9: NMOS LC-VCO topologies.

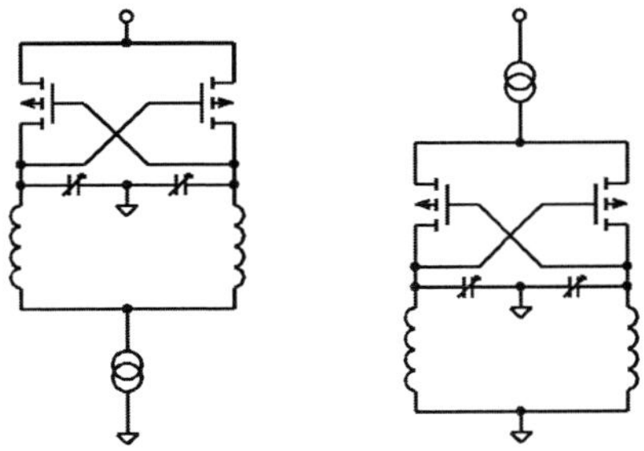

Fig. 10: PMOS LC-VCO topologies.

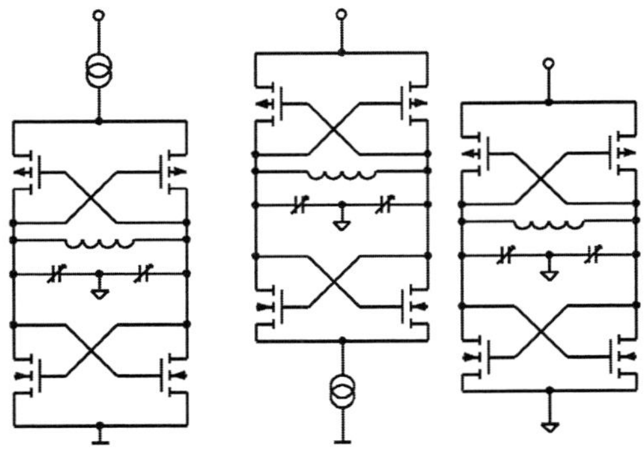

Fig. 11: NMOS/PMOS current reusing LC-VCO topologies.

NMOS- or PMOS-topology choice is driven mainly from technology (speed or f_T) and flicker noise aspects. PMOS-transistor has a lower flicker noise coefficient in most technologies, although it is no longer a buried channel transistor in modern sub-μm technologies. The combination of NMOS and PMOS transistors is very well suited for low power designs as tail current is reused.

Phasenoise mechanisms in VCO's have been treated extensively in open literature [8,9,10,11]. An empirical expression was published by Leeson in 1966 [14], which describes the phasenoise performance of VCO's effectively:

$$\mathscr{L}(f)=\frac{F\,kT}{2P_{sig}}\left[1+\left(\frac{f_0}{2\,Q\Delta f}\right)^2\right] \tag{2}$$

With Q the quality factor of the LC-tank, f_0 the carrier frequency, P_{sig} is the signal power of the oscillator, □f is the frequency offset, F the noise factor for the active devices, k Boltzmann constant and T absolute temperature. (2) shows clearly that the most effective parameter to improve phasenoise is the quality factor Q of the LC-tank. Furthermore (2) shows that, at constant phasenoise, a higher Q allows a reduction of signal P_{sig} or a lower power consumption, concluding that optimal LC-tank design enables low noise design as well as low power design.

Among many definitions for Q [13], some important insights result from the bandpass bandwith definition of the complete LC-tank. Physically the energy defined Q, or from the oscillator application, the phase stability defined Q should be more appropriate for the VCO-tank Q-definition. From the comparison in [13], qualitatively there is no significant difference, furthermore bandpass Q is more easy to handle from simulation and designers-insight viewpoints. The bandwith Q definition leads to 2 main conclusions. First, the inductor is used closer to its self resonant frequency (defined by Im(Z)=0), whereas conventional Q=Im(Z)/Re(Z) equals to zero at self resonance. Second conclusion arises from the bandpass bandwith properties of an

LC parallel connection: the ratio of L/C has to be maximized for highest Q. Best Q of the tank will be obtained at maximal inductance, maintaining just enough capacitance to assure the required tuning range.
The integrated inductors can be laid out as two single separate inductors, or as one fully symmetrical structure with middle tap. The coil structure with middle tap has the advantage to increase the effective inductance by a couple factor of about 0.8. It is also preferred from mixed-mode aspects, as substrate noise will be coupled into both nodes of the oscillator as a common mode signal. In the case of the coil with middle tap, self-resonance frequency, bandpass bandwidth and Q have to be redefined using differential voltages for differential operation in a fully balanced configuration. Looking at the simple lumped model of fig. 12, it is obvious that the capacitance at the middle tap of the coil is at common mode, resulting in increased differential self resonance frequency and differential bandpass Q.

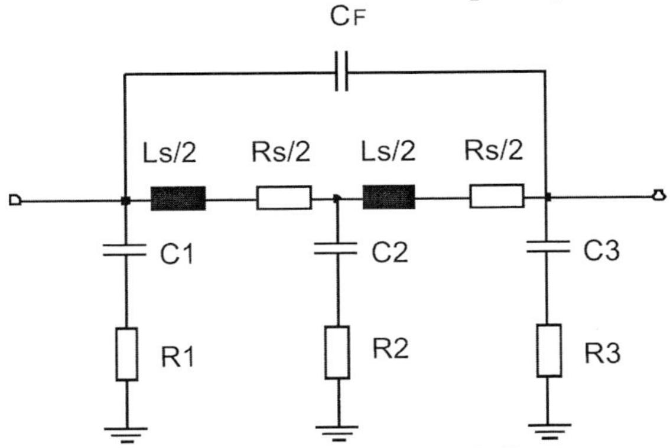

Fig. 12: lumped model for fully symmetrical coils with middle tap.

This effect can further be exploited through a special layout of the coil as shown in fig. 13. The middle tap of the coil is the outer winding in the structure. This winding is laid out wider to reduce its ohmic resistance, without capacitive penalty as this capacitor is at common mode. The inner windings are thinner, this decreases the capacitive load at the RF-nodes of the oscillator and thereby increases the differential self resonance frequency.

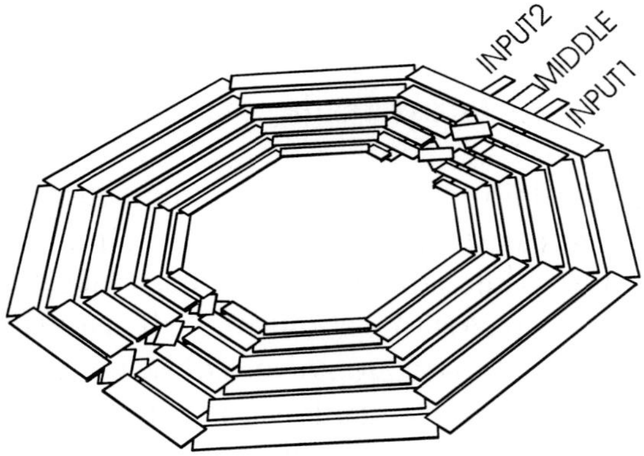

Fig. 13: fully symmetrical coil with middle tap

To conclude the design of the LC-tank, a varactor is needed. In CMOS, parasitic junction diodes (e.g. n+/p-well or p+/n-well) or MOS-capacitors (NMOS, PMOS, PMOS in p-well or NMOS in n-well) are available [16]. Through layout folding, MOS-based varactors can be designed to have minimal series resistance. Also the ratio of maximal to minimal capacitance is higher for the MOS-varactors as for junction-diodes, especially when going to low-voltage supplies.
A big disadvantage of the MOS-varactors is the very steep C-V-characteristic as shown by simulation results in fig. 14.

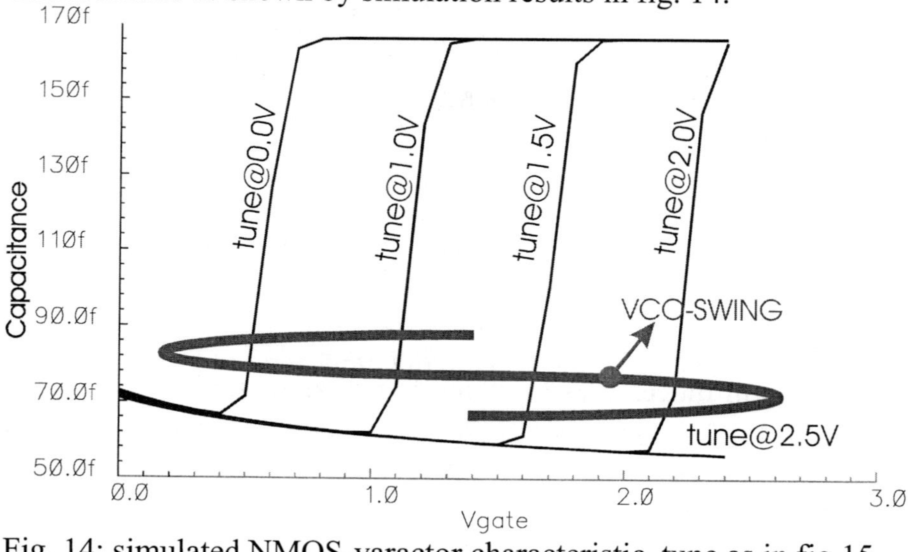

Fig. 14: simulated NMOS-varactor characteristic, tune as in fig.15.

The output of the loop filter represents the tuning input of the VCO in PLL's. Any noise at the tuning input of the VCO in a PLL, is converted up to phasenoise with VCO gain [MHz/V] squared. A VCO with such a steep characteristic or high gain would put impossible specifications on the loop filter.
This problem however, can be circumvented by directly coupling the varactor to the oscillator nodes as shown in fig. 15 and tuning the oscillator through the backgate of the MOS-varactors. This way the varactor capacitance is averaged over the oscillator swing as indicated in fig.14 and VCO tuning characteristic is effectively linearized as demonstrated by the measured designs in the following sections of this paper.

4. A 0.35μm low power 1.8GHz CMOS VCO for DECT

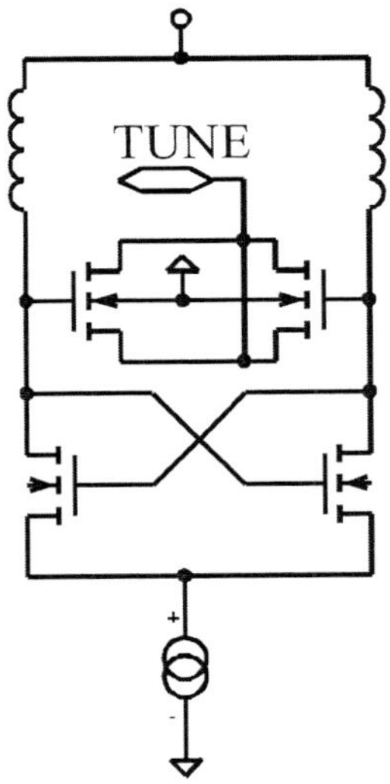

Fig. 15: VCO schematic.

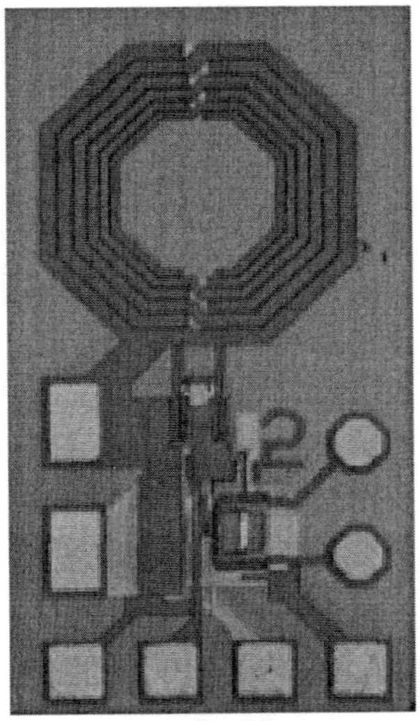

Fig. 16: Chipfoto.

A fully integrated CMOS VCO was designed in INFINEON standard CMOS 0.35μm process. VCO-tuning is realized with backgate tuned NMOS-varactors as described in the previous section. VCO-schematic is presented in fig. 15, the chipfoto (fig. 16) shows the fully symmetrical coil layout with middle tap.

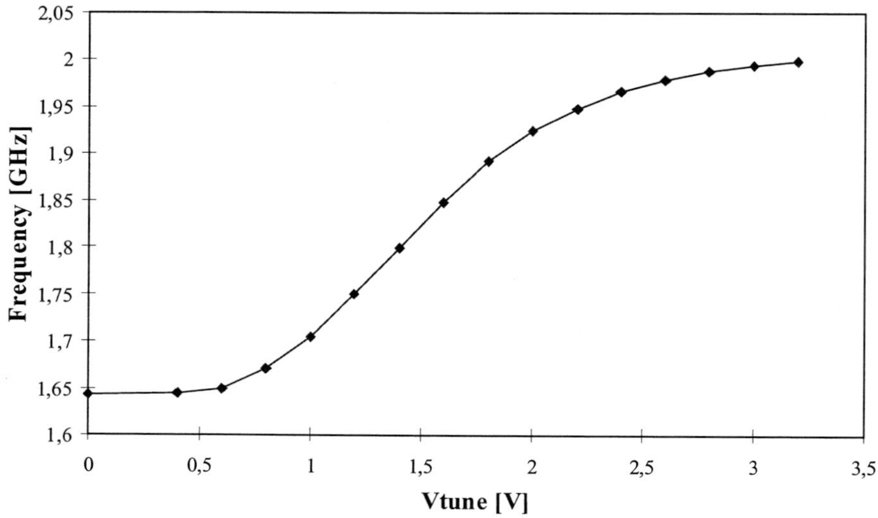

Fig.17: measured VCO tuning.

The process offers 3 metal levels (Al) with a thickness of respectively 0.4μm, 0,4μm and 0,6μm for metal1,2,3. Inductor windings were laid out with Metal2 and Metal3 in parallel ,cross connections in metal1. The tuning curve of the VCO is shown in fig. 17. Phasenoise was measured using Europtest PN9000 equipment, the measured phasenoise over oscillator core current is presented in fig. 18.

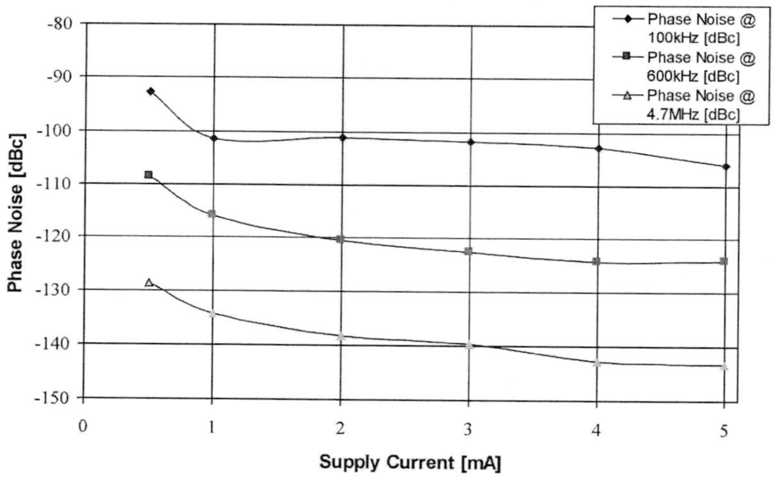

Fig. 18: VCO measured phasenoise performance (VDD=2V).

5. A 0.35µm 1.2GHz CMOS VCO for GSM

The toughest phasenoise specification for the local oscillator in a heterodyne GSM receiver comes from the blocker specifications at 3MHz offset. Assuming a minimal SNR of 9 dB at the input of the baseband part, combined with GSM channel bandwidth of 200kHz, a phasenoise performance of at least –141dBc/Hz@3MHz can be calculated (fig. 19).

The design uses exactly the same structure and process as the design discussed in the previous section. The same coil was reused, dimensions were linearly scaled by 1.4 to increase the inductance, at the cost of some tuning range. Varactor and MOS layout were extremely folded to reduce parasitic resistances. The measured tuning curve is shown in fig. 21, measured phasenoise in fig. 20. Although this design fulfills GSM phasenoise specifications, the power consumption (45mW) is too high to be considered as an alternative for the external VCO-modules used today (typically 15-20mW [22]).

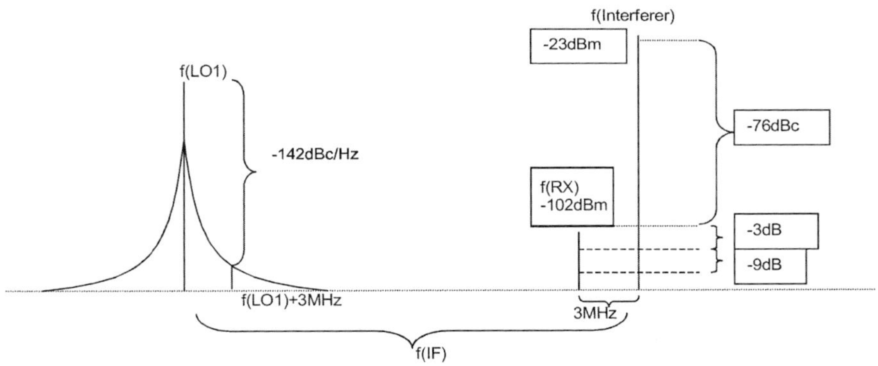

Fig.19: GSM blocker phasenoise specifications for LO1.

Frequency Synthesis for Integrated Transceivers

Jan-Wim Eikenbroek[1] and Sven Mattisson[2]

[1]Ericsson Radio Systems B.V., Emmen, The Netherlands
[2]Ericsson Mobile Communications AB, Lund, Sweden

Abstract

With the recent increase in radio transceiver integration levels it has become feasible to use more complex frequency synthesizer structures. The use of a fractional-N synthesizer decouples the channel spacing from the choice of synthesizer reference frequency which enables much faster frequency hopping. Static moduli synthesizers have required complex spurious compensation methods, but recently, $\Delta\Sigma$-dithered moduli fractional-N synthesizers have attracted a lot of attention as no analog spurious-compensation circuits are needed. Design equations for the analog parts of a $\Delta\Sigma$-based synthesizer are presented together with a discussion on practical implementation aspects.

R. J. van de Plassche et al. (eds.), Analog Circuit Design, 339–355.

1 Introduction

Regular integer-divide phase locked loops, see figure 1, are governed by the equations, [1, 2, 3],

$$
\begin{aligned}
i_e(s) &= K_P \phi_\epsilon = K_P \left(\phi_R(s) - \frac{\phi_o(s)}{N} \right) \\
\phi_o(s) &= i_e(s) Z(s) \frac{K_V}{s},
\end{aligned}
\tag{1}
$$

where s, K_P, $Z(s)$, and K_V are the complex frequency, phase detector gain, loop-filter trans-impedance, and VCO gain, respectively, and $\phi_R(s)$, $\phi_o(s)$, ϕ_ϵ, and $i_e(s)$ are the reference phase (or frequency as $2\pi f = s \cdot \phi(s)$), the VCO phase, the phase error, and the phase-detector error current, respectively.

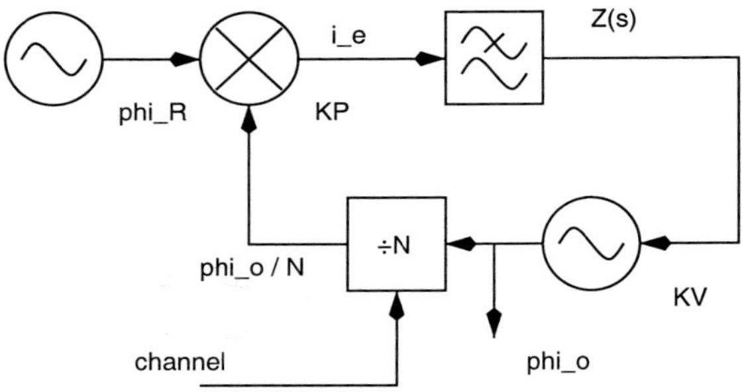

Figure 1: Basic integer-N PLL.

Solving the above equations for ϕ_o yields the well-known result that $f_o = N \cdot f_R$, that is, the VCO frequency is an integer multiple of the reference frequency.

In general it is desirable to design the loop to be a second- or third-order type 2 loop [4] (i.e. a loop with two integrators). This can be accomplished by the loop filter depicted in figure 2.

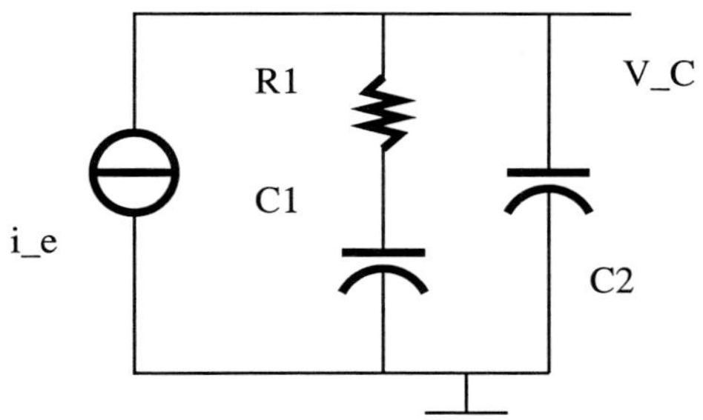

Figure 2: Typical passive loop filter.

If the PLL phase detector is of the charge-pump type we get the VCO control voltage $V_C(s)$ to, [1],

$$\begin{aligned} V_C(s) &= i_e(s)Z(s) \\ Z(s) &= \frac{1 + sR_1C_1}{s(C_1 + C_2) + s^2R_1C_1C_2}, \end{aligned} \tag{2}$$

where $Z(s)$ is the loop-filter impedance. The PLL transfer function can now be evaluated as

$$H(s) = \frac{\phi_o(s)}{\phi_R(s)} = N\frac{K_VK_PZ(s)}{s \cdot N + K_VK_PZ(s)}, \tag{3}$$

and after expanding $Z(s)$ and simplifications we get

$$H(s) = N\frac{1 + sR_1C_1}{1 + sR_1C_1 + s^2\frac{N(C_1+C_2)}{K_PK_V} + s^3\frac{NR_1C_1C_2}{K_PK_V}}. \tag{4}$$

In general performance will be good if the $H(s)$ poles coincide, that is to write the numerator as $(1 + s\tau)^3$. This pole placement results in high loop gain while maintaining good stability properties. It should be noted, though, that should the loop gain vary too much, parasitics may cause the loop to be unstable and a second-order loop may be preferable [4].

Applying the third-order pole placement to (4) and identifying terms we get

$$\begin{aligned} H(s) &= \frac{1+3s\tau}{(1+s\tau)^3} \\ R_1C_1 &= 3\tau = \frac{3}{\omega_N} \\ C_1 &= 8C_2 \\ \omega_N &= \sqrt{\frac{K_P K_V}{3NC_2}}. \end{aligned} \tag{5}$$

A PLL with this transfer function will be able to track a fixed reference frequency (i.e. $2\pi f_R = s \cdot \phi_R(s)$) without errors but will have both phase and frequency transient errors when subject to a frequency step (e.g. when switching channels by setting a different divider ratio N). The VCO frequency will over shoot by 25 % of the frequency step. Thus, the VCO must have a wider tuning range, not only due to production tolerances, but also due to over shoot when switching between the extreme channels. For example, a 80 MHz tuning range requires a 120 MHz tuning range to avoid saturating $V_C(s)$.

The phase detector has a limited range. Typically charge-pump detectors, with a single state, will have a 2π linear range (i.e. $||\phi_\epsilon|| < 2\pi$) [1]. More states will extend the range but will be more complex. If the PLL has been unlocked, the phase detector states must be zeroed, or the PLL has to "unwind" the states before locking is possible and this will slow down lock-in. By evaluating (5) with a frequency step $\Delta\omega$, in the time domain, it can be shown that the maximum phase error will be

$$||\phi_\epsilon||_{\max} \approx 0.84 \frac{\Delta\omega}{N\omega_N}. \tag{6}$$

With a given output frequency range and reference frequency (i.e. $\Delta\omega$ and N) we can calculate the minimum ω_N required for a given maximum phase error. Furthermore, depending on the required frequency accuracy, the PLL typically needs several reference frequency cycles (i.e. samples) to settle after a frequency step. To get within 0.1 % of the frequency step, some 12 cycles may be needed (i.e. for a 80 MHz step the residual error will be less that 80 kHz after some 12 cycles). Since, in an integer-N PLL, the reference frequency has to be equal to the channel spacing (or an integer thereof) we will have conflicting requirements between settling time, loop bandwidth and frequency resolution (channel spacing).

2 Fractional-N techniques

If non-integer division ratios could be provided, fast switching synthesizers with high resolution could be designed. The fractional-N dividers address this issue [1, 3].

Suppose our divider in figure 1 may divide either by N or $N+1$, that is a variable modulus divider, and that we divide by $N+1$ every K VCO cycle and with N otherwise. We then get

$$\begin{aligned} \hat{N} &= (N+1)\frac{1}{K} + N\left(1 - \frac{1}{K}\right) = N + \frac{1}{K} \\ f_o &= \hat{N} f_R = \left(N + \frac{1}{K}\right) f_R. \end{aligned} \tag{7}$$

With this divider the synthesizer would have a resolution of f_R/K which is smaller than the reference frequency. To generate, for example, 2407 MHz from a 13 MHz reference, a divider ratio of 185+ 2/13 is needed. Thus, if we, for every 13 cycles of the reference, divide two times by 186 and 11 times by 185 we get the desired average ratio. However, we have a cyclic error that will generate ripple on the VCO control voltage and, hence, VCO spurs. The loop filter will suppress this ripple somewhat, but as the whole objective was to increase the loop bandwidth relative to that of an integer-N PLL, we will typically not get sufficient suppression.

The V_C ripple is determined by the PLL parameters and the divider modulus selection. Since all these parameters are known it is possible to cancel the VCO control voltage ripple. Typically an accurate D/A converter is needed to inject an anti-phase sawtooth signal at the phase detector output. This cancellation has been successfully applied in many applications, for example laboratory frequency synthesizer, but requires tight control of parameters to be effective. In integrated synthesizers applications it will often times be to expensive, if at all possible, with such compensation schemes.

2.1 $\Delta\Sigma$ techniques

If there would be no cyclic behavior in the fractional-N divider scheme, as described above, no VCO spuriouses would be generated. This may be accomplished by letting the divider modulus control appear to be random but with the desired average value. In A/D converters the $\Delta\Sigma$ modulator has been used for quite some time for this purpose. By oversampling and by noise shaping a high resolution with little low-frequency noise is accomplished. A first-order $\Delta\Sigma$ modulator is depicted in figure 3.

Unlike the $\Delta\Sigma$ A/D converter, the fractional-N $\Delta\Sigma$ modulator only consist of digital blocks. The channel selection (input signal) as well as the modulus control (output signal) are digital and, hence, no analog blocks are needed. Thus, this technique of fractional-N synthesis is very well suited for integration (see e.g. [5]). The modulator can be modelled with an output signal that has the desired output plus quantization noise. Let $\phi_d = \phi_o/N$ represent the fractional-N divider output phase, then the power spectral density of the phase error of the modulator, as seen at the divider output,

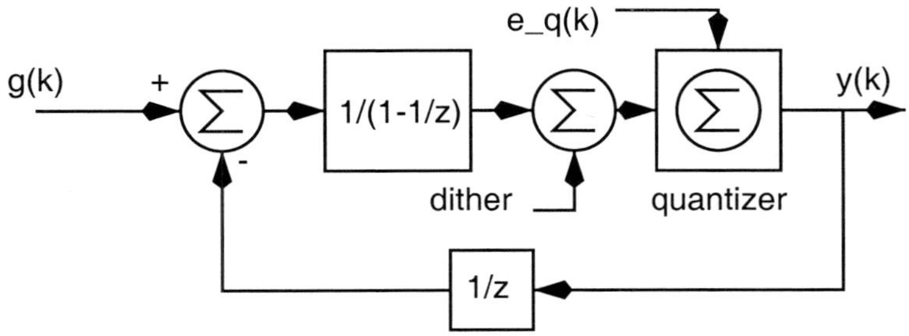

Figure 3: First-order $\Delta\Sigma$ modulator.

can be depicted as

$$S_{\Delta\phi_d}(z) = \frac{4\pi^2}{N^2}\left|\frac{Q(z)}{1-z^{-1}}\right|^2, \tag{8}$$

where $Q(z)$ is the $\Delta\Sigma$-modulator noise transfer function and z^{-1} is the delay operator. Assuming one-bit quantization and white quantization noise we can approximate $Q(z)$ to [3]

$$|Q(z)|^2 \approx \frac{2}{12f_R}\left|1-z^{-1}\right|^{2m}, \tag{9}$$

with m being the modulator order. Combining (8) and (9) we get the modulator phase-noise power spectral density to

$$S_{\Delta\phi_d}(z) \approx \frac{2}{12f_R}\frac{4\pi^2}{N^2}\left|1-z^{-1}\right|^{2(m-1)}. \tag{10}$$

If the loop bandwidth is small compared to the sampling rate f_R, we can approximate the modulator noise in the frequency domain by

$$S_{\Delta\phi_d}(f) \approx \frac{2}{12f_R}\frac{4\pi^2}{N^2}\left[2\sin\left(\frac{\pi f}{f_R}\right)\right]^{2(m-1)}, \tag{11}$$

where f represents the frequency offset from the carrier. This noise distribution is sketched in figure 4, where a residual thermal noise floor (e.g. thermal noise due to the divider circuitry) is also included.

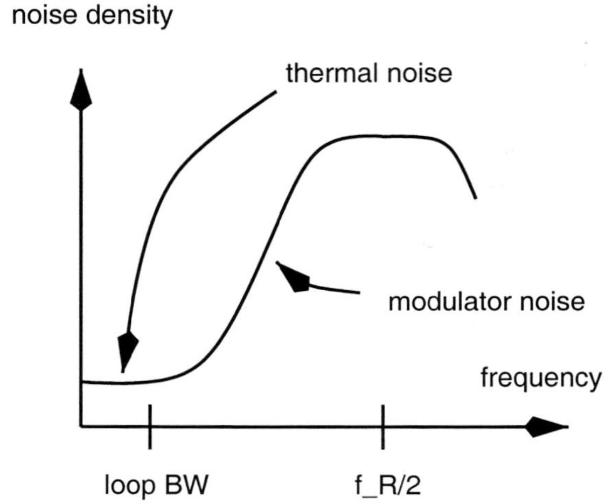

Figure 4: $\Delta\Sigma$-modulator noise density.

When the $\Delta\Sigma$ modulator is used in frequency synthesizer applications, it sees a constant input signal (i.e. the channel). In the absence of noise or varying input signal, the modulator will generate limit cycles, just like the above traditional fractional-N divider. By introducing a random dither signal with zero mean, the $\Delta\Sigma$-modulator limit cycle will be converted to noise. This dither signal may add extra "quantization" noise, depending on the dither injection technique and dither magnitude, but the expression in (10) may still be used. Depending on the exact $\Delta\Sigma$ modulator topology, and order, we may have to modify (9) slightly to correctly predict the noise density.

To evaluate the impact of the modulator noise on the PLL, we add the $\Delta\Sigma$-modulator noise power from (11) to the PLL equations (1).

After simplification the PLL error phase becomes

$$\phi_\epsilon(s) = \left(\phi_R(s) - \frac{\phi_o(s)}{N} + \sqrt{S_{\Delta\phi_d}(s)}\right), \tag{12}$$

with $\phi_\epsilon(s) = i_e(s)/K_P$. By recasting the above equation, and using 3, we may calculate the PLL output phase, including the modulator noise, to

$$\phi_o(s) = H(s)\left(\phi_R(s) + \sqrt{S_{\Delta\phi_d}(s)}\right), \tag{13}$$

and the $\Delta\Sigma$ fractional-N synthesizer single-sided output phase noise spectrum will then be

$$\mathcal{L}(f) \approx \frac{1}{2} S_{\Delta\phi_d}(f)\,|H(f)|^2. \tag{14}$$

The modulator noise will pass through the loop and be directly visible at the output, see figure 5. This may seem undesirable but the loop bandwidth is, typically, much smaller than the reference frequency. At very low frequencies, the modulator noise is proportional to $(f/f_R)^{2(m-1)}$, and for a third-order modulator this noise (i.e. the sinusoidal term) is some 60 dB down two decades below f_R.

3 Implementation aspects

Conceptually the $\Delta\Sigma$ fractional-N synthesizer is very similar to a regular PLL with a variable modulus [1] divider. The only change is that the selection of the divider modulus is done dynamically via a $\Delta\Sigma$ modulator. As a consequence the divider will have to be able to divide by at least all integers in the range $f_{o\,min}/f_R, \ldots, f_{o\,max}/f_R$; normally only two moduli (i.e. $N, N{+}1$) are required. The extended

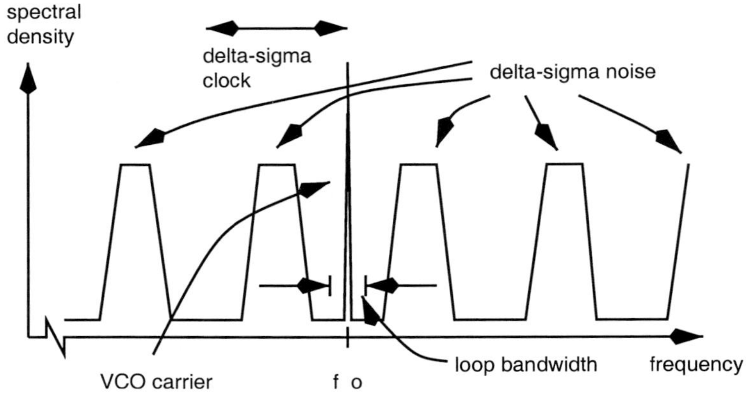

Figure 5: Open-loop $\Delta\Sigma$ fractional-N synthesizer output noise. The loop will suppress all components outside the loop bandwidth when locked.

range of moduli will add to the complexity of the divider and also its power consumption.

The reference frequency is, typically, much higher in a $\Delta\Sigma$ fractional-N than an integer-N synthesizer (after all that was the objective). This is more demanding for the phase detector, and especially if it is a charge-pump based detector. The detector will now have to switch at several tens of MHz rather than below the MHz range. Again, this has a complexity and power consumption impact.

Both the divider and the phase detector do, however, benefit from IC technology advances. The reference frequency to device transit frequency quotient (i.e. f_R/f_T) will decrease as IC feature sizes shrink. Todays BiCMOS technology typically achieve peak f_T values on the order of 20–50 GHz for bipolar as well as MOSFET devices which offers some design margin also for high reference frequencies. Still there are some subtle problems with the $\Delta\Sigma$ fractional-N synthesizer that need proper attention.

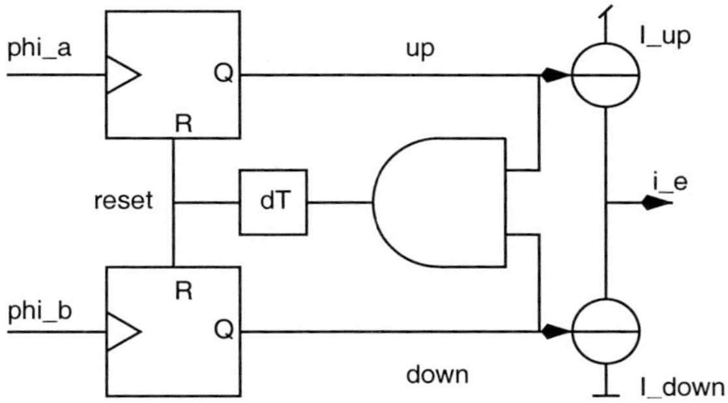

Figure 6: Block diagram of a dead-zone compensated sequential phase detector. The extra delay, δt, guarantees that both charge pumps will have enough time to switch on, hence preserving gain at near-zero phase error.

3.1 Phase-frequency detector signals

In a regular PLL (integer-N or static modulus) the phase detector signal, ϕ_ϵ, asymptotically approach zero as the loop settles (assuming a reasonably clean VCO signal). Only when the reference signal or the divider ratio is changed will the phase error increase. The phase detector performance is, thus, usually optimized for small error signals during tracking (e.g. the dead-zone compensation due to Egan [6]), and for a wide error range during frequency hopping (e.g. sequential phase detectors with multiple states [1]).

Due to the quantization error in the $\Delta\Sigma$ modulator noise is injected into the loop, as explained by equations (8) and (11) and in figure 4. The dithering of the modulator makes the divider modulus vary more than between N and $N+1$ during steady state[1]; typically a modulus variation in the range $N-M, \ldots, N+M$, as indicated in figure 7, where M may be three or even larger, may be required.

[1]The steady-state modulus range is centered around N, whereas the divider has to implement all integers needed to cover the frequency band of interest.

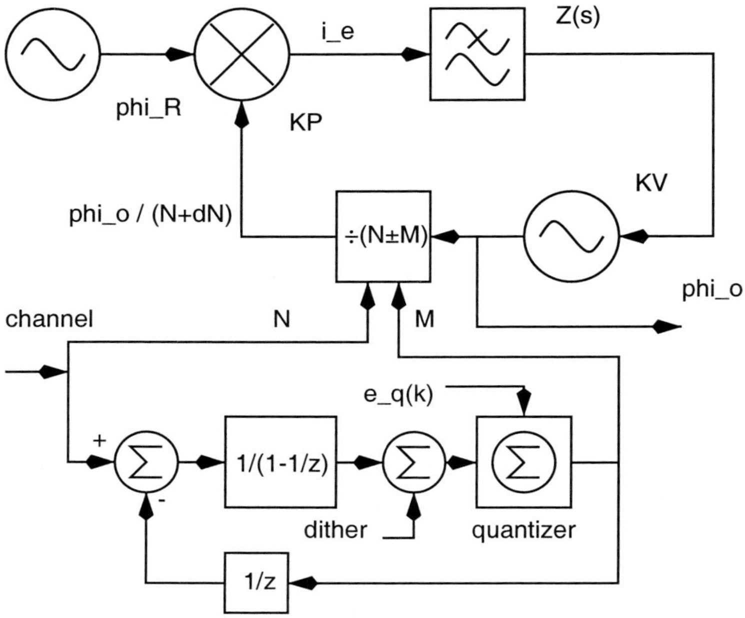

Figure 7: Block diagram of first-order noise shaping $\Delta\Sigma$ fractional-N synthesizer.

Both the quantization noise and the extended modulus range result in a larger phase error as seen by the phase detector. This is true even if the loop has settled, when the mean error is zero but the instantaneous error may be large due to high-frequency components.

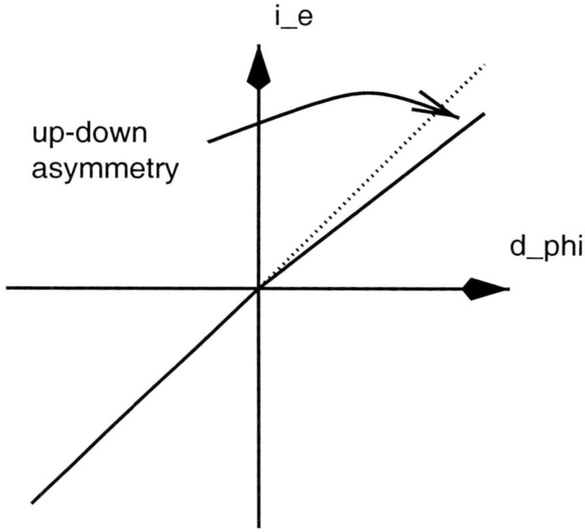

Figure 8: Charge-pump output current vs. phase error.

If the phase detector response is not symmetric we will get rectification of the error signal. This asymmetry is indicated in figure 8, and it can be modelled by a square term in the phase detector transfer function, as depicted in (15),

$$i_e(s) = K_P\left(\phi_\epsilon(s) + \alpha|\phi_\epsilon(s)|\right) \approx K_P\left(\phi_\epsilon(s) + \beta\phi_\epsilon^2(s)\right); \qquad (15)$$

the resulting spectrum folding is sketched in figure 9. This noise folding problem is similar to the zero-IF receiver DC-offset problem [2], and has to be avoided or else excessive noise levels will result. The advantage of the $\Delta\Sigma$ modulator was that noise shaping moved the quantization noise up in frequency such that the loop filter could remove it. Now with this asymmetry, the quantization noise folds back inside the loop bandwidth and cannot be removed. Up-down charge-pump current matching better than 1 % may be required for optimum performance.

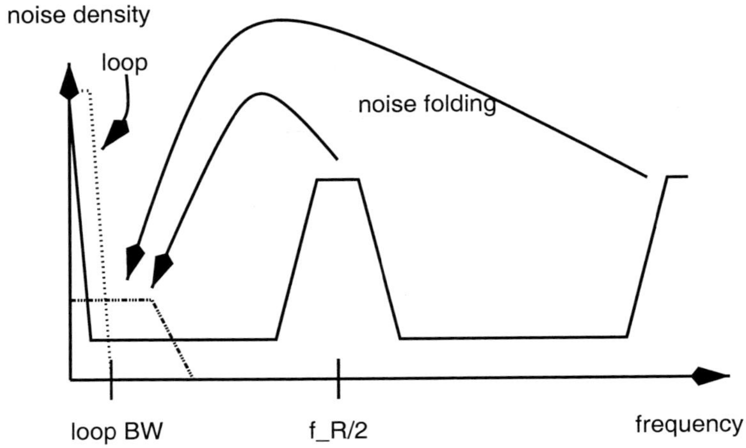

Figure 9: Open-loop noise components at phase-detector output. Closed-loop transfer is indicated by the dotted line and down-converted noise by the dot-dashed line.

3.2 Spurious generation

In an integer-N PLL, the VCO frequency is divided down to the reference frequency. Hence, any mixing products will be at DC or at twice the reference frequency. In the first case, the loop will compensate for the DC offset by justifying the phase of the VCO (not frequency). In the second, high frequency, case the loop filter will suppress most of the mixing products. In addition, phase-detector or loop-filter leakage currents will also try and force the VCO off the steady-state frequency. During closed-loop conditions, however, this leakage current will be compensated for by a VCO phase adjustment, just like in the previous case, and be rather harmless (in a frequency synthesizer; other applications may require zero phase error).

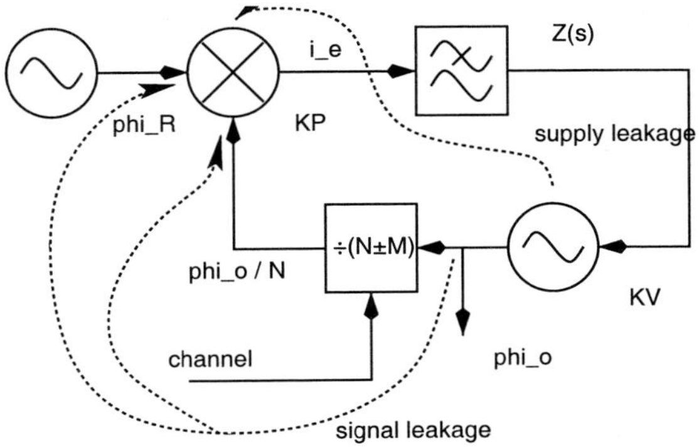

Figure 10: Fractional-N VCO leakage paths.

A fractional-N synthesizer typically has a VCO frequency that is not an integer multiple of the reference frequency. Thus, if the VCO frequency leaks into either phase detector input, see figure 10, low-frequency mixing products may result. Typically, the phase detector inputs are connected to latches, see figure 6, which are very wideband. When the VCO signal leaks into any of the latch clocks a timing jitter will result. This jitter will cause mixing of the VCO with the closest harmonic of the reference, and if the difference between the two is small enough two in-band spurious tones will appear on either side of the VCO. Careful layout is mandatory and low-pass filtering of the phase detector input signals may be required. Balanced circuitry will help to suppress even harmonics, but the odd ones will not benefit from a high common mode rejection. The VCO will also couple signals to the supply and ground rails as well as into the substrate. The traces of the VCO signal, or its second harmonic, may be picked up by circuits with poor power-supply rejection ratio (PSRR) and cause unwanted mixing.

4 Conclusion

The design of $\Delta\Sigma$-modulator-based fractional-N synthesizers has been discussed. This synthesizer offers better design flexibility with respect to settling time and channel spacing than a standard frequency synthesizer does. The changes relative to a regular variable modulus PLL is a frequency divider with a wider modulus range, a more linear phase detector, and the addition of an all digital $\Delta\Sigma$ modulator. The power consumption of the $\Delta\Sigma$ fractional-N synthesizer will be higher than that of the integer-N PLL, due to the higher reference frequency. Thus, in cases when settling time or channel spacing is not an issue, an integer-N loop can be used and there is no advantage in using a $\Delta\Sigma$ loop.

A $\Delta\Sigma$ fractional-N synthesizer aggravates the linearity requirements of the phase detector and is more sensitive to mixing between reference frequency harmonics and the VCO signal. The likely injection paths has been indicated.

References

[1] T. H. Lee. *The Design of CMOS Radio-Frequency Integrated Circuits.* Cambridge Univerity Press, Cambridge, 1998.

[2] B. Razavi. *RF Microelectronics.* Prentice-Hall, Upper Saddle River, 1998.

[3] James A. Crawford. *Frequency Synthesizer Design Handbook.* Artech House, Boston, 1994.

[4] Donald R. Stephens. *Phase-Locked Loops for Wireless Communications – digital and analog implementations.* Kluwer, Boston, 1998.

[5] Michael H. Perrot, Theodore L tewksbury III, and Charles G. Sodini. A 27-mW CMOS Fractional-N Synthesizer Using Digital Compensation for 2.5-Mb/s GFSK Modulation. *IEEE Journal of Solid State Circuits*, 32(12):2048–2060, December 1997.

[6] W. F. Egan. *Frequency Synthesis by Phase Lock.* John Wiley, New York, 1981.

PLL Frequency Synthesizers: Phase Noise Issues and Wide Band Loops

Marina de Queiroz Tavares
marina.tavares@philips.com
Philips Semiconductors

ABSTRACT:

New DVB standards impose very tight requirements for phase deviation in frontend tuners. Combining these requirements with higher integration levels are the challenges of current mixer-oscillator and PLL circuits, where behavioural models are essential to understand and guide the design work. Linear loop models and algorithms for filter calculation are developed to discuss stability and noise issues for the PLL synthesizer. The limitations of the models in the analysis of wide band loops are studied and theoretical limits are calculated. The accuracy of the models is verified in the implementation of testchips for multi-loop wide band PLLs.

1. INTRODUCTION

The central issue of this work is the stability and noise performance of PLL frequency synthesizers. Frequency synthesizers are a common block of the frontend of RF telecommunication systems. In particular, PLL synthesizers are extensively used for their programming flexibility, ease of integration and low production cost.

We focus on the context of TV broadcasting tuners, where the new standards of digital modulation broadcasting (DVB) which are appearing, and the continuous trend for higher integration levels, are bringing new issues for IC design and application.

We are mostly concerned with models: calculations and behavioural simulation tools, which were developed to support the activities of design and engineering for the integrated circuits in frequency synthesizers.

R. J. van de Plassche et al. (eds.), Analog Circuit Design, 357–382.

Issues arising in the implementation of wide band loops are examined such as: maximum loop bandwidth and maximum comparison frequency for tri-state frequency and phase detectors. The design of a monolithic mixer-oscillator and PLL synthesizer is presented and used as a practical example to compare the simulations and calculation tools with measurement results.

This presentation is derived from a thesis work developed between March 1996 and June 1999, by the author. The thesis was part of a collaboration contract between Philips Semiconductors Caen and the laboratory CEGELY of the INSA de Lyon.

2. CONTEXT

Let us introduce the context of the TV tuner and the current tendencies in architecture and IC requirements. These tendencies point to low phase noise synthesizers, implemented in very monolithic architectures with integrated oscillators.

The denomination tuner is mostly used in terrestrial TV receivers referring to the part of the frontend between the antenna and the output of the IF stage which is sent to the demodulator. Inside a tuner we find several stages of filtering and amplification and at least one block of frequency conversion. The latter is constituted by a mixer and controllable oscillator, driven by a programmable PLL. Nowadays, tuners often have one single IC, including the PLL, mixer-oscillator and IF amplifier blocks.

Terrestrial TV broadcasting allocates three RF-bands covering roughly the spectrum between 40MHz and 900MHz. This wide input frequency range often requires the first filtering and mixing stages to be doubled, having one set to cover the VHF bands (40MHz to 400MHz) and another for the UHF band (400MHz to 900MHz). In most standards the channel bandwidth varies between 6 and 8MHz, and the intermediate frequency between 39 and 55MHz. The filters preceding the frequency conversion have to reject the image channel by about 60 dB. Rejection of the adjacent channels is mostly made after down mixing and is in the same order of magnitude. The dynamics for amplitude gain are between 40 and 50 dB, with another 60 dB controllable amplification capacity in the demodulator. For

analog standards, the minimum SNR at the IF output is in the order of 55 dB, to start causing visible errors in the video reception.
Satellite TV is broadcast in the Ku-band: 10.7 GHz to 12.75 GHz. A first frequency conversion (by a fixed oscillator close to the antenna) brings the signal down to the band L (950 MHz to 2150 MHz) before it enters the tuner and demodulator blocks. Older analog standards use FM modulated channels with a bandwidth between 27 and 36 MHz. More recent digital norms have different channel compositions, using multiplexing in frequency and time domain and also sharing a spectrum window between an analog and a digital modulated signals. The tuner architecture evolved from heterodyne receivers with two IF stages (at about 470MHz and 70 MHz) to monodyne or near-zero IF receivers with increasing filtering capabilities in the digital circuits (after A to D conversion in the base band stage). AGC dynamics in the tuner is also in the order of 50 dB, but adjacent channels rejection is less constrained than in terrestrial TV receivers since satellite transmitted channels all have the same power levels, as they come from a common source.

Satellite DVB (digital video broadcasting) was one of the early digital standards being offered in large commercial scales. It is based on a single carrier modulated in QPSK, which transmits either a package of TDM channels with data rates up to 55 Mbps, or single channels in narrow bandwidths with rates down to a couple of Mbps. The minimum SNR at the base band output depends on the maximum BER (bit error rate) that can be corrected by the signal decoder. A maximum BER of 10^{-4} is usually accepted for most decoders, and it implies a minimum SNR of 11.4 dB for QPSK modulated data.
The relationship between the SNR and the CNR (carrier to noise ratio) of local oscillator (LO) used in the down conversion, depends on the channel characteristics (bandwidth, rejection on selectivity stages and power mismatch with respect to interfering channels) and in the underlying modulation principles. For terrestrial transmissions with strong adjacent channels, that can be up to 60 dB above the wanted channel, the influence of the spectral purity of the LO can be visualised in figure 1.

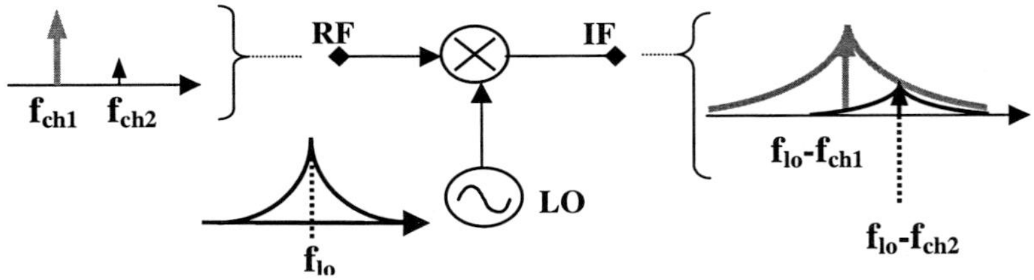

Figure 1 *Local Oscillator Spectral Purity X SNR*

If we take the example of FM signals for analog standards, we would be concerned about the LO noise in the $1/f^2$ power sidebands, which appears after demodulation as a white noise added to the channel signal. Therefore in the FM context, noise specifications are often bound to the free running, or out of loop, carrier spectrum, transmitted by the VCO intrinsic noise.
DVB standards have QPSK and QAM as basic modulation principles. Cable and satellite norms use a single carrier in contrast to terrestrial and DAB (digital audio broadcasting) which apply multi-carrier techniques to increase their noise robustness. Bandwidth efficiency is specially high in cable transmission where SNR requirements will ramp to 30.2 dB for a carrier modulated in QAM256. Terrestrial DVB uses a complex modulation scheme called OFDM (orthogonal frequency division multiplexing), which combines up to 6817 subcarriers, modulated in QAM16 or 64, with a frequency spacing down to 1.12 kHz. In both cases, the price to be paid for the increasing bandwidth efficiency and modulation complexity, is the narrowing margin for signal degradation in the tx-rx path.
In our study we concentrate on the losses related to the spectral purity of the local oscillator used in the frequency conversion block. DVB standards define a margin of 0.2 dB of implementation loss due to the LO noise. This noise can also be define as a residual phase deviation in an oscillator locked by a phase controller. For phase modulation principles the phase noise of the LO can be seen as an interference blurring the output constellation and increasing the BER. For multi-carrier standards it is also common to define a noise threshold for the level of the LO sidebands, for offsets that are comparable to the frequency spacing between subcarriers.

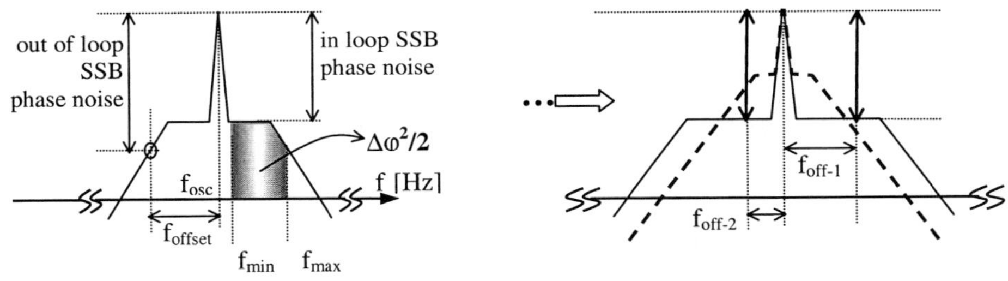

Figure 2 ***Phase Noise requirements***

The phase deviation can be calculated integrating the sidebands of the LO spectrum, as shown in figure 2, left side. The lower and upper limits of the integral are determined by the demodulator and channel bandwidth parameters. The phase noise specification may concern both the intrinsic behaviour of the oscillator (out of loop phase noise) and the PLL blocks (in loop phase noise).

In figure 2 (right side), we represent two carrier spectra and a noise threshold requirement related to two offset frequencies. The dotted spectrum has a better free running oscillator performance but a narrower PLL bandwidth, which limits its noise performance for low offsets. The solid line spectrum shows an option where the in loop (PLL related) noise performance is adapted to the CNR specification at both offsets. In practice this last spectrum conforms to two contexts: requirements for multi-carrier standards and applications with fully integrated oscillators for near-zero IF receivers.

The second situation send us back to the trend for higher integration levels. Currently, most LOs are based on a LC loaded amplifier with an external resonator. The large frequency range of the TV applications limits the possibility of integrating the resonant circuit, as occurs in narrow band reception systems, like mobile telephones. Therefore other oscillator structures, like ring or relaxation, have to be tried. The drawbacks of these other structures are: their poorer phase noise performance as compared to LC resonators with high quality factors, and the impossibility to track the LC matched filters in the input stages of the tuner. The advantages appear mostly in the zero-IF configurations, where a totally integrated oscillator, with no LC resonator, increases the robustness to RF interference.

Therefore the integration tendency forces architectural modifications in the tuner. The absence of external tracking filters can be more easily coped with in satellite receivers, where the uniform input level enables a feasible compromise between selectivity and linearity requirements.

Direct conversion schemes have new constraints related to the suppression of the IF stage. The AGC dynamics in the RF and BB parts have to replace the previous IF dynamics while preserving the linearity and noise figure properties. Coupling interactions between the local oscillator and the RF input signal (now in the same frequency), have to be controlled to reduce the signal degeneration by "self-reception" or "self-demodulation". These constraints brought an additional interest to a completely integrated oscillator suffering from less external coupling problems.

The use of an integrated oscillator covering a large tuning range often brings an inherent degradation of the oscillator spectral purity. Thus achieving strict phase noise requirements becomes obligatory for the PLL circuitry.

Figure 2 showed that the noise requirement imposes a compromise between the PLL and the VCO noise performances. The variable parameter adapting these performances is the loop bandwidth, which unfortunately is not independent of other parameters such as loop gain, comparison frequency, minimum tuning step and DC tuning range. Therefore as we optimise the noise performance of the LO we have to keep in mind the stability constraints of the feedback loop. A linear time invariable model is used to discuss these issues on sections 3 and 4, which focus in a careful fitting of the loop filter. Section 5 discusses the limitations of this idealised model with respect to the behaviour of discrete or sampled systems, as digital phase detectors and dividers. Section 6 discusses calculations and simulation methods for phase noise in PLLs and finally testchip implementations are presented in section 7.

3. PLL Phase Model and Loop Filter Calculation

The phase representation for the PLL frequency synthesiser (figure 3) contains the following blocks: programmable dividers (in the reference and feedback paths, called respectively reference and main

dividers), phase and frequency detectors based on D flip-flops, a tri-state charge pump, which is loaded by the loop filter, and a voltage controlled oscillator (VCO). Throughout the following sections we name PLL all the blocks in the synthesiser excepting the VCO.
A linear time invariant model can efficiently and simply describe the system behaviour around the locked condition. The input and output of the model are phase variations measured with respect to reference and VCO phases. In this manner the phase detector and charge pump can be very shortly represented by an adder and a gain block with the sensitivity: average output current over input phase error, $K_\varphi = I_{cp}/2\pi$ *A/rad*. The dividers in the phase domain are directly gain blocks with the magnitude of the division ratio. The VCO becomes a base band equivalent as an integrator with a gain accounting for the tuning sensitivity, $K_o = K_{VCO}.2\pi$ rad.Hz/V.

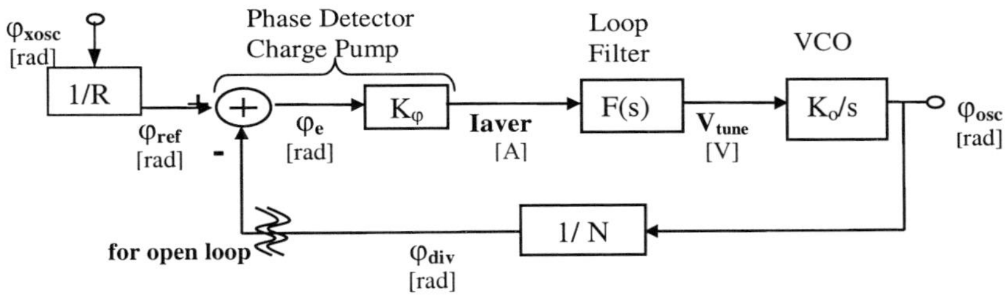

Figure 3 ***Linear Phase Model for a PLL***

In order to examine the stability conditions of the loop we need to introduce the loop filter impedance F(s) and the open loop frequency response H(s).

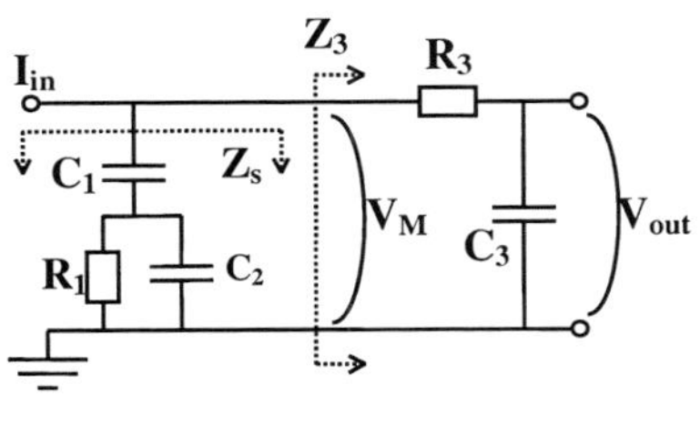

$$H(s) = \frac{\alpha \cdot (1 + s \cdot T_{z1})}{s^2 \cdot C_1 \cdot (1 + s \cdot T_{p2}) \cdot (1 + s \cdot T_{p3})}$$

$$T_{z1} = R_1 C_1 \; ; \; T_{p2} = R_1 C_2 \; ; \; T_{p3} = R_3 C_3$$

$$\alpha = \frac{I_{cp} K_{vco}}{N} \; ; \; C_1 >> C_2 \; ; \; T_{p3} << T_{p2}$$

Figure 4 ***Loop Filter Impedance and Open Loop Transfer***

H(s) is a factored approximation to the filter impedance which is accurate enough when the zeros and poles frequencies are far from each other. The variable α ensembles the variable loop gain parameters. Therefore the maximum and minimum loop gain values are referred as: α_{max} and α_{min} . We also define two frequency ratios, r_{21} and r_{31} , and an average gain value, α_n , which help us to centre the gain range around the maximum phase margin point.

$$r_{21} = \frac{f_{p2}}{f_{z1}} = \frac{T_{z1}}{T_{p2}} \quad ; \quad r_{31} = \frac{f_{p3}}{f_{z1}} = \frac{T_{z1}}{T_{p3}} \quad ; \quad \alpha_n = \sqrt{\alpha_{\min} \cdot \alpha_{\max}}$$

The open loop zero cross frequency or bandwidth is noted w_{ol} . For $r_{31} >> r_{21}$, the open loop bandwidth for a maximum phase margin w_{oln} is equidistant to f_{z1} and f_{p2} . The filter components can then be calculated as:

$$\left|H(jw_{\text{oln}})\right|\Big|_{\alpha=\alpha_n} = \frac{\alpha_n}{w_{\text{oln}}^{\ 2} \cdot C_1} \cdot \frac{\sqrt{1+r_{21}}}{\sqrt{1+1/r_{21}} \cdot \sqrt{1+r_{21}/(r_{31})^2}} \xrightarrow[\substack{r_{21}>>1 \\ r_{31}>>1}]{\text{supposing}} \frac{\alpha_n}{w_{\text{oln}}^{\ 2} \cdot C_1} \cdot \sqrt{r_{21}} = 1$$

$$\Rightarrow \begin{cases} C_1 = \dfrac{\alpha_n \cdot \sqrt{r_{21}}}{w_{\text{oln}}^{\ 2}} = \dfrac{\alpha_n}{w_{z1} \cdot w_{\text{oln}}} \quad ; \quad R_1 = \dfrac{T_{z1}}{C_1} = \dfrac{1}{w_{z1} \cdot C_1} = \dfrac{w_{\text{oln}}}{\alpha_n} \quad ; \\ C_2 = \dfrac{T_{p2}}{R_1} = \dfrac{T_{p2}}{T_{z1}} \cdot C_1 = \dfrac{C_1}{r_{21}} \quad ; \quad T_{p3} = R_3 \cdot C_3 = \dfrac{\sqrt{r_{21}}}{r_{31} \cdot w_{\text{oln}}} \quad . \end{cases}$$

The gain loop variation can be related to the open loop bandwidth by an arithmetic fitting function:

$$\frac{w_{ol}}{w_{o\ln}} = \left(\frac{\alpha}{\alpha_n}\right)^{\frac{1}{1+\sqrt{r_{21}}}}$$

This function is then used to calculate the maximum α variation that can be accommodate within convenient phase margin values. The results are: for a minimum r_{21} and r_{31} values such as, $r_{21}>10$ and $r_{31}>1.6*r_{21}$, a gain variation of $2*r_{21}$ can be fit in a frequency interval with phase margin equal or larger to 30°.

For smaller ratios between r_{31} and r_{21} (close to the 1.6 limit), the nominal gain value and open loop bandwidth, α_n and w_{oln} , need to be repositioned to compensate for the phase margin loss due to the 3rd pole f_{p3} or post-filter. In this case the initial chosen values are noted, α_{npf} and w_{olnpf} , and the nominal values are derived as:

for $\alpha_{npf} = \sqrt{\alpha_{max} \cdot \alpha_{min}}$ and $r_{pf} = \left[\frac{r_{31} - r_{21}}{r_{31}}\right]$:

$$w_{oln} = \frac{w_{olnpf}}{\sqrt{r_{pf}}} \quad \text{and} \quad \alpha_n = \alpha_{npf} \cdot \left(\frac{1}{r_{pf}}\right)^{\left(\frac{1+\sqrt{1/r_{21}}}{2}\right)}$$

Finally, the filter components can be calculate as previously using the compensated w_{oln} and α_n values.

4. OPTIMUM CLOSE LOOP BANDWIDTH

The choice of the centred open loop bandwidth, w_{olnpf} , is tied to several parameters as: rise and settling time, attenuation of the reference breakthrough and adequacy to the noise performance of the VCO. The latter issue relates to the optimisation of the phase noise in the spectrum of an oscillator driven by a PLL.

The linear phase model is used to calculate the transfer of two noise contributions: the global PLL noise gathered at the phase detector input, N_{pll} in rad/sqrt(Hz), (accounting for the crystal oscillator, the dividers, the phase detector and the charge pump) and the intrinsic VCO noise, V_{nvco} in V/sqrt(Hz) .

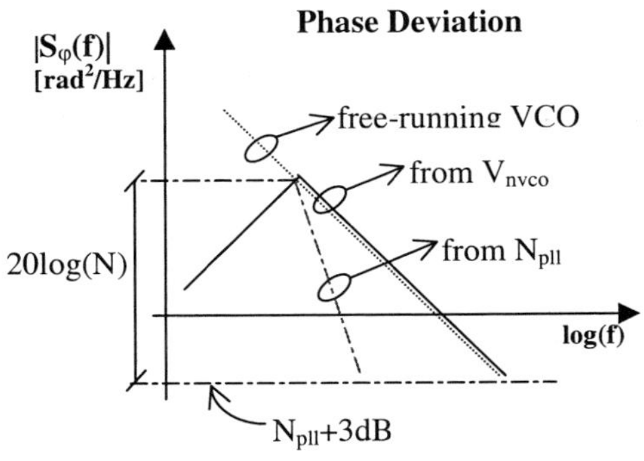

Figure 5 Spectrum of a VCO locked by a PLL

Figure 5 shows a base band representation of the output spectrum. The magnitude, $S_\varphi(f)$, represents the phase deviation which is equal to the level of the sidebands compared to an unitary carrier level. The closed loop transfer functions from N_{pll} and V_{nvco} equal:

$$B(s) = \frac{\varphi_{osc}}{\varphi_{ref}} = N \cdot \frac{H(s)}{1 + H(s)} \approx \frac{N}{(1 + s \cdot T_{p3}) \cdot \left(\frac{s^2}{w_n^2} + \frac{2\xi}{w_n} \cdot s + 1 \right)}$$

$$B_{vco}(s) = \frac{\varphi_{osc}}{V_{tune}} = \frac{B(s)}{K_\varphi \cdot F(s)} \approx \frac{K_o \cdot s \cdot C_1}{\left(\frac{s^2}{w_n^2} + \frac{2\xi}{w_n} \cdot s + 1 \right) \cdot \alpha}$$

In the equations above, the right most terms are approximated forms which put in evidence the low pass filtering behaviour of B(s) and the band pass filtering behaviour of $B_{vco}(s)$. These expressions both have a second order polynomial in the denominator, written in a standard ξ and w_n form. We choose this common notation to indicate similar roots in the two functions.

Figure 5 shows an ideally smooth intersection between the two zones of the spectrum, the in-loop one and the out-of-loop one. Nevertheless, the dominant noise in each of these zones originates from independent noise sources, and in practice the feedback bandwidth and gain determine whether the intersection is smooth or bumpy. In order to minimize the noise of the output spectrum, we need to match the PLL closed loop bandwidth with the intersection frequency, where the noise contributions from N_{pll} and V_{nvco} cross each other. Mismatches result in additional peaking or excessive PLL noise as drafted in figure 6.

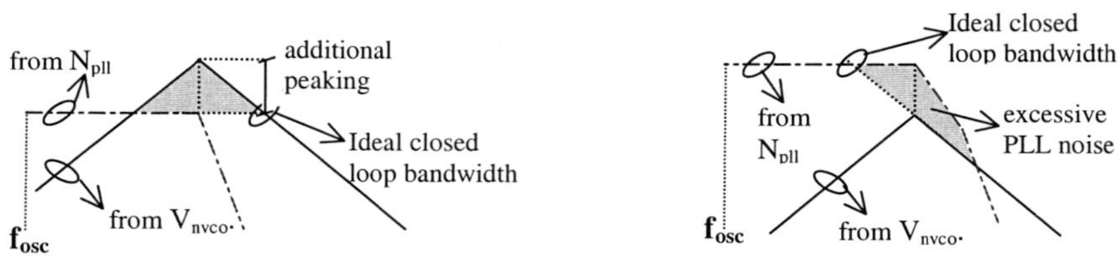

Figure 6 ***Peaking X Optimum Closed Loop bandwidth***

The notation introduced in section 3 is used to look for a relationship between the open and closed loop PLL bandwidths, so that we can minimise the phase noise and still guarantee the stability of the loop.

We choose the 3 dB attenuation point of B(s) as the pll closed loop bandwidth, w_{3dB} . Polynomial reductions can be used to find a relationship between open and closed loop bandwidths, w_{ol} and w_{3dB} . Numerically, for typical values of r_{21} and r_{31} and for a centred gain variation of $2*r_{21}$ around w_{olnpf} , the ratio w_{3dB}/w_{ol} is contained in a limited range.

$$\left.\begin{array}{l} r_{21} \in [10 \;,\; 50] \\ r_{31} \in [16 \;,\; \infty] \\ \wedge \;\; \dfrac{r_{21}}{r_{31}} \geq 1.6 \end{array}\right\rangle \Rightarrow \frac{w_{3dB}}{w_{ol}} = 1.63 \pm 0.28$$

The limiting intervals for r_{21} and r_{31} include the typical values encountered in synthesizer applications.

A numerical example can illustrate the utility of the algorithms discussed above. Let us consider the example of a satellite tuner with the following characteristics:

- PLL noise: N_{pll} = -154 dBc/Hz for comparison frequency, f_{cp} = 1 MHz ;
- Main divider ratio: N ∈ [950, 2150] ;
- VCO phase noise: L_{vco}(100KHz)=-100dBc/Hz ;
- Tuning sensitivity: K_{vco} ∈ [4.5, 100] MHz/V .

For a fixed charge pump current, the ratio $\alpha_{max}/\alpha_{min}$ equals to 50[1]. So the minimum value for r_{21} is 25, for the stability criterion. We take r_{21} equal to 25 and r_{31} equal to 50. The intersection frequency for the PLL and VCO noise can be calculated with the following expression:

$$f_i = f_{offset} \cdot 10^{-\left[\frac{N_{pll} + 20\cdot\log(N) - L_{vco}(f_{offset})}{20}\right]}$$

The centred open loop bandwidth, w_{olnpf} , is placed such as to have f_i as the geommetrical average between w_{ol} and w_{olnpf} .
Figure 7 shows the output spectra for several gain values and the corresponding phase deviation or jitter. The jitter is calculated

[1] A gain variation of 50 is a typical value for satellite applications, whether in terrestrial it can go up to 100, in which case it is convenient to use different charge pump currents to reduce this interval.

integrating upper and lower sidebands, in a large frequency offset interval (from $f_{z1}/500$ up to $40{*}f_{p3}$).

For small α values the spectra is dominated by the VCO intrinsic noise, and for large α by the PLL noise. Nevertheless the total gain range is stable and presents a maximum phase deviation of 1.8°. It is not rare to find a maximum phase jitter specification in the order of 2° for low noise satellite synthesizers.

The curves from left to right correspond to the gain values:

a) $\alpha_{npf} \cdot (2.r_{21})^{-0.5}$
b) $\alpha_{npf} \cdot (2.r_{21})^{-0.25}$
c) α_{npf}
d) $\alpha_{npf} \cdot (2.r_{21})^{+0.25}$
e) $\alpha_{npf} \cdot (2.r_{21})^{+0.5}$

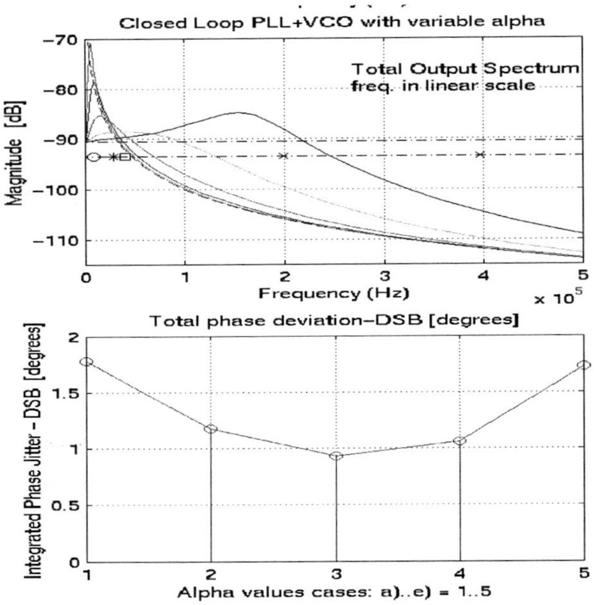

Figure 7 *Output Spectra and Phase Deviation for Satellite tuner example*

The linear loop model can also be used to calculate reference breakthrough and the influence of the AC characteristics and noise of an active loop filter.[2] Limitations of this model appear when examining the acquisition mode of the PLL and when treating wide band loops, as examined in the following section.

---

[2]Reference [Tava99] chapters 3 and 4.

5. Limitations of the Linear-Time-Invariant (LTI) Phase Model

Phase noise constraints, and even more integrated oscillator architectures, demand increasing bandwidths in PLL synthesizers. As the PLL bandwidth increases the comparison frequency (w_{cp}) needs to increase as well to keep the system stable.
In fact, design and stability constraints will appear to limit the values of both w_{ol} and w_{cp}. They concern the maximum feedback bandwidth for a loop that is partially discrete, and the maximum comparison frequency that still guarantees an unlimited tracking range for the tri-state phase detector. These limitations are not contained in the LTI model discussed so far, but they can be evaluated and/or added with additional considerations.

♦ *Minimum Phase Deviation Range*

We consider the classical digital phase detector implementation with D flip-flops and an asynchronous reset commanded by an and-gate and a delay to avoid dead-zone problems. This detector has an important characteristics, essential for frequency synthesizers, to be able to track frequency and phase errors. However, as the comparison frequency increases, its period becomes comparable to the reset delay, and we start reducing the continuous input range of the phase detector transfer, which eventually disables the loop to track frequency errors.

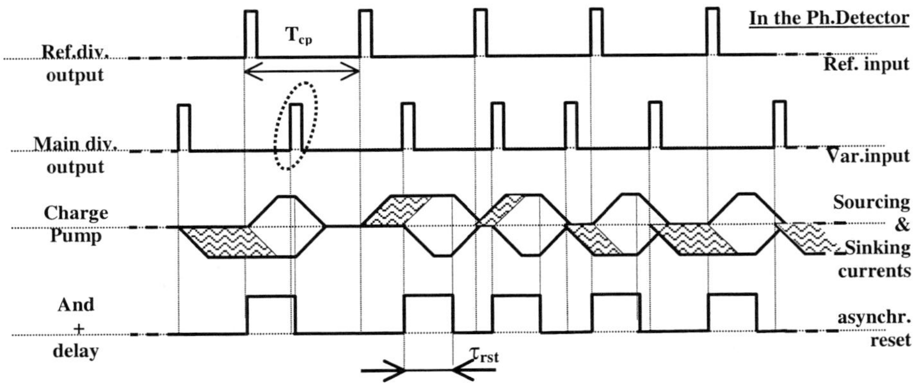

Figure 8 ***Maximum Phase Detection Range & Cycle slips***

Figure 8 shows a VCO varying towards lock. The VCO is initially at a good frequency but it has a phase advance of $\Delta\varphi_1$. The reset delay is large enough to hide the following front of the variable input, and consequently the next phase deviation is measured with respect to the reference input. The phase detector has slipped one cycle.

These cycle slips may be represented in the transfer function $I_{average}/\Delta\varphi_{in}$. They appear as a decrease in the linear portion; in reality, the transfer is not linear up to $\pm 2\pi$, but only up to $\pm 2\pi.(1-\tau_{rst}/T_{cp})$. The resulting transfer is shown in figure 9 for $\frac{\tau_{rst}}{T_{cp}} = \frac{1}{2}$.

The unlimited tracking range is assured by a monotonously increasing or decreasing average charge injected in the loop filter, for input signals with a positive or negative frequency difference.

In figure 9, we observe that τ_{rst} equals $T_{cp}/2$, is the limiting value for which the accumulated charge has the same sign as the derivative of the phase difference. Therefore to guarantee an unlimited frequency tracking range, f_{cp} is limited to: $f_{cp} < \frac{1}{2 \cdot \tau_{rst}}$.

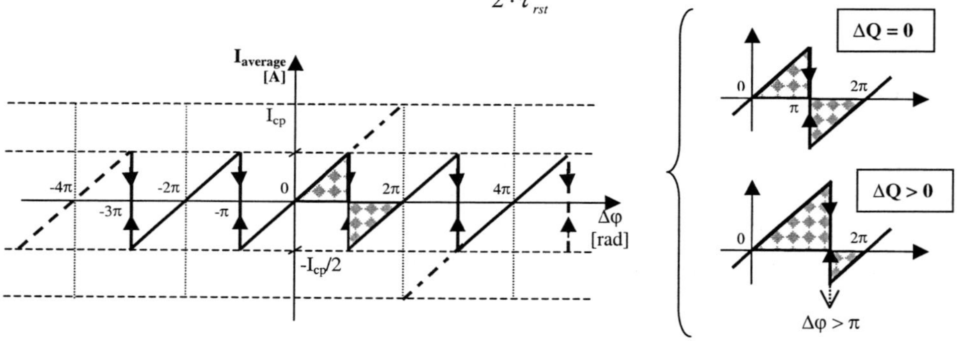

Figure 9 Condition for unlimited frequency tracking range

Another way to derive the minimum range of the linear portion, is to seek a convergence condition for the phase deviation values.

Let us consider a discrete variable $\Delta\varphi_n$, representing the phase deviation of the nth comparison period. Close to lock the phase deviation sequence should decrease towards zero: $|\Delta\varphi_{n+1}| < |\Delta\varphi_n|$.

This degressive sequence can only be obtained over a cycle slip, if the linear portion of the transfer covers the range [$-\pi$, $+\pi$]. Otherwise the module of the phase deviation would increase after each cycle slip, avoiding the convergence towards the lock condition. Thus we confirm the boundary proposed by the average charge approach.

♦ *Lock convergence in acquisition mode*

The phase deviation sequence, as a discrete variable in the time domain, can be used to calculate the maximum ratio between loop bandwidth and comparison frequency, which still guarantees the convergence from tracking towards locked condition.
The expressions derived in the time domain for a 3rd order loop (without post-filter), are dependent of 4 variables: w_{oln} / w_{cp} , $\Delta\varphi_{initial}$, r_{21} and α. The smallest ratio, w_{oln} / w_{cp} , is imposed for $\Delta\varphi_{initial}$ tending to zero. Since we express the maximum bandwidth ratio with respect to the nominal open loop bandwidth w_{oln}, the variable gain, α, accounts for the bandwidth variation. Let us consider a numerical example: for a given r_{21} equal to 25 and a nominal gain value $\alpha=\alpha_n$, we find $\max\{w_{oln}/w_{cp}\}=0.24$. Including a gain variation of $\alpha_{max}/\alpha_n=\mathrm{sqrt}(2.r_{21})=7.07$, determine a much lower ratio of $\max\{w_{oln}/w_{cp}\}=0.05$.
This maximum ratio shows the limit of increasing the open loop bandwidth in single loops synthesizers. The stability problem deriving from an increasing feedback bandwidth, can be represented in the LTI phase model as a delay with the magnitude of half the period of the comparison frequency.

6. Phase Noise: Theoretical to Practical Approach

Phase Noise is a convenient parameter to quantify unwanted phase variation in a periodic signal. Phase variation can be caused by a linear phenomenon such as signal addition and also by non-linear phenomena such as angular modulation.
In the PLL synthesizer we consider two sources of periodic signals, which are disturbed by phase noise: the reference oscillator and the voltage controlled oscillator. The disturbances are either intrinsic to

the periodic sources, or are accumulated as their outputs propagate through the PLL blocks.

Phase noise is represented in many different notations, which are chosen with respect to the origin of the phase deviation, or to the measurement tools. The intrinsic noise of the VCO is often described as a modulation of the noise contributions inside the oscillation loop, combined to an amplitude limitation mechanism, which attenuates the amplitude modulated sidebands, and transmits the angular modulated ones. On the other hand, time deviations that are added to periodic signals which are transmitted through switching stages, are more conveniently represented by the notation below.

♦ *Superposed noise in switching stages or Slope approach*

The result of noise simulations in analog circuits is usually given as a voltage noise density at a specific node. In order to evaluate the influence of such noise in the output spectrum, we need to convert it into amplitude and or phase disturbance.

Let us consider a logical or switching stage that has two output values, low and high. The instants where the signal crosses the threshold are called zero-crossings. The interval between two successive zero-crossings is the period of the signal driving the stage. The variations of this period that are due to additional voltage noise are called time jitter.

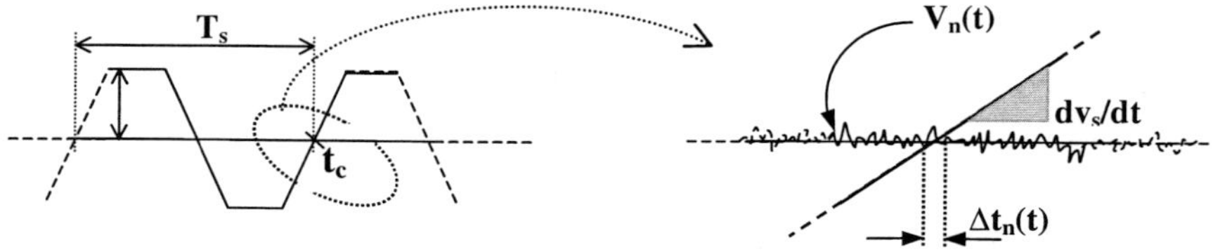

Figure 10 ***Slope approach: voltage & time deviations***

The noise voltage $V_n(t)$ is calculated by a small signal noise simulation around a zero-crossing instant. The result is usually presented as a voltage noise density $\delta v_{n\text{-}rms}(f)$ in $[V/\sqrt{Hz}]$. The rms amplitude equals the square root of the power spectral density for the unitary impedance. The time deviation is represented by similar functions in the time and frequency domain: $\Delta t_n(t)$ and $\delta t_{n\text{-rms}}(f)$ in $\lfloor s/\sqrt{Hz} \rfloor$.

The relationship between the voltage and time deviations is given by the voltage slope of the large signal driving the stage. We name $v_s(t)$ the output signal and t_c the zero-crossing time instant; and we start looking at a single tone portion of $V_n(t)$ that we call $v_n(t)$, or as a frequency function: $v_n(t) \leftrightarrow \delta v_{n-rms}(f_n)$

Superposed noise into a periodic signal can be represented by two orthogonal contributions causing amplitude or phase modulation. The error caused by this superposed sideband at the zero-crossing instants, can not be represented by an amplitude error, therefore it is necessarily a phase or time[3] error. The modulating tone appear as a sideband with an equivalent base band frequency equal to the offset frequency with respect to the carrier. Thus the time deviation that is caused by the single tone component $\delta v_{n\text{-rms}}(f_n)$ becomes:

$$\delta t_{n-rms}(f_n - f_c) = \frac{\delta v_{n-rms}(f_n)}{\dfrac{dv_s(t_c)}{dt}} = \delta t_{n-rms}(f_{no}) = \frac{\delta v_{n-rms}(f_{no} + f_c)}{\dfrac{dv_s(t_c)}{dt}} \qquad \left[\frac{s}{\sqrt{Hz}}\right]$$

for f_{no} an offset frequency defined as: $f_n = f_{no} + f_c$.

This is the time deviation due to a single side band (SSB) superposed noise at a frequency offset f_{no} from the carrier. If the voltage noise density $\delta v_{n\text{-rms}}(f)$ has the same amplitude for the frequencies f_c+f_{no} and f_c-f_{no} the time deviation due to a DSB superposed noise becomes:

$$\delta t_{n-rms}(f_{no}) = \frac{\sqrt{\delta v_{n-rms}^2(f_{no} + f_c) + \delta v_{n-rms}^2(f_{no} - f_c)}}{\dfrac{dv_s(t_c)}{dt}} = \frac{\sqrt{2} \cdot \delta v_{n-rms}(f_{no} + f_c)}{\dfrac{dv_s(t_c)}{dt}}$$

Finally the phase deviation due to a time deviation is:

$$\delta \varphi_{n-rms}(f_{offset}) = \frac{2\pi}{T_s} \cdot \delta t_{n-rms}(f_{offset}) \qquad \left[\frac{rad}{\sqrt{Hz}}\right]$$

---

[3] Phase and time deviations are related through the period of the large signal which is being modulated. Furthermore we are supposing that the crossing moment is determined by the large signal.

where T_s is the period of the signal, and we indicate the independent parameter as the frequency offset to remember that the voltage noise that originates this time deviation is found at $f_c \pm f_{offset}$. The phase deviation relates the time jitter to the SSB phase noise of the output signal. It follows that:

$$L_{dB}\left(f_{offset}\right) = 20 \cdot \log\left(\frac{\Delta\varphi_p\left(f_{offset}\right)}{2}\right) = 20 \cdot \log\left(\frac{\Delta\varphi_{rms}\left(f_{offset}\right)}{\sqrt{2}}\right) = 20 \cdot \log\left(\frac{\sqrt{2} \cdot \pi \cdot \delta t_{n-rms}\left(f_{offset}\right)}{T_s}\right)$$

The equation above shows the degradation of a periodic signal due to a time deviation. It also shows that the phase noise is inversely proportional to the period of the signal.

- *Large Signal Linearization*

The slope approach can also be described as a linear time variable transfer function[4]. The transfer function for a small disturbance signal, that is transmitted by a block which is driven by a large input signal.[5]

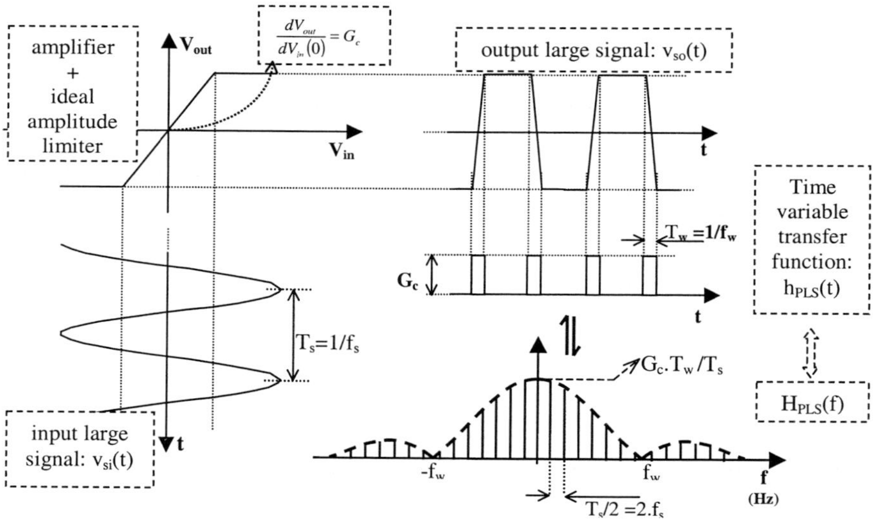

Figure 11 ***Periodic Transfer determined by a large signal***

---

[4] These ideas are based on the convolution transfer discussed in reference [Boon89]. A similar discussion focused on oscillators noise can be found in [Haji98].

[5] The supposition of a linear transfer holds for small signals whose amplitude does not disturb significantly the periodic large signal transfer.

Let us consider the transfer function of a voltage amplifier that has an ideal limiting output.
Figure 11 shows the transfer of a sinusoidal input signal $v_{si}(t)$ that overdrives the ideal limiting amplifier. The output signal $v_{so}(t)$ has a fundamental harmonic at the same frequency as the input, but it also has higher harmonics that are generated by the non-linear clipping of the limiter. The transfer function $v_{so}(t)/v_{si}(t)$ is time variable, and it may be represented in both time and frequency domains. We call it the periodic large signal (PLS) transfer.
The transfer of a small signal that is added to $v_{si}(t)$ may be calculated making a 1st order development of the periodic transfer around the steady-state that is driven by $v_{si}(t)$. If the small signal is represented by a noise component $v_n(t)$, it becomes:

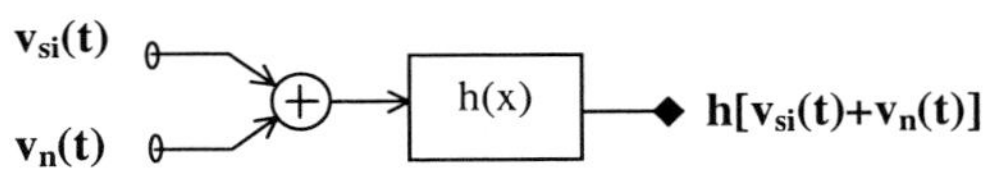

$$h[v_{si}(t)+v_n(t)] \approx h[v_{si}(t)] + \left.\frac{dh(x)}{dx}\right|_{x=v_{si}(t)} \cdot v_n(t) = v_{so}(t) + h_{PLS}(t)\cdot v_n(t)$$

where $h_{PLS}(t)$ is the transfer function for a small signal that is added to the large input signal. The Fourier transform of this time transfer is denoted as $H_{PLS}(f)$, and we use it to define the convolution transfer of the small signal in the frequency domain:

for
$$\left.\begin{array}{l} v_n(t) \leftrightarrow \delta v_{n-rms}(f_n) \\ h_{PLS}(t) \leftrightarrow H_{PLS}(f) \end{array}\right\rangle \quad v_n(t)\cdot h_{PLS}(t) \leftrightarrow \delta v_{n-rms}(f_n) \otimes H_{PLS}(f)$$

It is important to notice that the time variable characteristic of this transfer causes frequency translation of the input signals. For broadband noise contributions the frequency translation also causes aliasing or folding.

- *Phase Noise in the PLL Context*

The aliasing effect can be quantitively understood in a transient simulation combined to a FFT of the periodic inputs and outputs. Such simulations can provide a better understanding of the noise generation mechanisms and can help to isolate the dominant noise contributors. More precise numerical evaluations can be done using one of the recently RF simulators which have implemented periodic transfer functionality. Nevertheless the kernel design parts have to be isolated, and the surrounding parts replaced by behavioural models, since the periodic transfer tools are often limited in number of components or maximum complexity that can be resolved.

The transfer functions and phase noise generation mechanisms discussed in the previous sections, can be used to distinguish in measurements the dominant noise contributors in a synthesizer implementation.[6] For instance, the charge pump noise contribution, translated in a phase deviation (part of N_{pll}), is sensible to I_{cp} , and it is a narrow band noise contribution (filtered by the loop filter) which is not aliased. Therefore its level is not dependent of the comparison frequency. These characteristics are quite distinct from the noise contribution of the reference oscillator, which is often a large bandwidth contribution, which is strongly aliased by the phase detector. In this manner the contribution of the reference oscillator (in N_{pll}) often presents a sensitivity to the comparison frequency to the order of 3dB/octave of f_{cp}.[7]

The behavioural model of the PLL can be extended to include the carrier and clock recovery loops inside a digital demodulator. This larger model can then be used to translate the implementation loss specification into a maximum jitter requirement.[8]

7. TESTCHIP IMPLEMENTATIONS

In this section two testchip designs are briefly presented. They contain a PLL and a monolithic Gm-C oscillator that covers the satellite band L, 950MHz to 2150MHz. Testchip TC2 is part of a double loop synthesizer with a comparison frequency that goes up to 330MHz, with an in-loop noise in the order of –108dBc/Hz. Testchip TC3

---

[6] Reference [Tava99] chapter 7 presents several calculation examples.

[7] The calculations of the N_{pll} contribution suppose a phase detector switching slope that is limited to the maximum slew rate of the design, for given biasing and loading conditions.

[8] Vide references [Sinde98] and [Tava99]

explores the maximum bandwidth of a single loop PLL and confirms the theoretical approach of section five.
The testchips were realized in a bipolar process that is derived from a BiCMOS process. The stripped bipolar process kept the gate oxide of the CMOS components for the capacitors. This enables us to compose a native PMOS, which gives us a bipolar+PMOS process. The peak value of the ft of the NPN transistors equals 13GHz. The maximum ft of the lateral PNP equals 200MHz. There are three levels of metallization, with a pitch of 2.4μm.
We start describing the results of the Gm-C oscillator, which is a common block in the two testchips. A fuller description of the double loop structure and the Gm-C oscillator can be found in references [Vauc98] and [Tang97].

♦ *Gm-C oscillator*

A fully integrated oscillator becomes quite interesting in monodyne receivers where the radiation of the input RF signal may significantly deviate a LC externally-coupled oscillator. In terrestrial and satellite tuners the usual range of the tuning voltage is 30V. This high voltage supply can be suppressed if the LO can be tuned under a 5V range. The integrated Gm-C oscillator has a range divided into 4 bands that are tuned in a 5V range.
The Gm-C oscillator is a ring structure with two integrator stages and an inverting feedback. The two stages have outputs with an equal frequency, and phases that are shifted by 90° with respect to each other.[9] The oscillating frequency depends on the value of the capacitors and on the transconductance Gm. The frequency tuning is made by varying the biasing current of the transconductance stages. The resistive load of the transconductance are compensated by an active negative resistor (another transconductance connected in positive feedback), which also enables the amplitude to be controlled. However for large amplitude values, well above the linear range of the transconductances, amplitude and frequency tuning are no longer

[9] Hence the oscillator is also called a QCCO: quadrature current controlled oscillator. These quadrature outputs are very convenient for a receiver with a monodyne structure. A monodyne receiver needs to provide two outputs, in quadrature to each other, so that the demodulator can distinguish the channel from its mirror image, which is the proper wanted channel for a zero-IF receiver.

independent. Therefore, meeting the constraints of amplitude, range coverage, maximum power consumption, and robustness to process spread, become a tricky compromise. The current design was optimised for an uniform power consumption, tuning sensitivity and noise performance throughout the tuning range. The results are: peak amplitude value between 200 and 300mV for a constant control input, with 26mA under 5V supply; four tunable bands, each with 425MHz ±10%, which cover the complete band-L, with overlaping intervals between bands; almost constant tuning sensitivity 125MHz/V ±5% , and also uniform phase noise performance of L(100kHz)=-76.5dBc/Hz ±1 for $f_{QCCO} \in$ [850MHz, 2250MHz].
Nevertheless the phase noise of the integrated QCCO under 5V, is in average 20dB worse than a LC oscillator covering the same range with a 30V tuning range. The solution, to cope with the degradation of the phase noise, is to increase the closed loop bandwidth. In order to respond to both the specifications of a maximum tuning step and a minimum closed loop bandwidth, a multi-loop structure is needed.

♦ *Double Loop PLL Synthesizer*

Testchip TC2 is an implementation developed in collaboration with Nat.Lab. the research laboratory of Philips. It is a double-loop PLL synthesizer. The first loop drives an oscillator in the VHF band, which is used as the input reference for the second loop which drives the Gm-C oscillator. The two oscillators are tuned in a 5V range.
The testchip TC2 contains the higher-frequency loop of a double loop synthesizer. It is a mixer-oscillator and PLL circuit, with a dual mixer, a QCCO and a PLL working with comparison frequencies of 210MHz up to 330MHz. The RF path is dimensioned for a monodyne receiver, with single ended base band outputs. Figure 12 shows a block diagram of the chip.
The testchip is basically divided into two parts, analog and digital, that interact through interface blocks. The ensemble of blocks is programmed by a 3-wire bus. The bus has an additional acknowledge block which is included to test the sensibility to bus cross-talk.

There are 4 supply pins, a pair for the analog part and another for the digital one. The total consumption is 60mA under 5V, and the active layout area equals 1.2mm^2 . The total layout area is 2.1mm^2 , which includes the 20 input/output pins.
The symmetry of the layout of the analog part is stressed to guarantee the quadrature characteristics of the I and Q branches. Figure 13 shows a photo of a testchip TC2. On the left side there are the digital blocks (bus, main divider and phase-detector /charge pump, from the higher to the lower corner); and on the right side, the analog part (QCCO, mixer, regulator, input and output buffers).

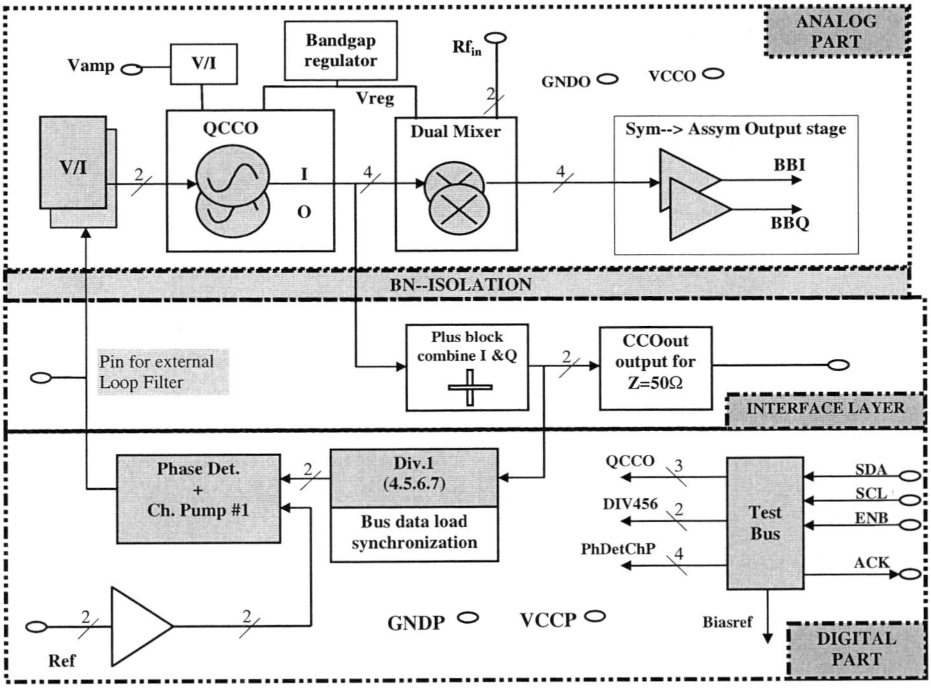

Figure 12 *Block diagram of testchip TC2*

Although the bus acknowledge block pulses currents up to 7mA, there was no visible interference on the LO spectrum for continuous programming sequences. The oscillator output is also very little affected by the RF power, and the first visible interference are only seen for RF input power equal to –10dBm (10 dB above the maximum power previewed for the design). The quadrature of the LO outputs has an error below 2° for oscillator amplitudes above 200mV_p. For the worst case, of the smallest amplitude, the quadrature error has a maximum of 3.5°. The spurious rays at $\pm f_{cp1}$ are lower than –62 dBc, for a loop filter with a closed bandwidth around 2MHz.

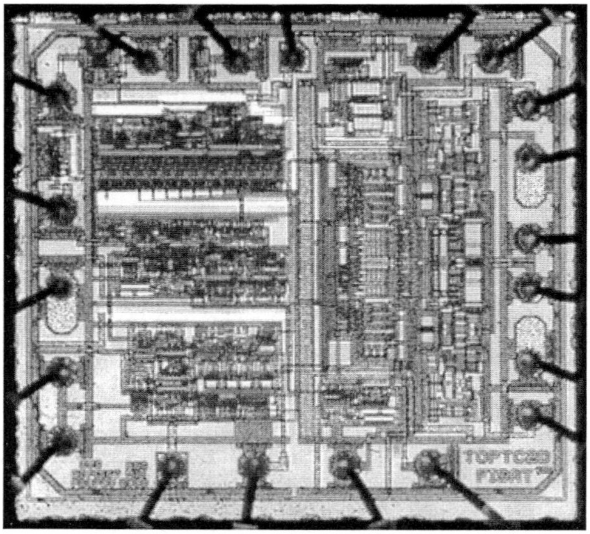

Figure 13 ***Photo of a open die of testchip TC2***

The comparator (phase detector and charge pump) is able to retrieve frequency differences for a maximum f_{cp} equal to 450MHz, with no loss in its sensibility K_φ (no dead zone). The N_{pll}, or synthesizer noise floor, for a f_{cp} of 300MHz is measured as –124dBc/Hz.[10]
Figure 14 shows the in loop zone of the QCCO spectrum. It is a measurement in a small span of 250kHz, for an f_{QCCO} of 2.1GHz. The L(f) is indicated in the plot.

$$L(25kHz)\big|_{f_{cp1}=300MHz} = -107\ {}^{dBc}\!/_{Hz} \quad\Rightarrow\quad N_{pll}(300MHz) = -107 - 20\log(N1)\big|_{N1=7} = -123.9\ {}^{dBc}\!/_{Hz}$$

[10] This result can be very accurately matched to noise simulations of the charge pump, which is the dominant noise contributor in this implementation.

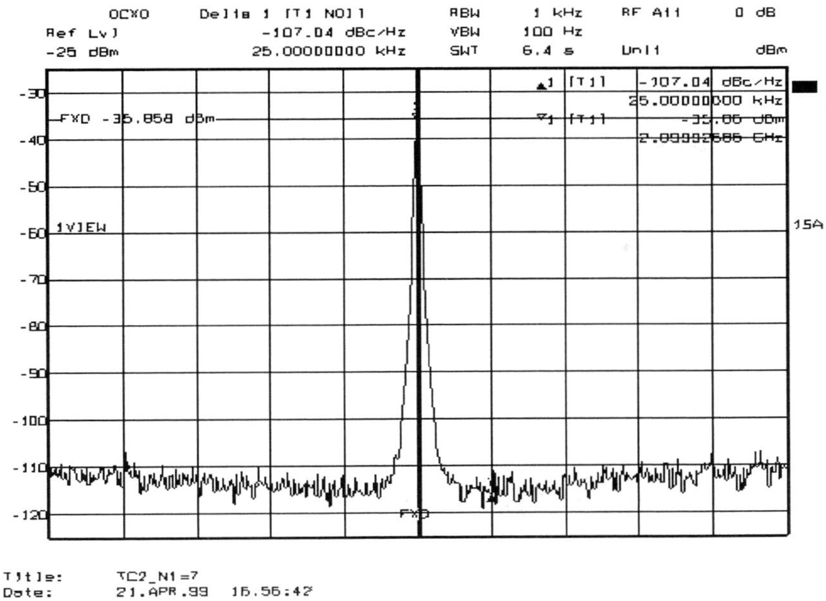

Figure 14 *TC2 _ in-loop spectrum for N=7 and f_{cp}=300MHz*

♦ *Single loop synthesizer plus QCCO*

Testchip TC3 confirms the theoretical study of the maximum feedback bandwidth of section 5. On a single loop PLL working with a comparison frequency of 1MHz, we can measure stable ouputs (with no excessive PLL peaking) for open loop frequencies up to 165kHz, and closed loop 3dB bandwidth of 280kHz.

8. CONCLUSIONS

New communication standards are very demanding for tuner specifications. Therefore behavioural system analysis becomes a more and more relevant step, in evaluating and dimensioning circuit and block requirements.

The PLL synthesizer was presented as a control system, in order to study the influence of different parameters using a simple and flexible model. This representation was used to examine some issues around the application and specification of the PLL: controlling the feedback bandwidth, working with larger comparison frequencies, dealing with phase noise and stability. The theoretical concepts were exemplified in the implementation of testchips with wide band loops.

9. REFERENCES

[Boon89] C.A.M.Boon, *Design of High-Performance Negative-Feedback Oscillators*, PhD thesis, Delft, University of Technology, 1989.

[Haji98] A.Hajimiri and T.H.Lee, "A General Theory of Phase Noise in Electrical Oscillators", IEEE Journal of Solid-State Circuits, Vol.33, No.2, pp. 179-194, Feb. 1998.

[Robi91] W.P.Robins, *Phase Noise in Signal Sources (Theory and Applications)*, Peter Peregrinus, London, UK, 2nd edition reprinted 1991. ISBN 0-86341-026-X

[Sinde98] Sinderen, J. v. *Phase Noise Requirements of Some Digital Modulated Systems.* Eindhoven (NL): Philips Research Laboratory Nat.Lab., 1998. 6 p. Internal study report.

[Tang97] J.D.van der Tang and D.Kasperkovitz, "A 0.9 – 2.2GHz monolithic quadrature mixer oscillator for direct-conversion satellite receivers", in Digest of ISSCC, San Francisco, USA, February 1997, pp. 88-89.

[Tava99] Tavares, M.Q. *PLL Synthesizers: Phase Noise Issues and Wide Band Loops.* PhD thesis: INSA de Lyon, 1999. http://philweb1.natlab.research.philips.com/PWWReports/

[Vauc98] C.Vaucher and D.Kasperkovitz, "A Wide-Band Tuning System for Fully Integrated Satellite Receivers", IEEE Journal of Solid States, Vol.33, No.7, pp. 987-997, July 1998.

LOW POWER CIRCUITS FOR RF-FREQUENCY SYNTHESIZERS IN THE LOW GHz RANGE

Dirk Pfaff, Qiuting Huang
Integrated Systems Laboratory
Swiss Federal Institute of Technology Zurich
CH-8092 Zurich

Abstract

The voltage controlled oscillator and the prescaler are building blocks which dominate power consumption of RF frequency synthesizers. This paper explores the low power aspects of CMOS VCO and prescaler design. Different resonator circuits are analyzed and compared. Measurement results on a 1GHz oscillator, a quadrature signal generator, and a dual modulus prescaler, all with very low power consumption, are presented.

1. Introduction

The last decade has seen an explosive growth of wireless handheld personal communication devices. The design of the latter is governed by two objectives: First, it needs to satisfy type approval requirements of the targeted standard with a respectable margin to achieve a high yield in production. Second, low cost is endeavored. The total cost of a handset is composed of the cost for battery, chipset, discrete and mechanical components as well as assembling cost. It is worth mentioning, that the battery takes 25% of the overall cost, while the chipset (RF and baseband) takes another 25% [1]. Passive components, including filters, cause relatively little cost. The latter figures demonstrate, that low power translates directly into low cost. A smaller, and therefore less expensive battery can be used, thereby further reducing size and weight, which are both strongly related to the battery.

Thus, the focus in wireless tranceiver design must be on low power. While transceivers are traditionally built using bipolar technologies, CMOS designs have been reported in the last few year [2] [3]. Consuming about 60-100mA, most CMOS receivers realized to date are not considered sufficiently competitive to

R. J. van de Plassche et al. (eds.), Analog Circuit Design, 383–400.

BJT implementations, which typically consume 50mA in receive mode [4] [5]. The potential cost advantage of CMOS is hence outweighed by the increased power consumption.

Recently, however, it has been demonstrated that with 0.25μm CMOS and careful design, it is feasible to realize RF receivers for cellular applications such as GSM, at less than half the current consumption of today's typical BJT receivers [6].

Progresses made in CMOS transceiver design shift the burden to find low power solutions to the designer of RF frequency synthesizers. The latter's power consumption is dominated by the local oscillator and the prescaler which tend to consume 2-5mA each. Relative to the consumption of the receiver, the frequency synthesizer consumption is high and needs significant improvement.

The low power aspects of CMOS oscillator and prescaler design are explored in this paper. First, the relation of oscillator amplitude as well as phase noise performance to the design parameters is given. Second, various resonator circuits are proposed and their quality factors are analyzed. A low power voltage controlled oscillator (VCO) is presented, together with measurement results. Some receiver topologies need local oscillator providing quadrature outputs. Two different approaches suited for the generation of quadrature outputs are compared and a 1.5GHz low power solution is presented. Last, a dual modulus prescaler, aimed to GSM applications, is presented with very low power consumption.

2. Low power VCO design consideration

The aim of voltage controlled oscillators is to generate a periodic signal with a large enough amplitude to drive a mixer and prescaler circuits. On the one hand, a weak oscillator amplitude may deteriorate the mixer's noise figure and may further reduce its conversion gain. On the other hand, an excessively large amplitude will lower the maximum operating frequency of the prescaler. Hence, the oscillator amplitude must be well controlled. Another important oscillator requirement is its phase stability. The linear phase trajectory of an ideal oscillator in a practical implementation is more or less covered with noise causing a noise skirt around the carrier. This noise skirt has tremendous impact on receiver and transmitter: Inband noise can directly destroy information contained in the phase modulated signal, while oscillator phase noise far from the carrier will degrade the signal to noise ratio in the IF section of a receiver due to reciprocal mixing.

Apart from amplitude and phase stability, tuning characteristic is another crucial requirement for a voltage controlled oscillator. The oscillator must be tunable to all required channels. The frequency tuning range should be significantly larger than the effectively needed frequencies to compensate for process tolerances.

Otherwise trimming may be necessary, a technique which is not cost effective and should therefore be avoided.

It is essential for low power design to gain insight how amplitude and phase noise are related to the design parameters, such as drain current, transistor transconductance and resonator quality factor. Because of its simple topology, the Clapp or Colpitts oscillator, shown in figure 1a, is a suitable test vehicle to explore the latter relations. These oscillators are widely employed in fully discrete designs, since they can be built from a single bipolar transistor. Its steady state amplitude was found [7] to be

$$V_m = 2 \cdot \frac{I_B}{g_{mc}} \cdot \frac{I_1\left(\frac{V_m}{U_t}\right)}{I_0\left(\frac{V_m}{U_t}\right)} \approx 1.9 \cdot \frac{I_B}{g_{mc}} \tag{1}$$

with I_B the drain current, g_{mc} the critical transconductance, and I_0 and I_1 modified Bessel functions. Notice the steady state amplitude is dependent on the drain current and the tank loss, but independent of the transistor transconductance. The steady state amplitude of a MOS realization has been derived recently [8]

$$V_m = \frac{I_B}{g_{mc}} \cdot f\left(\frac{g_{mo}}{g_{mc}}\right)\Bigg|_{\frac{g_{mo}}{g_{mc}} = 3} \approx 1.75 \cdot \frac{I_B}{g_{mc}} \tag{2}$$

with g_{mo} being the transistor transconductance. Equation (1) and (2) demonstrate that CMOS can be regarded as a serious competitor for bipolar oscillators since the amplitude of the MOS oscillator is virtually identical to the BJT implementation. The amplitude is strongly related to the ratio of bias current and critical transconductance, but is only a weak function of MOS transistor transconductance. This outcome is plausible since the transistor conducts current just for a short fraction of the full period and behave more like an ideal switch. Indeed, if a device such as an ideal switch would exist, the oscillator built with the latter device would generate an amplitude equal to $V_m = 2 \cdot I_B / g_{mc}$. The outcome of equation (1) and (2), which is very close to the latter result, indicates, that the active device in an oscillator may be simplified as a switch, at least for steady state amplitude calculations.

So far, the relation between oscillator amplitude and the design parameters was outlined. Doing the same for oscillator phase noise is more onerous. Most designer relied in the past on Leeson's observation, which predicted phase noise as

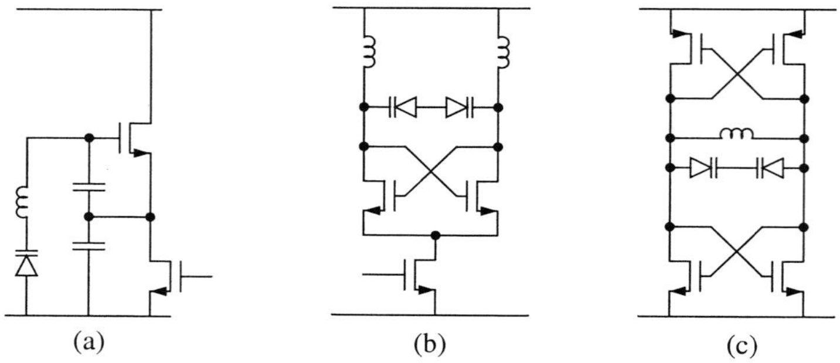

Figure 1: Oscillator topologies: (a) Clapp oscillator (b) differential oscillator with a single cross coupled pair (c) differential double cross coupled pair oscillator as proposed in [12]

$$L(\Delta\omega) = \frac{kT \cdot F}{I_B^2 \cdot Q^2} \cdot \left(\frac{\omega_0}{\Delta\omega}\right)^2 \tag{3}$$

with ω_0 the angular center frequency, $\Delta\omega$ the angular offset frequency from the carrier and Q the resonator quality factor. Leeson's observation proved very useful, especially for the design of discrete bipolar oscillators where the transconductance of the active device was not under control of the designer. However, the contribution of the noise due to the active device remains mysterious and was expressed as the so called oscillator noise figure F, which needs to be evaluated experimentally.

Recently reported oscillator phase noise performance varies considerably even though drain current and quality factor are comparable. Hence, a more accurate analysis of oscillator phase noise was highly desirable, which could be found for the CMOS Clapp oscillator as [9]

$$L(\Delta\omega) = \frac{kT}{4I_B^2 Q^2} \cdot \left(\frac{\omega_0}{\Delta\omega}\right)^2 \cdot \left[\frac{2}{3}g_{mB} + g_{mc} + \frac{2}{3}g_{mo}\sqrt{\alpha}\right] \tag{4}$$

with g_{mB} the bias current source transconductance and α the duty cycle of the active device. Referring to Leeson's formula, it is widely believed, that doubling the quality factor improves phase noise performance by 6dB. Indeed, referring to equation (4), this is true if in the brackets the last term dominates. However, since the transconductance is under control of the designer, g_{mo} can be kept moderate. If the g_{mo}/g_{mc}-ratio is set to 3, large enough to guarantee startup, and g_{mB} to 20% of g_{mo}, equation (4) simplifies to

$$L(\Delta\omega) \cong \frac{3}{4} \cdot \frac{kT}{I_B^2 Q^2} \cdot \left(\frac{\omega_0}{\Delta\omega}\right)^2 \cdot g_{mc} \tag{5}$$

An interesting relation can be given further, since g_{mc} is proportional to *1/Q*:

$$L(\Delta\omega) \propto p, \frac{kT}{I_B^2 Q^3} \cdot \left(\frac{\omega_0}{\Delta\omega}\right)^2 \tag{6}$$

Equation (6) indicates that doubling the quality factor can improve phase noise performance by 9dB if the doubled quality factor is accompanied by halved transconductances!

Equations (1)-(6), although based on analysis on a particular oscillator, allow us to draw some general design guidelines for low power tuned oscillators:

- Use the best quality resonator available. Both amplitude and phase noise performance profit from high Q. Doubling the quality factor allows the drain current to be lowered by a factor of 2.8, if the transistors are sized correspondingly.
- The choice of BJT as active device is not an advantage a priori since the oscillation amplitude is only a weak function of the transistor transconductance. CMOS offers the ability to control its transconductance by choosing the W/L-ratio to minimize phase noise.
- The transconductance of the active device should be kept moderate since a large transconductance does not increase the amplitude but injects more noise current into the tank. A g_{mo}/g_{mc}-ratio of 3 is appropriate if the tank quality factor is fairly well controlled.
- As will be shown in the next section, a large inductor, combined with a small capacitor, can increase the quality factor. However, a trade off between tuning range and high-Q exists.

3. The choice of the resonator circuit

The previous chapter demonstrated that the LC resonator not only serves as the frequency setting circuit, but also has tremendous influence on the oscillator phase noise performance. This section explores three different realizations of resonators.

3.1. Fully integrated resonators

Although highly desirable, fully integrated resonators are rarely used in current low phase noise wireless applications such as cellular, since they suffer from a

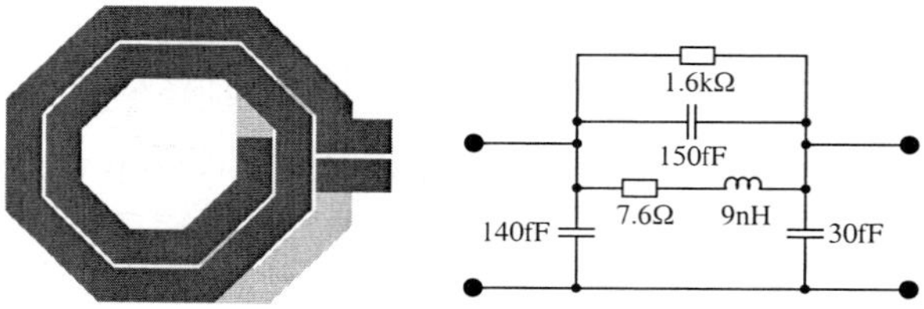

Figure 2: Layout and equivalent circuit of a planar spiral inductor integrated in a 6 metal layer CMOS process

very low quality factor. Their inherent advantages, such as less components, less cost for assembling and package, are outweighed by higher current consumption. Nevertheless, considerable effort has been spent to improve the situation. Implementing integrated planar spiral inductors in standard CMOS in the 10nH region with quality factors of 5 to 6 is quite straight forward. Figure 2 shows a layout and the equivalent circuit of a planar spiral inductor with 8 windings designed in a 6 metal layer CMOS process. The values of the equivalent circuit are extracted from a network analyzer connected to a wafer probe station. The Q was measured as 5.9 at 1GHz. However, no CMOS integrated inductors with significantly improved quality factors are reported. The main limitations lie in the finite resistivity of the aluminium tracks which must be designed relatively narrow to keep substrate loss moderate. An increase of 33% in quality factor was reported [10] by introducing a shielding technique to reduce the latter loss.

Apart from the inductor, an integrated solution for a varactor must be found. Some implementation of integrated resonators use reverse biased pn junctions as varactors [11] [12]. These pn junctions again suffer from low quality factor, which is around 7 at 1GHz for typical CMOS processes [13]. The quality factor of the resonator, simply found as $1/Q_{res} = 1/Q_{ind} + 1/Q_{var}$, is no more dominated by the inductor Q alone. Gate capacitances of MOS transistors are alternatively utilized as varactors [13]. Their quality factor is around 15, but the tuning characteristic is very nonlinear, and even worse, the spread of quality factor can be as high as ten over the full tuning range.

We can conclude, that the Q of integrated resonators is affected by loss due to the inductor as well as due to the varactor. Resonators built in CMOS technology tend to have Q's in the region of 3 to 5. The tuning range is limited due to strong alterations in loss in the integrated varactor under different tuning voltages.

3.2. Resonator built from surface mounted devices

The limitations of integrated resonators, low Q of integrated spiral inductor and limited tuning range of the MOS varactor, can be overcome by using external surface mounted components. Indeed, virtually all commercial VCO's for portable wireless applications use external resonators [4], [14], [15].

Low cost, small size 10nH inductors with quality factor of 80 at 1GHz are available [16] while hyperabrupt junction varactors offer a large tuning range [17]. The only remaining question is which L-C combination for a given frequency is the best choice. It can be observed, that the Q of inductors increases slightly with increasing inductance value while the Q of the junction varactors increases with smaller capacitance values. The first phenomenon is explained that the inductance is proportional to the number of turns squared, while the effective series resistance scales only linearly with the number of turns. Hence, the inductor Q increases with larger inductances.

This leads to the conclusion that the tank resonator quality can be maximized by involving a large inductor, combined with a small varactor. The lower limit for the varactor capacitance is determined by parasitic capacitances (package, ESD protection and the active devices) which should not dominate the resonator capacitance, the upper limit of the inductor is given by its self resonance.

A resonator built with a 10nH inductor and two common cathode varactors, housed in a single package, has been measured with a network analyzer connected to a test fixture. The measured quality factor at 1GHz is 20, slightly increasing for frequencies above 1GHz. This is a factor 4 to 5 above its integrated equivalent, which allows, according to equation (6), the drain current to be reduced by one decade while maintaining phase noise performance.

3.3. Tuned microstrip resonators

Resonators built from surface mounted inductors and varactors are the preferred choice for frequencies up to 2GHz. However, parasitic inductances in the region of 1 to 2nH, caused by the packages, limit their use at higher frequencies. As an alternative, open ended transmission lines of λ/2 or λ/4 length exhibit parallel and series resonances respectively at their input. These devices can therefore replace a discrete tank circuit at higher frequencies. Transmission lines, implemented as microstrip lines, can be integrated onto the printed circuit board without any assembling cost. The guided wavelength, given as

$$\lambda_g = \frac{30cm}{\sqrt{\varepsilon_r} \cdot F} \tag{7}$$

with ε_r the relative dielectric constant, which is between 3 and 4 for standard printed circuit board substrates, and F the frequency in GHz, determines the physical length of λ/2 and λ/4 resonators to be in the acceptably small range of 1 to 2cm at 4GHz.

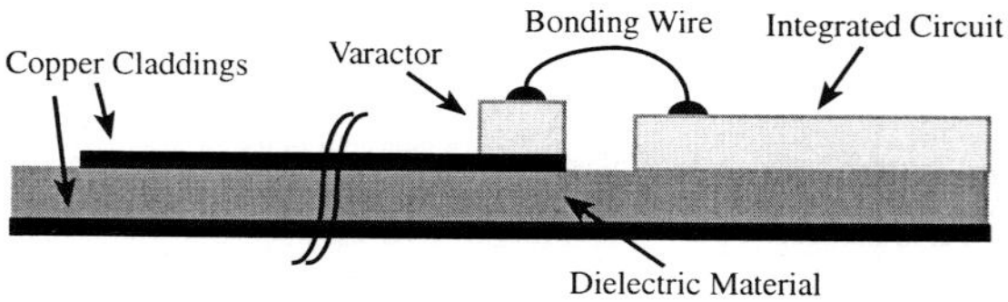

Figure 3: Cross section of a varactor tuned microstrip resonator built from a double layer printed circuit board

The microstrip resonator frequency can be tuned with a varactor, assembled as naked die onto the microstrip line to avoid package inductance. Figure 3 shows a cross section of such a resonator structure.

Although microstrip lines generally display extraordinarily low loss, they must be carefully designed to maximize quality factor [19]. Three main contributors of loss exist in the low GHz range which determine the quality factor:

- **Conductor loss**: The finite resistivity of the copper tracks decays a signal along the microstrip line. Wide tracks, which corresponds to a small characteristic impedance, minimize this loss.
- **Dielectric loss**: Further energy is dissipated in the dielectric material. The effect is quantified with tanδ, a material constant. Dielectric loss is nearly independent of characteristic impedance, as well as frequency.
- **Radiation loss**: Each discontinuity radiates energy into the air, especially the beginning and the end of a transmission line. The radiated energy can be modeled with a shunt resistance [20]. Radiation is minimized by the use of narrow tracks and a thin substrate. Furthermore, radiation can become the dominant loss at higher frequencies since it scales with f^2.

The total quality factor of $\lambda/2$ or $\lambda/4$ microstrip resonators, given by the conductor quality factor Q_C, the dielectric quality factor Q_D and the radiation quality factor Q_R as

$$\frac{1}{Q_0} = \frac{1}{Q_C} + \frac{1}{Q_D} + \frac{1}{Q_R} \tag{8}$$

is described in depth in the literature [21] by powerful design formulas. A plot of the individual Q's versus characteristic impedance is shown in figure 4a for a commercial low cost, low ε_r substrate material [18]. Figure 4a shows that a well defined optimum characteristic impedance exists. While conductor loss dominates at large characteristic impedance, radiation loss dominates at lower characteristic impedance. The optimum Z_0 of 51Ω for the latter substrate, which is frequency dependant, results in a peak quality factor of 150.

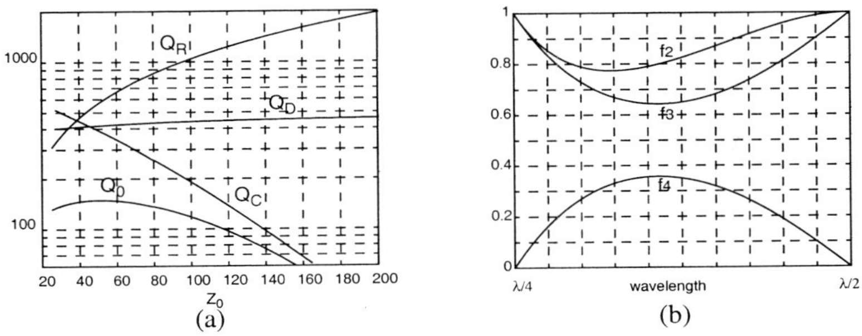

Figure 4: (a) Radiation Q_R, dielectric Q_D, conductor Q_C and total Q_0 quality factor vs. characteristic impedance of RO4003 substrate material at 4GHz. The substrate thickness is 0.8mm (b) Trigonometric functions according to equation (12)

Does the latter result means that tunable microstrip resonators can have a quality factor as high as 150? The answer is negative due to two reasons. First, a lossy tuning device is connected in series to the microstrip line which will deteriorate the overall quality factor. Remarkably that the overall quality factor changes even if the tuning device were free of any loss, since the energy stored in the resonator is altered. Second, the above result is valid only for λ/2 and λ/4 microstrip lines. Since a capacitive tuning device is added to the microstrip line, the electrical length θ must be between λ/2 and λ/4 according to

$$\tan\theta = -2\pi \cdot f \cdot C_{var} \cdot Z_0 \tag{9}$$

with C_{VAR} the varactor capacitance. Hence, the overall quality factor of a tunable microstrip resonators cannot be given as straightforward as is the case for LC tanks. It depends not only on the loss of the tuning device, but also on the microstrip line Q, which in turn is dependent on the electrical length. To the author's knowledge, no rigorous analysis is reported in the literature which could predict the overall quality factor.

We have recently derived the Q of a varactor tuned microstrip resonator including the influence of the electrical length as well as the loss of the tuning circuit, which tends to dominate the latter quality factor. A short outline of the analysis and the outcomes are presented here:

The overall quality factor is given by the physical definition of quality factor, which is given as the ratio of the peak stored energy and the energy dissipated in one period

$$Q_{tot} = 2\pi \cdot \frac{E_{pkms} + E_{pkvar}}{E_{lssms} + E_{lssvar}} \tag{10}$$

with E_{pkms} and E_{pkvar} the peak stored energy in the microstrip line and in the varactor respectively, E_{lssms} and E_{lssvar} the energy dissipated in the microstrip and the varactor respectively. Equation (10) can be rewritten as

$$\frac{1}{Q_{tot}} = \frac{1}{Q_{ms}} \cdot \frac{E_{pkms}}{E_{lssms} + E_{lssvar}} + \frac{1}{Q_{var}} \cdot \frac{E_{pkvar}}{E_{lssms} + E_{lssvar}} \tag{11}$$

with Q_{ms} and Q_{var} the microstrip and the varactor Q's respectively. The individual stored energies and energy losses, both functions of the electrical length, can be evaluated. Leaving out this tedious part of the analysis, the final result is given here directly as

$$\frac{1}{Q_{tot}} = \frac{1}{Q_C} + \frac{1}{Q_D} \cdot f_2(\theta) + \frac{1}{Q_R} \cdot f_3(\theta) + \frac{1}{Q_{var}} \cdot f_4(\theta) \tag{12}$$

with $f_2(\theta)$ to $f_4(\theta)$ trigonometric functions. A plot of these function is presented in figure 4b. It can be emphasized that the overall quality factor can be up to three times larger than the varactor quality factor, if the microstrip line Q is significantly larger than the latter, which is normally the case. Indeed, in a practical 4GHz resonator using a 0.35λ microstrip line built from the earlier mentioned substrate material and a 1.6pF varactor with a Q of 20, the overall quality factor can be calculated as 40. However, it will be hard in a practical implementation to achieve a so large quality since the latter will be prone to other losses, such as bonding wire coupling and on chip parasitic resistances.

4. A very low power, 900MHz GSM Local Oscillator

This section describes the design and measurement results of a voltage controlled oscillator which meets specifications of the 900MHz GSM standard. Recently published oscillators with below -100dBc/Hz phase noise at 100kHz offset frequency tend to consume 2 to 5mA [12] [11]. An external tank is employed in this design since the design goal was to reduce the latter drain current by roughly one order of magnitude. First, an appropriate oscillator topology has been chosen. A differential topology shown in figure 1c has been used. This oscillator produces at the same drain current and with the same LC tank a larger steady state amplitude than its more conventional counterpart in figure 1b. The topology uses a double cross coupled pair, and since both cross coupled pairs provide negative g_m, the W/L-ratio can be designed slightly smaller than in single cross coupled pair oscillators. The steady state amplitude can be calculated, assuming a square wave flowing through the PMOS transistors and deep class C operation of the NMOS transistors, as

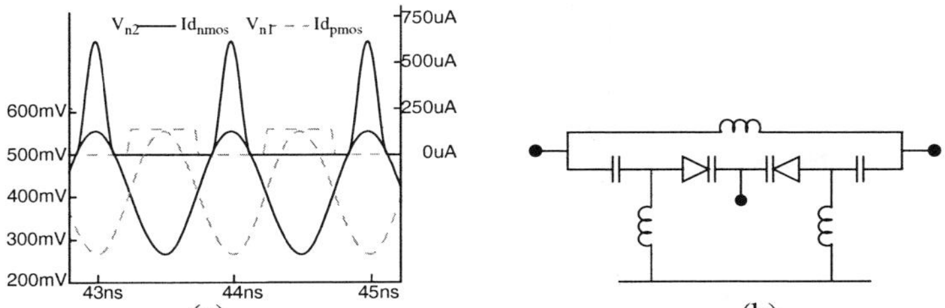

(a) (b)

Figure 5: (a) Voltages and currents in the double cross coupled pair oscillator (b) Tank decoupling circuit

$$V_m = \left(\frac{2}{\pi} + 1\right) \cdot I_D \cdot \sqrt{L/C} \cdot Q \tag{13}$$

Figure 5a visualizes the voltages across the tank as well as the NMOS and PMOS currents.

The oscillator is laid out in a 0.25μm standard CMOS technology and incorporates an output buffer for measurement purposes. Drain current is 250μA from a 2.5V battery, excluding the buffer. The chip size is 170μmx80μm without pads, which are ESD protected. The die is bonded directly onto the printed circuit board containing the tank with a measured quality factor of 20.

Figure 6 summarizes the measured performance. Phase noise was measured as low as -112dBc/Hz at 100kHz offset from the 1GHz carrier and as -118dBc/Hz at 200kHz offset, indicating that flicker noise is not significant. The oscillator can be tuned linearely from 1020MHz to 850MHz with only 2V tuning voltage. Phase noise is below -110dBc/Hz at 100kHz offset over the full tuning range, as illustrated in figure 6c. An increase in phase noise below 850MHz is observed due to forward biasing of the varactors.

Due to the large VCO gain of 85MHz/V and the low phase noise level, the oscillator is susceptible to low frequency alterations in the DC common mode level, which generate additional FM noise around the carrier. To overcome this limitation, the anodes of the varactors are decoupled from the oscillator and two RF chokes provide a low ohmic path to ground, as shown in figure 5b. However, this increases the number of external components from 2 to 6.

The topology shown in figure 1b was considered as an alternative to reduce the number of components despite its larger current consumption. The varactor cathodes are directly connected to AC ground through the inductors without additional network. An oscillator was built in the same manner as before, but using only 3 external components. Drain current was increased to 400μA to achieve the same amplitude as in the double cross coupled pair topology. The measured tun-

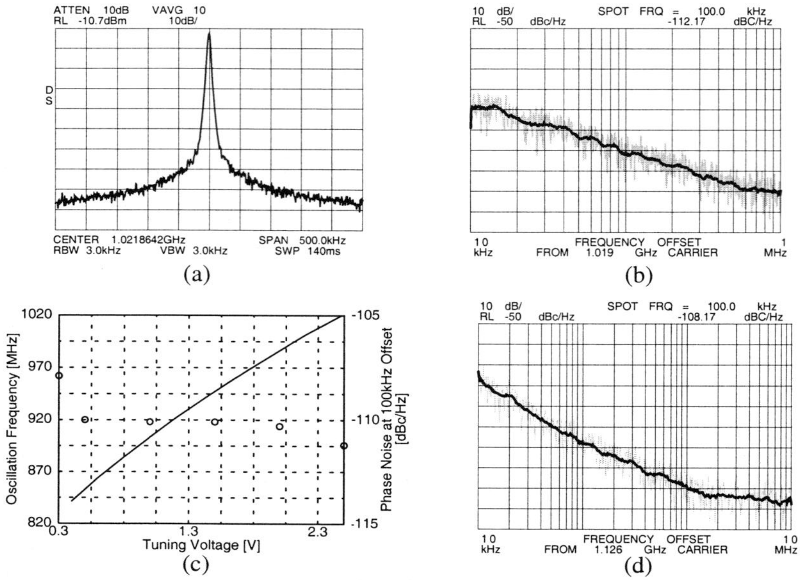

Figure 6: 1GHz CMOS oscillator measurements: (a) oscillator spectrum, (b) oscillator phase noise, (c) oscillator tuning characteristic and phase noise at 100kHz offset vs. tuning voltage (d) phase noise of the alternative oscillator topology with single cross coupled pair

ing range remains the same and phase noise is measured as low as -108dBc/Hz at 100kHz offset, as shown in figure 6d.

5. Quadrature signal generation

Image reject mixers as well as direct conversion receivers demand local oscillators providing quadrature outputs. Excellent phase match is required to achieve a strong image reject ratio. To preserve the phase information in the base band of a direct conversion receiver, the image rejection ratio (IRR) must be below -30dB. However, the latter is not only dependant on the phase mismatch, but also on the amplitude mismatch of the quadrature local oscillator signals, given as

$$IRR = 20\log\sqrt{\sin^2(\Delta\varphi/2) + \Delta A^2\cos^2(\Delta\varphi/2)} \tag{14}$$

Equation (14) forces phase and amplitude mismatch below a burdensome 2° and 2% respectively to achieve the required performance. Among different circuit techniques suited for quadrature output generation, we discuss the most promising, which are the RC-CR phase shift network and the digital divider approach:

- **RC-CR network**: This technique uses a combination of a low pass and a high pass filters which provide two signals with 90° phase difference as shown in

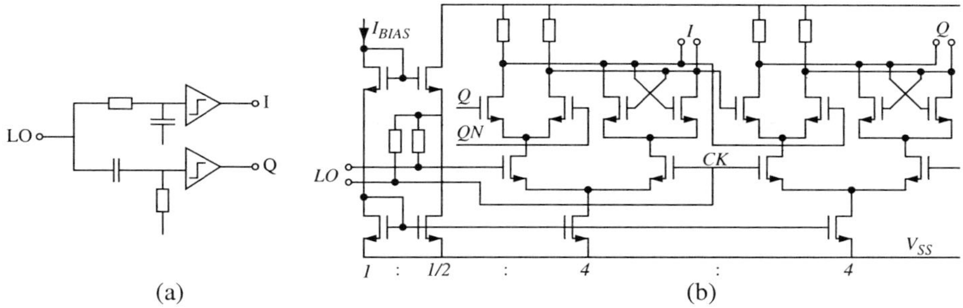

Figure 7: (a) Quadrature signal generation by a RC-CR network phase shifter (b) Quadrature signal generation with a digital master slave flip-flop

figure 7a. Phase mismatch, dependent only on the relative mismatch of the resistors and capacitors but not on frequency, can be as low as 1 to 2%, if matched integrated polysilicon resistors and matched MIM capacitors are used. On the other hand, amplitude mismatch is frequency dependent. Hence, limiting amplifiers must be used to equalize the amplitudes. The necessity of these limiters, which are power hungry, is the major drawback of the phase shifter approach. Indeed, a solution presented for 900MHz GSM consumes 12mA [6]. Another drawback of this approach is the low impedance input of the circuit, requiring another buffer to isolate the local oscillator from the phase shifter network.

- **Digital divider**: A digital master slave flip-flop, operated as a divider and driven from a local oscillator with doubled frequency, provides broadband, accurate quadrature signals. The phase mismatch does not rely on any passive components values. The only prerequisite required is a 50% duty cycle local oscillator signal, which can be easily achieved with a fully differential oscillator and divider. The advantage of the divider approach lies in the fact that no buffers are required, either at the output, or at the input, which shows a high impedance. The lack of buffers easily outweighs the increased power budget in the local oscillator, which must run at doubled frequency.

Despite the potential difficulties in oscillator design at doubled frequencies, we prefer the digital divider to the phase shifter solution, since the overall current consumption is lower. A prototype divider has been designed and laid out in a 0.25μm standard CMOS technology. Current mode topology was chosen due to various advantages, such as differential operation, low supply as well as substrate noise and low logic swing, allowing the oscillator to be connected directly to the flip-flop. Special attention was paid to a symmetric layout, in order to minimize phase and amplitude mismatch. The prototype chip contains further a 3GHz local oscillator and two mixers, added to the chip to downconvert the quadrature sig-

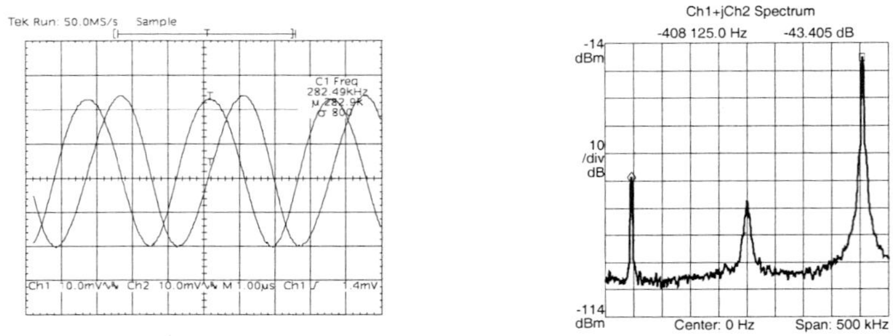

(a) (b)

Figure 8: Measurements of quadrature signals (a) time domain (b) complex spectrum

nals to a lower frequency, which eases image rejection ratio measurements. The flip-flop can be driven from the 3GHz oscillator or, alternatively from an RF signal generator. Figure 7b shows a schematic of the divider flip-flop.

The flip-flop, driven from the RF signal generator, consumes 2.8mA with a 8GHz input and 1.8mA with a 4GHz input, both from a 2.5V battery. Image rejection was measured with the flip-flop driven from the 3GHz local oscillator, with the flip-flop current set to the former 1.8mA at 2.5V supply voltage. Figure 8 shows the measured quadrature signals, downconverted to somewhat 200kHz. The measured image rejection ratio is -43dB, giving a comfortable margin to the required -30dB.

6. A low power 64/65 1GHz CMOS prescaler

Dual modulus prescalers dominate, together with the local oscillator, the power consumption of RF frequency synthesizers. They must therefore be optimized carefully for low power. While prescalers in the low GHz range are traditionally built in bipolar technologies, newer designs are reported using CMOS [11] [12] [22] [23], but they consume significantly more current than their bipolar counterparts. However, the situation changed with the introduction of 0.25μm CMOS. We outline in this section the implementation of a 1GHz 64/65 dual modulus prescaler which falls even below the power consumption of commercial prescalers in advanced bipolar technologies.

Figure 10a shows a block diagram of the circuit. Flip-flop 1 to 3 form a synchronous 4/5 divider. They are designed for maximum speed, while the asynchronous divider chain, formed by flip-flop 4 to 7, is designed for lower speed to save current. The NAND and OR gates of the synchronous part are built into flip-flop 1 and 3 for further current reduction. Figure 10b shows the schematic of a flip-flop, which are all implemented in current mode as is the flip-flop for quadrature generation. Current mode was chosen for the same reasons as in the quadrature flip-

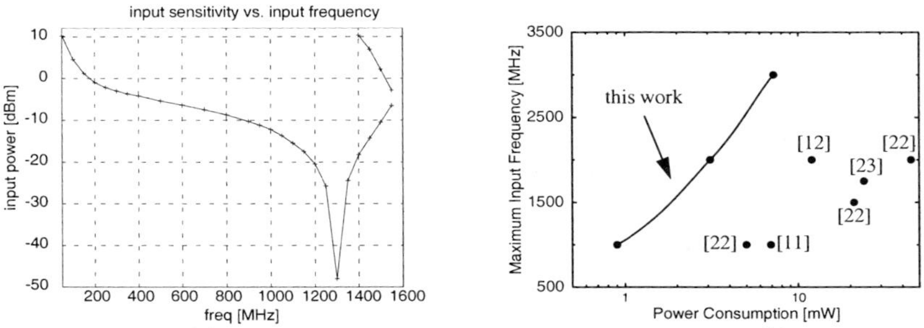

(a) (b)

Figure 9: (a) Prescaler minimum/maximum input amplitude with 900μA current consumption at 2.5V battery voltage (b) Prescaler maximum input frequency vs. power consumption. Comparison with other recently published CMOS prescaler

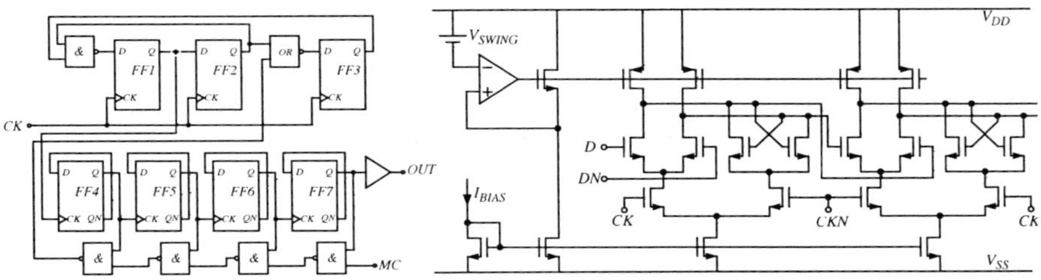

(a) (b)

Figure 10: (a) Block diagram of the 64/65 dual modulus prescaler (b) Schematic of a current mode flip-flop and replica circuit

flop design. However, triode region biased transistors are employed as load instead of resistors, since no accurate polysilicon resistors were available in this process. The PMOS triodes are biased with a replica circuit, which allows the logic swing and the triode resistance to be controlled independently.

The prescaler has been integrated in 0.25μm CMOS. The current consumption for a 1GHz input is 500μA at a 1.7V power supply, excluding the output buffer. Although intended for a 1GHz input frequency, the prescaler operates at 2GHz with 1.25mA and at 3GHz with 2.1mA, while the power supply is increased to 2.5V and 3.5V respectively. These numbers are best measured performance. To ensure a high yield in production, the prescaler has been designed to be biased at 900μA for a 1GHz input. This is still less current consumption compared to commercial products built in an advanced bipolar technology which consume 1.6 to 2.5mA [24]. Figure 9a shows the required input amplitude versus input frequencies. The minimum input amplitude is -10dBm in the 800 to 1500MHz band. Figure 9b presents the maximum operating frequency versus power consumption.

Furthermore, this work is compared with other recently published CMOS prescalers.

7. Conclusions

Local oscillator and prescaler are the most challenging part of an RF frequency synthesizer. Although these parts are traditionally built in bipolar processes, it has been demonstrated, that 0.25μm CMOS solutions can be found with below bipolar power consumption. It has been outlined that high quality factor resonators are mandatory for low power, low phase noise oscillators. Phase noise performance of CMOS oscillators can be improved by 9dB with a doubled quality factor, if the transconductances are sized accordingly. Apart from the oscillator analysis, different tank implementation were described. Fully integrated resonators yield quality factors between 3 to 5 in the 1GHz range, while external surface mounted resonators exhibit quality factors around 20. This allows oscillators to be biased with one tenth of the current of fully integrated oscillators, while maintaining phase noise performance. Beyond 2GHz, tuned microstrip resonators are attractive and not too bulky. Very high quality factors around 40 are expected. A very low power 1GHz oscillator with 18% tuning range has been presented. Phase noise below -110dBc/Hz at 100kHz offset over the full tuning range was achieved with only 0.5mW power consumption. Quadrature signals can be generated from a divider flip-flop in a power consumption effective way, since no buffers are needed. A 0.25μm CMOS flip-flop, consuming 1.8mA from a 2.5V battery, driven from a 3GHz oscillator, shows a image reject ratio of -43dB. The flip-flop operates with slightly increased power consumption up to 8GHz. Beyond the oscillator, a 1GHz 64/65 dual modulus prescaler has been presented with very low current consumption of 900μA.

References

[1] *When less is more*, Mobile Europe, Vol. 7, No. 2, February 1997

[2] J. Rudell et al., *A 1.9-GHz Wide-Band IF Double Conversion CMOS Receiver for Cordless Telephone Applications*, IEEE J. of Solid State Circuits, Vol. 32, Dec. 1997, pp. 2071-2088

[3] A. Roufougaran, G. Chang, J. Rael et al., *A Single-Chip 900MHz Spread-Spectrum Wireless Transceiver in 1mm CMOS, Part I-II*, IEEE J. of Solid State Circuits, Vol. 33, April 1998, pp. 515-547

[4] T. Stetzler et al, *A 2.7-4.5V single chip GSM transceiver RF integrated circuit*, IEEE J. of Solid State Circuits, Vol. 30, pp. 1421-1429

[5] C. Marshall et al, *A 2.7V GSM transceiver IC with on-chip filtering*, IEEE ISSCC Dig. Tech. Papers, Feb. 1995, pp. 148-149

[6] P. Orsatti, F. Piazza, Q. Huang, *A 20mA-Receive, 55mA-Transmit, Single-Chip GSM Transceiver in 0.25μm CMOS*, IEEE. J. of Solid State Circuits, Vol. 34, No. 12, December 1999

[7] D. Pederson and K. Mayaram, *Analog Integrated Circuits for Communications*, Kluwer Academic Publishers, 1991

[8] Q. Huang, *Power Consumption vs. LO Amplitude for CMOS Colpitts Oscillators*, Proc. CICC 1997, pp. 255-258

[9] Q. Huang, F. Piazza, P. Orsatti, T. Ohguro, *The Impact of Scaling Down to Deep Submicron on CMOS RF Circuits*, IEEE J. of Solid State Circuits, Vol. 33, No. 7, July 1998

[10] C. Patrick, S. Wong, *On-Chip Spiral Inductors with Patterned Ground Shields for Si-Based RF IC's*, IEEE J. of Solid State Circuits, Vol. 33, No. 5, May 1998

[11] B. Park, P. Allen, *A 1GHz, Low -Phase-Noise CMOS Frequency Synthesizer with Integrated LC VCO For Wireless Communications*, Proc. of the Custom Integrated Circuits Conference 1998, pp. 567-570

[12] J. Craninckx, M. Steyaert, H. Miyakawa, *A Fully Integrated Spiral-LC CMOS VCO Set with Prescaler for GSM and DCS-1800 Systems*, Proc. of the Custom Integrated Circuits Conference 1997, pp. 403-406

[13] T. Soorapanath, et al., *Analysis and Optimization of Accumulation-Mode Varactor for RF ICs*, Digest of VLSI Symposium, 1998, pp. 32-33

[14] S. Lo, C. Olgaard, D. Rose, *A 1.8V/3.5mA 1.1GHz/300MHz CMOS Dual PLL Frequency Synthesizer IC for RF Communications*, Proc. of the Custom Integrated Circuits Conference 1998, pp. 571-574

[15] J. Tham et al., *A 2.7V 900MHz/1.9GHz Dual-Band Transceiver IC for Digital Wireless Communication*, Proc. of the Custom Integrated Circuits Conference 1998, pp. 559-562

[16] Datasheet, *Chip Inductors 0805HQ Series*, Coilcraft Inc.

[17] Datasheet, *Hyperabrupt Tuning Varactors SMV1232-37*, Alpha Industries

[18] Datasheet, *RO4000 Series High Frequency Circuits Materials*, Rogers microwave products, 1998

[19] A. Gopinath, *Maximum Q-Factors of Microstrip Resonators,* IEEE T. on Microwave Theory and Techniques, Vol. 29, No. 2, February 1981

[20] J. James, A. Henderson, *High-frequency behavior in microstrip open-circuit termination*, IEE J. on Microwaves, Optics and Acoustics, Vol 3, No. 5, 1979

[21] T. Edwards, *Foundations for Microstrip Circuit Design*, John Wiley, 1992

[22] H. Cong et al., *Multigigahertz CMOS Dual-Modulus Prescaler IC*, IEEE J. of Solid State Circuits, Vol. 23, 1988, Oct. 1988, pp. 1189-1193

[23] J. Craninckx, M. Steyaert, *A 1.75-GHz/3V Dual-Modulus Divide-by-128/129 Prescaler in 0.7mm CMOS*, IEEE J. of Solid State Circuits, Vol. 31, July 1996, pp. 890-897

[24] Motorola Semiconductor Data Sheet *MC12053A*, 1997